T0311522

Sustainable Entrepreneurship, Renewable Energy-Based Projects, and Digitalization

Sustainable Entrepreneurship, Renewable Energy-Based Projects, and Digitalization

Edited by
Amina Omrane
Khalil Kassmi
Muhammad Wasim Akram
Ashish Khanna
Md Imtiaz Mostafiz

CRC Press
Taylor & Francis Group
Boca Raton London New York

CRC Press is an imprint of the
Taylor & Francis Group, an **informa** business

First edition published 2021
by CRC Press
6000 Broken Sound Parkway NW, Suite 300, Boca Raton, FL 33487-2742

and by CRC Press
2 Park Square, Milton Park, Abingdon, Oxon, OX14 4RN

© 2021 Taylor & Francis Group, LLC

CRC Press is an imprint of Taylor & Francis Group, LLC

Library of Congress Cataloging-in-Publication Data
Names: Omrane, Amina, editor.
Title: Sustainable entrepreneurship, renewable energy-based projects, and digitalization / edited by Amina Omrane, Khalil Kassmi, Muhammad Wasim Akram, Ashish Khanna, and Md Imtiaz Mostafiz.
Description: Boca Raton, FL : CRC Press/ Taylor & Francis Group, LLC, 2021.
| Includes bibliographical references and index.
Identifiers: LCCN 2020032461 (print) | LCCN 2020032462 (ebook) | ISBN 9780367468378 (hardback) | ISBN 9781003097921 (ebook)
Subjects: LCSH: Sustainable engineering. | Industries--Environmental aspects. |
Social responsibility of business. | Sustainable development.
Classification: LCC TA170 .S875 2021 (print) | LCC TA170 (ebook) | DDC 338.9/27--dc23
LC record available at https://lccn.loc.gov/2020032461
LC ebook record available at https://lccn.loc.gov/2020032462

ISBN: 978-0-367-46837-8 (hbk)
ISBN: 978-1-003-09792-1 (ebk)

Typeset in Times
by SPi Global, India

Contents

Preface

Sustainable Entrepreneurship, Renewable Energy-Based Projects, and Digitalization presents recent studies and new techniques and tools proposed for scholars, leaders, entrepreneurs, and/or practitioners who are concerned and/or responsible for taking strategic decisions in their companies, and who aim for sustainable development. This book highlights the use of new business systems, techniques, and models that can be employed by organizations and researchers to save millions of dollars, to enhance economic growth, as well as to resolve many environmental and social issues such as renewal energy distribution, green entrepreneurship, and solar energy.

On the one hand, the book considers sustainable entrepreneurship as a discipline at the crossroads of many others, such as social sciences, economics, management, and organizational science. It deals with many topics that aim at tackling a variety of problems related to unemployment, economic growth and development, disparities between urban cities and rural areas, pollution, lack of resources, and so on. Therefore, it helps not only scholars to disseminate their knowledge, but also practitioners and business leaders to strengthen their abilities to succeed in a globally inclusive economy.

On the other hand, the book highlights the importance of developing renewable energy-based projects. Indeed, it attempts to provide sustainable solutions to society, which widely suffers from energy consumption, scarcity of natural resources, global warming, and many other environmental problems related to the atmosphere. Those solutions are mainly based on optimized energy consumption, renewable energy applications, and photovoltaic solutions. New sustainable systems will help managers and entrepreneurs to improve local energy delivery, enhance performance, save costs, and therefore generate revenues.

The book also serves as a walk through of the emerging trends in digitalization that can support and help in the application of sustainable solutions in various functional domains. The book provides a comprehensive discussion of practical techniques and tools, such as Bitcoin, robotics, and digital currency. Such instruments can help decision makers in their everyday life and enable them to grow their businesses. The focus will be on how to benefit from such techniques to develop sustainable projects as well as digitalized new ventures.

Sustainable Entrepreneurship, Renewable Energy-Based Projects, and Digitalization has 17 chapters providing insights into sustainable development, renewable energy, digitalization, and entrepreneurship as a whole.

Chapter 1 analyzes the role and importance of digital education against the backdrop of epidemics and pandemics, which cause lockdowns and the conversion of schools and offices into human quarantine areas across the globe. In particular, it underlines the challenges of delivering quality educational services and programs, based on technological platforms, to diverse groups of learners in this digitalized world.

Chapter 2 analyzes the required tools for the transition towards a green economy based on a sustainable entrepreneurship. Relying on a literature review, it aims at solving problems related to resource scarcity and climate disruption. For this purpose, a model of sustainable development is proposed, emphasizing to what extent "sustainability" and "greenness" may occur spontaneously at lower costs.

Chapter 3 explores the effects of green innovation-oriented practices and managerial environmental concerns on improving environmental performance in the Chinese context. A cross-sectional study based on a Partial Least Square (PLS) approach is conducted, targeting 234 entrepreneurs and top managers of petrochemical firms located in Guangdong province, China. Findings support that green business strategies lead to business growth and environmental sustainability.

Chapter 4 proposes a framework of actions for the promotion of sustainable social entrepreneurship. It aims at meeting the challenges of sustainable development by nurturing inclusion and the impact of social entrepreneurship in Morocco.

Chapter 5 highlights the cross-sectional roles of chemistry in transforming the energy system and responding to "energy" and "climate" crises by offering non-nuclear alternative energy processes. It provides a synopsis of key findings, and theoretical and technological advances directed towards entrepreneurial sustainable development through chemistry, materials, renewable resources, and clean energy technologies.

Chapter 6 deals with marine fisheries' management in Pakistan. It emphasizes the role of such management, including fishing grounds, fish landing sites, and fish consumption/production and exports, in promoting sustainable development and marine biodiversity in Pakistan.

Chapter 7 identifies the factors that may lead customers to choose either the digital or the traditional marketing approach while making their purchases of goods and services. An artificial neural network (a multilayer perceptron) applied in Kalkota (India) revealed that customers select the purchase process based on seven identified factors, including a "fast process" of marketing followed by "technical orientation".

Chapter 8 evaluates the effects of the main motivational factors behind the participation of the individual in digital crowdsourcing platforms through the moderation of trust. The self-determination theory was used to support the study conducted on 209 employees working in the transportation service sector. Results show that different motivations have different effects on the participation efforts of the individual in crowdsourcing.

Chapter 9 sheds light on Bitcoin as a novel digital currency system. It investigates whether a relationship exists between trading volumes of the Bitcoin currency and the volume of queries addressed to search engines. The findings of the study revealed that web-search media activities could be used by professional investors in order to analyze search volumes and the power of forecasting trading volumes related to the Bitcoin currency.

Chapter 10 presents the structure, the functioning, the numerical modeling, and the first results concerning the experimental heating of the multi-stage solar desalination station still being carried out at Douar Al Hamri in the province of Berkane (Morocco).

Chapter 11, which deals with the design and realization of a pilot solar desalination plant in Douar El Hamri in the province of Berkane (Morocco), aims at making such plants more reliable by creating a system management block. This acquisition, control, and supervision system guarantees the production of fresh water and allows remote intervention in the event of a failure or malfunction.

Chapter 12 sheds light on developing innovative solar cooking equipment in the Moroccan context. The structure, the functioning, the modeling, and the experimentation of such a prototype cooker are presented for this purpose.

Chapter 13 investigates the thermal modeling of photovoltaic (PV) ovens and hotplates. It advocates the design and realization of innovative cookers running on PV energy, adaptable to the needs of the inhabitants, in or out of the homes of the rural or urban world, in the Moroccan context.

Chapter 14 proposes the development and testing of autonomous cooking plates, operated by photovoltaic energy in the Moroccan context.

Chapter 15 concerns autonomous photovoltaic installations and the injection of electricity into the network at the Technopole of Oujda (Morocco). It analyzes the measurement results for two proposed and realized PV installations: an autonomous one with storage in batteries (10 kW, 48 V) and one connected to the grid (10 kW, 50 Hz, 230 V). All measurements obtained are in accordance with the specifications and electricity requirements of the buildings of the University Campus at the Technopole of Oujda.

Chapter 16 explores the sustainable and strategic sources of livelihood among workers who have migrated from rural areas of Uttar Pradesh to its capital city Lucknow for employment purposes. It examines the impacts of migration on the socio-economic conditions of work in states like Pradesh and Bihar.

Chapter 17 underlines the significance of need-based financial products and a more focused inclusive policy for the urban destitute and other marginal sections of society, known as "slum dwellers", by banks and microfinance institutions for poverty alleviation in Lucknow, India.

Notes on the Editors

Amina Omrane is currently an associate- professor (HDR), authorized to supervise researches in management science in the department of Management Science at the University of Sfax in Tunisia. She has more than 10 years of academic experience in Management Science and Entrepreneurship, and teaches post-graduate and masters' courses related to Research methods, Strategic Management, Business Plan, Entrepreneurship, as well as Human/Personal development.

She obtained a research master degree in Management and Strategy as well as a professional master degree in Management by the goals, before earning her PhD in management science from the University of Jean Moulin Lyon III (France) and IHEC-Carthage (Tunisia).

Dr. Amina Omrane has presented numerous research papers in international events and conferences. She has written many other papers that are published in international journals, such as IJBG, IJESM, IJBE, RIPME, and JAB. Besides, she is usually serving as a reviewer for many other international journals. She also authored five considerable books in management science, entrepreneurship and human development, as well many book chapters in the area Series, CRC Press, Taylor and Francis Group, as well as IGI Global. Finally, she is ensuring the role of guest editor for other international journals.

Khalil Kassmi born in Casablanca (Morocco) in 1963, made all the higher education in the University Paul Sabatier (UPS) of Toulouse (France) from 1983 until 1991. In 1991, he obtained his PhD degree in Electronics from UPS in France. In 1992, he was recruited at SGS Thomson-Casablanca (Engineer). In 1993, he integrated the Mohammed 1st University in Oujda, Morocco, as Professor-researcher, and he obtained his PhD degree (Doctorat d'état) in Electronics in 1996. He is responsible for cooperation's national and international projects in the field of the renewable energies. He has two patents of invention on photovoltaic applications, and serves as an expert at the CNRST (in Morocco).

Muhammad Wasim Akram earned a PhD from Universiti Teknologi Malaysia in 2019. He is currently serving as managing director of Scientia Academia Malaysia and also working as Assistant Professor at University of Sialkot Pakistan. He has published numerous research papers and chaired 5 five international conferences. He is also serving as editor of three international journals.

Ashish Khanna has expertise in Teaching, Entrepreneurship, and Research & Development with specialization in Computer Science Engineering Subjects. He received his Ph.D. degree from National Institute of Technology, Kurukshetra in March 2017. He is serving the research and academics as a teacher, researcher, keynote speaker, consultant, book author & editor, and project consultant. He has numerous accepted and published research papers and book chapters in reputed SCI, Scopus journals, conferences and book series, including many papers in SCI indexed Journals of Springer, Elsevier, IEEE Transaction, Wiley Journals. He also has published a Patent. Additionally, he has authored and edited numerous books for publishers like Springer, Wiley, and Elsevier. He has worked as Guest Editor, Associate Editor in various journals (Springer, Wiley, Bentham Science) and he is Series Editor in De Gruyter house publishing (Germany) of "Intelligent Biomedical Data Analysis" series and Consulting Editor of Elsevier Book series. Furthermore, he has served the research field as a Faculty Resource Person/Session Chair/Reviewer/TPC member in various conferences and journals. His main research interest includes image processing, distributed systems and its variants, machine learning, evolutionary computing, and many more. He is currently working at the Department of Computer Science and Engineering, at Maharaja Agrasen Institute of Technology, under GGSIPU, Delhi, India. He is also originator of a research unit under the banner of "Universal Innovator". He has also played the key role in promoting Smart India Hackathon at MAIT.

Md Imtiaz Mostafiz is a Lecturer of International Business and Strategy at Sheffield Business School, Sheffield Hallam University, England. Prior to joining Sheffield Hallam University, he served Taylor's Business School, Taylor's University, Malaysia. His scholarly works published in the Multinational Business Review, Journal of Business and Industrial Marketing, European Journal of International Management, International Review of Entrepreneurship, Journal for International Business and Entrepreneurship Development, Asia Pacific Journal of Business Administration, International Journal of Emerging Markets, and so forth. His research mostly comprises dynamic individual and firm-level capabilities, knowledge management, international entrepreneurship, and early internationalization using the quantitative mechanism.

Contributors

Hebatallah Adam
Jindal School of International Affairs (JSIA)
O.P. Jindal Global University
Sonipat, Haryana, India

Suhail Ahmad Bhat
Department of Economics
Babasaheb Bhimrao Ambedkar University
Lucknow, Uttar Pradesh, India

Naveed Ahmad Lone
Abdul Ahad Azad Memorial Degree College Bemina
Srinagar, Jammu & Kashmir, India

S. Alexopoulos
Solar-Institut Jülich of the FH Aachen
University of Applied Sciences
Julich, Germany

I. Atmane
Faculty of Science
Department of Physics
Laboratory of Electromagnetic, Signal Processing & Renewable Energy (LESPRE)
Team Electronic Materials & Renewable Energy (EMRE)
Mohamed First University
Oujda, Morocco

Adil Azzahidi
Department of Economics and Management
Laboratory of Studies and Research in Economics and Management (LEREG)
Faculty of Legal, Economic and Social Sciences
Ibn Zohr University
Agadir, Morocco

N. Bachiri
Association Humain and Environnement of Berkane (AHEB)
Berkane, Morocco

Sudin Bag
Department of Business Administration
Vidyasagar University
West Midnapur, West Bengal, India

Abdul Baset
Department of Zoology
Bacha Khan University Charsadda
Khyber Pakhtunkhwa, Pakistan

Hurmat Sumaiya Binti Bashir
Department of Economics
Islamic University of Science and Technology
Awantipora, Jammu & Kashmir, India

Mustapha Bengrich
Department of Economics and Management
Laboratory of Studies and Research in Economics and Management (LEREG)
Faculty of Legal, Economic and Social Sciences
Ibn Zohr University
Agadir, Morocco

H. Chayeb
Solar-Institut Jülich of the FH Aachen
University of Applied Sciences
Jülich, Germany

O. Deblecker
Polytech. Mons— Electrical Power Engineering Unit
University of Mons
Mons, Belgium

N. El Moussaoui
Faculty of Science
Department of Physics
Laboratory of Electromagnetic, Signal
    Processing & Renewable Energy
    (LESPRE)
Team Electronic Materials &
    Renewable Energy (EMRE)
Mohamed First University
Oujda, Morocco

K. Hirech
Faculty of Science
Department of Physics
Laboratory of Electromagnetic, Signal
    Processing & Renewable Energy
    (LESPRE)
Team Electronic Materials &
    Renewable Energy (EMRE)
Mohamed First University
Oujda, Morocco

Kamariah Ismail
Business School
Universiti Teknologi Brunei
Gadong, Brunei Darussalam

Yousfi Karima
Department of Economics
AbouBakrBelkaid University
Tlemcen, Algeria

K. Kassmi
Faculty of Science
Department of Physics
Laboratory of Electromagnetic, Signal
    Processing & Renewable Energy
    (LESPRE)
Team Electronic Materials &
    Renewable Energy (EMRE)
Mohamed First University
Oujda, Morocco
and
Association Humain and Environnement
    of Berkane (AHEB)
Berkane, Morocco

Mohammad Ayub Khan
Department of Management
UDEM Business School
Monterrey, Mexico

A. Lamkaddem
Faculty of Science
Department of Physics
Laboratory of Electromagnetic, Signal
    Processing & Renewable Energy
    (LESPRE)
Team Electronic Materials &
    Renewable Energy (EMRE)
Mohamed First University
Oujda, Morocco

Shaza Mahar
Azman Hashim International Business
    School
Universiti Teknologi Malaysia
Johor Bahru, Malaysia
and
Faculty of Management and
    Administrative Sciences
University of Sialkot
Sialkot, Pakistan

Z. Mahdi
Solar-Institut Jülich of the FH Aachen
University of Applied Sciences
Jülich, Germany

Lahcene Makhloufi
Department of Management
School of Technology Management and
    Logistics
Universiti Utara Malaysia
Sintok City, Malaysia

Firdous Ahmad Malik
Department of Economics
Babasaheb Bhimrao Ambedkar
    University
Lucknow, Uttar Pradesh, India

M. Melhaoui
Faculty of Science
Department of Physics
Laboratory of Electromagnetic, Signal
    Processing & Renewable Energy
    (LESPRE)
Team Electronic Materials &
    Renewable Energy (EMRE)
Mohamed First University
Oujda, Morocco

Amina Omrane
Department of Management
Faculty of Business and Economics
University of Sfax
Sfax, Tunisia
and
University of Carthage
Carthage, Tunisia

Nabil Ouarsafi
Faculty of Law, Economic and Social
    Sciences
CEDOC: Law, Economy and
    Management, Management and
    Development Research Laboratory
    (LRMD)
Hassan 1st University (UHP)
Settat, Morocco

Biswajit Roy
Department of MBA
Future Business School
Kolkata, West Bengal, India

Khalid Sadiqi
Hassan 1st University (UHP)
Faculty of Law, Economic and Social
    Sciences
CEDOC: Law, Economy and
    Management, Management and
    Development Research Laboratory
    (LRMD)
Settat, Morocco

P. Schmitz
Solar-Institut Jülich of the FH Aachen
University of Applied Sciences
Jülich, Germany

K. Schwarzer
Engineering Office of Energy and
    Environmental Technology (IBEU)
Jülich, Germany
and
Solar-Institut Jülich of the FH Aachen
University of Applied Sciences
Jülich, Germany

Aslan Amat Senin
Azman Hashim International Business
    School
Universiti Teknologi Malaysia
Johor Bahru, Malaysia

Isak Rajjak Shaikh
Razak Institution of Skills, Education
    and Research (RISER)
Malmö, Sweden
and
Kemistuga
Swedish Chemical Society
Stockholm, Sweden

Shambhavi Singh
Department of Economics
Babasaheb Bhimrao Ambedkar
    University
Lucknow, Uttar Pradesh, India

Mohamad Zamhari Bin Tahir
Department of Management
Universiti of Technology Sarawak
Sarawak City, Malaysia

Mohamed Talbi
Faculty of Law, Economic and Social
    Sciences
CEDOC: Law, Economy and
    Management, Management and
    Development Research Laboratory
    (LRMD)
Hassan 1st University (UHP)
Settat, Morocco

S. Talbi
Faculty of Science
Department of Physics
Laboratory of Electromagnetic, Signal
    Processing & Renewable Energy
    (LESPRE)
Team Electronic Materials &
    Renewable Energy (EMRE)
Mohamed First University
Oujda, Morocco

Meirun Tang
Department of Management
School of Management
Guizhou University
Guiyang City, China

Muhammad Wasim Akram
Scientia Academia Malaysia
Johor Bahru, Malaysia
and
Faculty of Management and
    Administrative Sciences
University of Sialkot
Sialkot, Pakistan

D. K. Yadav
Department of Economics
Babasaheb Bhimrao Ambedkar
    University
Lucknow, Uttar Pradesh, India

1 The Era of a Digitalized World

Pandemics and Academia

Mohammad Ayub Khan and Amina Omrane

CONTENTS

1.1 INTRODUCTION

In the contemporary world of diverse, complex, and globally scaled human phenomena, the role and importance of information and communication technologies (ICT) are gaining much attention from all walks of human life, support, and services. Whether it is about buying groceries, educational programs and services, health related facilities, job related training, social contacts, or production and transportation-related technological platforms, the subject of ICT has no substitute these days. Currently and for the last few months, we are experiencing the visible effects of the Covid-19 pandemic across the globe which has virtually shattered all nations regardless of their geographic sizes, or their economic and financial strength, or their rocksolid social fabric. In this state of affairs the only alternative to conventional and physical interaction among human beings is the various technological platforms such as emails, social media (e.g., Facebook, LinkedIn, Instagram, Twitter), as well as meeting and work technologies (Zoom, Google Hangouts, Skype, Microsoft Teams), to shortlist a few. These platforms are supplementing and complementing successfully our daily activities, whether they are professional or social. Without these, the potential economic and social loss of not being able to perform properly our quotidian functions would have been far greater and more devastating than what we see today.

Covid-19-like calamities have occurred in the past, such as Russian Flu in 1889, Spanish Flu in 1918, Asian Flu in 1957, HIV/AIDS in 1981, and SARS in 2003. There is also no guarantee that such catastrophic events will not occur again in the future. These pandemics and epidemics have caused human deaths and disabilities for the millions and have disrupted human activities to a considerable extent. The direct and indirect impacts of such pandemics in the short term and in the long run on health systems, education programs, businesses, supply and demand, services, human psychology and sociology are incalculable in many ways (United Nations, 2020).

Against such an unpredictable backdrop, the genius and the common people alike seek viable answers to a few pertinent questions:

- Have we learned from our past lack of preparation to confront and overcome such disasters and socio-economic paralysis?
- Are we working in a coordinated manner as global citizens to address such happenings?
- Are we prepared to tackle these and even more horrifying scenarios in the future?

1.2 THE DIGITALIZED WORLD: A PANACEA TO OVERCOME PANDEMIC EFFECTS

More than any other human activity (social, business, professional, economic, financial), the education related tasks and services from nursery schools to the institutions of higher education bear the brunt of effects from such unforeseen and unpredicted events. In the past, business, transportation, commercial activities, and educational institutions simply used to be locked down for several days or months. But today, due to the advances in ICTs and computer based technological platforms, business and economic activities, public services, and social interaction can be carried out in a normal manner. For example, while maintaining social distance among administrators, teachers, and students, educational services and programs are delivered at home via various digital tools and platforms.

1.2.1 THE DIGITALIZED WORLD AND THE EDUCATION SECTOR

Given the occurrences of environmental events and the advances in information, communication, and computation related technologies, the subject of digitalization both in the industry and the education sector has been a hotly debated subject for a long time now – to the point where even some ambitious academic experts propose Industry 4.0-like initiatives and online programs for the education sector as well. These experts consider digital skills as being one of the critical professional competencies required in the industry, in addition to saving money and time, to achieve flexibility and interactivity for learners. For example, Beckmann (2018), who is a member of the Executive Board at Merck (a German company founded in 1668, and dedicated to science and technology), explains:

> Our educational system and, in particular, the way we learn and teach must therefore be adapted for this era of digitalization. This is true for all educational levels – from primary school to vocational schools and universities to continuing education. A digital education will allow people to continue to participate in social life in a self-determined manner in the future. In addition, digital education will ensure that companies remain competitive. Simply put: Industry 4.0 needs education 4.0. With the digital transformation, we must radically alter both *how* we learn and *what* we learn. Grand-Clement (2017) is of the view that "the digital world is increasingly penetrating the education and skills domain, with technology gradually being used to deliver education, knowledge and skills in new and innovative ways".

1.2.2 THE PROVISION OF QUALITY EDUCATIONAL SERVICES AND PROGRAMS

In the era of disruptive and revolutionary technologies and in the context of the environmental crisis and its fallout, the provision of quality educational services and programs to current and potential learners is the responsibility of government and academic institutions, whether they are public or private. Whenever it comes to the delivery of educational services and programs both the national government and academic institutions should take the following issues seriously:

- The availability of these services to all interested parties and stakeholders. This is regardless of their geographic locations, time zones, and socio-economic standing.
- The quality of the services.
- The flexibility, affordability, and timeliness of the delivery of the services.
- Technical and technological (Internet, speed, capacity) support and assistance.
- Socio-emotional and psychological services.
- The security and privacy of the learners.

1.3 DIGITAL EDUCATION (OR EDUCATION 4.0): THE MAIN FOUNDATION, REQUIRED TOOLS, AND SKILLS

In the industrial world, the buzzword of the fourth industrial revolution (also known as Industry 4.0) is currently in the offing in both academia and industry. Industry 4.0 is simply about the digitalization of industrial activities from suppliers, to operation, to inventory, to the distribution of products and services, to the final consumer, and even to after sales services. Some writers on this subject describe Industry 4.0 as a connection between the real and the virtual world, leading to a cyber-physical operational system. In essence, this is a new industrial world order where new professional and social competences emerge, such as digital skills, information technology literacy, e-management, and mobile business management. For Roland Berger Consulting (2016), Industry 4.0 is about "a shift from rigid, efficiency-focused and manual manufacturing to a more dynamic, agile and automated manufacturing. It is about a shift from mass production to mass customization, production flexibility and shorter lead times. It reduces labor costs and decreases cost of complexity". Industry 4.0 is then about a transformational amelioration in the overall operational efficiency and effectiveness of business organizations. "The fourth Industrial Revolution is the stage in the development of knowledge in which the lines between physical, digital and biological spheres are being blurred" (Schwab, 2016).

1.3.1 REQUIRED SKILLS FOR A SUCCESSFUL DIGITAL EDUCATION

Industry 4.0 brought about demands for new professional and social competences, and thus expectations from the academic institutions to rethink their existing education models and accordingly enable their modus operandi to meet the demands and needs arising out of Industry 4.0. Educational technologies have been transformed already by this digital transformation, where rapid changes of knowledge reinforce the notion of designing a new model of education for current and future learners (Shahroom & Hussin, 2018). Besides, educational futurologists are predicting that there will be radical reforms in the ways of teaching and learning, including the content of teaching, and the roles of lecturers and students. The new education model will be customized to fix and fit the system to meet the needs of the learners (Shahroom & Hussin, 2018).

Whether it is a question of formal education, continuing education (lifelong learning) programs, consulting, or business advisory services, education 4.0 can "create a conducive environment for learners, academics and practitioners to break barriers, imagine, innovate, create and collaborate, stimulate greater human connectivity through the exchange of students and academic staff and through curriculum delivery" (Wahid Omar, 2017; cited in Shahroom & Hussin, 2018). Education 4.0 is not only about digitalizing the content and delivery mechanisms; it is also related to changes in the curriculum, pedagogical approaches, and teacher training. For example, because of the emerging automated manufacturing systems, robotics has become an integral part of the curriculum in many universities. Many others are establishing research and development centers, digital hubs, big data analytics laboratories, artificial intelligence centers, and cybernetic and robotic intelligence systems.

From this perspective, prior researchers argued that digital education or education 4.0 requires numerous skills, which revolve around abilities developed in order to handle easily digital tools and programs. In keeping with Omrane's (2020) assumptions, those skills encompass technology capacities, ICT competencies, and digital skills. Technology proficiency refer to practical or theoretical knowledge applied in methods, procedures, and experiences that enable persons to discover innovative outputs or resolutions (Zarefard & Cho, 2018; cited in Omrane, 2020). Beyond technological skills, learners are encouraged to develop digital competencies, including a wide range of high level professional capabilities that are not limited to merely technical and ICT skills (Ngoasong, 2018). They also involve broader organizational abilities, such as domain knowledge, strategic and operational management, as well as soft and behavioral capabilities.

Graduates should learn and be able to work with social media, social networks, and social technologies (e.g., Facebook, Twitter, Instagram, LinkedIn, and WhatsApp). How best to use these platforms to build social synergies, social networking, and communicate better? How to learn about netiquette (protocols, security, time zones, etc.)? Graduates should master professional technologies in terms of learning about them, and having the self-efficacy and the actual capability to work with them and use them to bring efficiency and effectiveness into the organizational setting. Professional technologies, including Zoom, Skype professional, Google meeting, and Microsoft work platforms, have saved the existing human-life support system from collapsing. Technological advances in the industrial sectors cannot be ignored or left to the future. Software and simulations related to the various value and supply chain management steps in the industrial setup are also gaining momentum and thus demanding from our graduates now and in the future to be able to work with such technological advances and capitalize on them. E-management, m-business, and virtual organizations demand knowledge and skills that are high-tech and which meet the demands of the industry.

1.3.2 Main Digital Education and Learning Tools

There are several digital education tools which allow students to self-learn, to improve the administrative system of the academic institutions, to develop collaboration among different actors and power domains, and to increase efficiency and effectiveness in the communication and feedback system between students/learners on the one hand and teachers/trainers on the other hand (Chauhan, 2018). From this perspective, diverse interactive tools are to be found in the existing literature related to the educational, professional, and business workplace (Chauhan, 2018), such as:

- Edmodo;
- Socrative;
- TED-Ed;
- Projeqt;
- ClassDojo;
- eduClipper;
- Storybird;
- Kahoot.

These various tools are briefly described below for readers to obtain a quick overview of them (Stoodnt Guest Author, 2018):

- **Edmodo** is an educational network that helps connect all the learning community with the relevant users and support resources needed to share and learn. It is a social and educational platform for teachers and students to communicate, interact, and share educational materials and information.
- **Socrative** is another educational platform which helps teachers to design a variety of exercises and games which students can carry out on different mobile devices such as tablets, mobile phones, and laptops. Teachers also use this platform to design exams and other learning measurement tools in their courses.
- **TED-Ed** is an educational platform that is helpful in the creation and dissemination of educational lessons by teachers, students, and animators; the platform allows full access to information sources for all interested learners and teachers.
- **Projeqt** helps users (students and teachers alike) to perform several learning and teaching activities such as developing multimedia presentations, and preparing and using dynamic slides using links, online quizzes, Twitter timelines, and videos.
- **ClassDojo** is helpful in encouraging and improving the behavior and attitude of the students. Teachers, for example, can provide instant feedbacks (comments or observations or points) to students on their activities and behavior which can change their attitude to the class positively.
- **eduClipper** is a technological tool that is very helpful for storing, exploring information, educational materials, and references, and sharing with the intended learners and users. For example, via eduClipper one can collect information (or sources of information) from the web and share it with the group of learners.
- **Storybird** is a tool to help improve the skills of writing and reading of students' through the technique of storytelling. Storytelling is an effective pedagogical approach for all categories of learners. The platform allows students and teachers to share information and interact as well as to facilitate project development, grading, and providing regular feedback to students.
- **Kahoot** is an educational-technological platform that is composed of games and basic questions. It helps create questions, answers, and graphs, and thus provides an environment for group discussion, arguments, and feedback on the spot in a classroom setting. Kahoot-based teaching and learning activities are also funny, interactive, social, and engaging for the students, especially for young learners.

Digital libraries and digital books are replacing continuously and in a revolutionary nature brick-and-mortar library systems and services, since most of the universities in developed nations, where technological platforms, services, and capacity, have reached a respectable level, have moved from traditional platforms to digital ones. Some universities are offering digitalized degrees to their graduating students. The day is not far off when graduation ceremonies will take place on digital platforms.

From another perspective, distance learning, v-learning, e-learning, mobile-learning, and digital learning are all about technology-based education models and systems. Our current and future education system and services are required to be aligned with emerging and future environmental trends and tendencies. Indeed, they should be so flexible as to adapt to: (1) technological advances, via social and professional platforms, as well as industrial operational systems based on artificial intelligence, the Internet of Things, cloud computing, Smart Data Analytics, cryptocurrencies, etc.; (2) alternative education models, via MOOC, SPOC, COOC, SPOOC, NOOC, blended, online, hybrid, and programs which can be affordable and convenient anywhere and at any time for anyone. Academic programs should then be student centered, just in time, and adjustable. Admission, graduation, certification, and all the other related administrative formalities can then be digitalized.

1.4 INNOVATION: A FUNDAMENTAL LEVER FOR IMPROVING THE QUALITY OF EDUCATION BEYOND DIGITALIZATION?

In order to design and deliver quality educational programs and services, academic institutions must make sure they have the following features in place:

- Create a supportive culture;
- Develop a facilitating operational system;
- Allocate sufficient financial resources;
- Hire qualified information technology experts in all key areas of the educational organization, regardless of their size, origin, and level of academic services;
- Motivate, train, and reskill academic leaders, professors, students, and support personnel on a regular basis;
- Invest in state-of-the-art technologies, both educational and professional – this can be achieved by building sustained synergistic alliances with technology firms and government regulators;
- Invest in innovation and related technology as applied to scientific research projects, such as technology-computer laboratories and digital hubs;
- Partner with other national and international academic counterparts to exchange ideas, models, and best practices across a variety of technology-pedagogy mergers.

All these key components, which make up the education environment, are in essence widely dependable on the larger context, which can have a moderating, mediating, or even devastating role in the design, delivery, and evaluation of educational programs and services. As shown in Figure 1.1, such a context encompasses three main elements:

- The national environment, which includes political leadership, financial and economic policies, educational programs, science and technology-related plans, national development orientations, and national safety and security strategies.

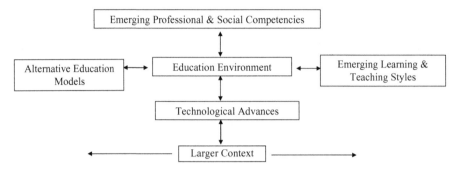

FIGURE 1.1   The larger context of education

- The global environment, which involves international relationships and col-laboration, wars and conflicts, diseases and poverty, international institutions, international norms and standards, international education, international economics and business.
- Human development needs and goals, which includes industry and community demand and the need for new skills, abilities, and values from graduating students. Such demands and expectations involve leadership skills, values and ethics, artificial intelligence, information intelligence, working with technological advances regardless of the discipline and field of expertise, social responsibility, and global understanding.

The key components of Figure 1.1 will be briefly explained for the general reader so as to become familiar with these concepts and theories. The figure shows how complex and dynamic the education industry is becoming, given the abrupt changes in the multi-dimensional larger context that surrounds the education system.

- Emerging professional and social competencies: Competencies are defined as knowledge, abilities, and values. Professional skills include decision-making, production management, marketing management, social capital leadership, change management, investment decisions, and other business-related factors. Social abilities are about people-related issues which involve conflict management, influencing, networking, communication, teamwork, empathy, and emotional intelligence.
- Alternative education models: Given all the technological advances, the globalization of educational programs and services, and diverse needs and demands for education and training, a variety of new education models (including design, delivery, and evaluation) are emerging on regular bases. Commonly known models are distance education, online education, virtual education, hybrid education, and mixed education models.
- Emerging learning and teaching styles: Students are working and learning these days. They have access to global education facilities and options. Educational programs of high quality and at affordable prices are available on their doorsteps.

In addition, students are technology savvy and have access to information and information sources all the time. They are fast, systematic learners, and they prefer more applied and practical knowledge. They also want flexibility in adopting teaching mechanisms, and to teach technology to fit their style. Similarly, faculty tends to adjust to these changing learning styles and modes of students.

- Technological advances: A variety of technological tools and platforms are being created these days to produce a teaching database, deliver it, and evaluate its impact on the learners' development. A teaching database means teaching materials and information available on technological platforms managed by educational institutions.

1.4.1 INNOVATION IN EDUCATION

The topic of innovation in education or innovative education is not rocket science. It is a matter of commonsense, and does not require a strategic guru. According to the OECD (2016), innovation in education has many advantages not only to learners, but also for the whole of society. In fact, it contributes to improving several key education performance indicators, such as learning outcomes or quality of education delivery, as well as to customize educational services and school management. It also helps to enhance equity in access and equality in learning outcomes. For this reason, academic leaders and policy makers alike should search for and collect comprehensive and enlightening explanations to the simple questions of:

- How to grow innovative and skilled graduates if the education model in itself is outdated?
- Is the academic institution obsolete in its processes, operation systems, and support programs?
- Are the academic programs out of date?
- Is the pedagogy obsolete?
- Is the academic leadership useless?
- Is the curriculum obsolete?
- Are the instructors behind the times?
- Is the university infrastructure ancient?

The OECD (2016) underlined also that academic institutions, even if using advances in educational technologies, are called upon to fully exploit the benefits of existing technological platforms. If they do, they will increase their productivity, improve their efficiency, increase quality, and foster equity, as compared to the industrial sector. While innovating educational models based on digital devices and the Internet, academic institutions must not neglect the severe implications/constraints associated with such new models, such as the lack of digital skills, the shortage of financial resources, and more importantly the need for a new organizational culture. Indeed, in many countries/territories of the world, some parties (including faculty and students) have no access to advanced technological resources. Hence, innovative digitalized education can create skill gaps and thus worsen socio-economic divisions in societies (OECD, 2016).

1.4.2 The Role of New Learning Methods and Approaches in the Education System

Since the 1990s, new learning models have appeared and prevailed to counter the myths and false proverbs about education and teaching. These new models support educational reform so as to make it more flexible, adaptable to adult learners' needs and aspirations, and make it more intelligible, meaningful, and creative. In this vein, the evaluation and competition system, which puts into perspective the alignment with the rating measures in accordance with vertically adopted training, has been re-examined.

Veldman (2004, cited in Omrane, 2019), the founder of haptonomy (the science of affectivity), opposes two approaches to teaching which should be in perfect alignment with the promoted learning styles: objective effectiveness and subjective affectivity. In what he calls "subjective affectivity", Veldman (2004, cited in Omrane, 2019) emphasizes that the instructor is asked to invite learners to learn by guiding, accompanying, and supporting them in the context of training programs or content requiring intuition, creativity, transversality, interdisciplinarity, and investigation. In another perspective, objective effectiveness emphasizes that the instructor is assimilated to a teacher who should impose (will), organize (duty), and sanction (power), within the framework of objects of more fundamental trainings, requiring the acquisition of knowledge, similar to concepts or objectives at a low taxonomic level (Belleau, 2015, p. 123).

The first spread learning co-constructivist vision based on "subjective affectivity" advocates a co-construction of knowledge. More precisely, it underlines the active participation of the learner even if he or she becomes the ambassador of new knowledge. Instead of confining his or her role to a vertical teaching process, which makes the learner a merely receptive but ignorant person in front of a scientific possessor of knowledge, the active approach is centered on the learner. In this context, Marcel Crahay (2015) states that the goal of education is learning to grow. In other words, every learner is invited to develop his knowledge by interacting with others, while building his own knowledge and prior learning. Therefore, he becomes the actor who participates in others' learning. Such an approach fosters curiosity, initiative, and trust. It also advocates for an active learning by placing the learner in an interactionist state, as well as empowering him to co-produce.

In this regard, it is worth mentioning that innovation in andragogy and learning is highly recommended. Unlike standard, prescribed, and well-crafted practices, innovative teaching practices would help to engage the learner in a more influential way. This could be explained in two ways: on the one hand, society as a whole evolves concomitantly with technological and digital evolution, which influences education that should be opened to international trends (inter-professionalism, connection to practice, etc.). On the other hand, over time, the learners change in terms of requirements, aspirations, and habits. For example, individuals who belong to Generation Z require more connectivity, creativity, and professional ethics than their Y or X generation predecessors. Currently, in order to innovate, three active learning models could be proposed and offered to instructors/trainers: problem-based learning, project-based learning, and cooperative learning.

- Problem based learning (PBL): Invented by the Faculty of Medicine at Hamilton University, Ontario, Canada, in 1970, this method involves bringing the learners together in groups in order to work collectively to solve a specific problem. According to Galaise (2001, cited in Tremblay, 2009, p. 18), problem-based learning is a pedagogical approach in which the problem serves as a learning motive. The learners are confronted with the problem, without prior study or presentation, neither of the topic nor of the problem. Then, they are called to determine what type of knowledge they have and what they lack to solve this problem. Therefore, each team is invited to understand and explain the phenomena underlying the problem, in order to solve it as part of a non-linear process. The teacher/trainer plays the role of a facilitator. Moreover, at the level of each group, three key persons join their efforts in order to ensure the same mission: a coordinator who supports the communication and exchanges between the different members of each group, a secretary who takes notes of what is stated by these members, and a manager who manages the time and division of tasks between the different actors. From this perspective, the seminal work of Barrows (1986) outlines two basic methods of problem-based learning: the case method, and discovery learning. According to Tremblay (2003, 2009), the case method deploys practical cases taken from real life in association with the lectures, while propelling the debates around the addressed topics. Regarding learning by discovery, this refers to Piaget's pioneering work (1972) which focuses on the intelligence and psychology of children. It refers to the renewal of the learner's understanding, which makes him or her become active, by asking questions as well as by reporting to other resourceful people in order to solve the target problem.
- Project based learning: This learning style is based on the theory of action that should become, in the sense of Nahas (1994), an integral part of university education. This theory emphasizes, among other things, that knowledge is simply the result of practice and theory in the context of a continuous dialectic. Moreover, Philippe Perrenoud (1999) claims that this learning style is based on a method that brings together a group of members in the context of a collective organization leading to a collective production. To do this, members have to get involved in a certain number of tasks, which would result in a project that promotes an identifiable learning. Perrenoud (1999) adds also that such an active approach will, in turn, meet the expectations of the different members. In fact, it will help them cultivate their learning motivation as well as mobilize and develop their faculties of research, analysis, reflection, experimentation, and imagination through a collective confrontation with concrete cases. The social interactions that derive from this approach will offer the various gathered individuals the opportunities to collectively build new skills and knowledge that enable them to tackle the studied problem situations and experience them. Those fruitful connections will also open up opportunities for the various individuals in order to develop further abilities collegially. As they realize projects, they will learn to reason better and overcome encountered obstacles through trial-and-error processes in order to achieve a better collective intelligence. Consequently, their actions will be

meaningful and their collective and individual identities will be strengthened by linking their different forms of skills and knowledge.

• Collaborative learning: This approach focuses on learning from each other, by putting the emphasis on the individual, as the actor of his or her own learning process. Indeed, according to this view, each trainee is able to participate in the development of his or her skills and knowledge, by cooperating effectively with the trainer. This style of learning presupposes putting the learner in a real action situation, where he can express, exploit, and exercise his capacities and mental faculties. This could result in appropriate answers to the various questions raised when creating individual and/or collective projects. In fact, it is by achieving a collegial and cooperative approach that the efforts and the different forms of knowledge and skills combine for the benefit of a constructive learning. Highlighting the advantages of cooperation goes back to the research undertaken by Piaget (1896–1980) who emphasized a part of the psychological approach which promotes the exchange of ideas, the questioning of representations, and any behavior that is likely to develop critical thinking, objectivity, and discursive reflection.

1.4.3 Toward a Holistic View of Digital Educational Quality: The Shared Responsibilities of Key Stakeholders

It has become an obvious and understood fact of life that, without technological platforms, modern educational programs and services will be difficult to design and deliver to the current as well as future learners in accordance with their demands and needs. It is important to note here that in order to provide such comprehensive and integrated educational programs and services, which encompass key factors of pedagogy, technology and content, to learners now and in the future, engaging the collective efforts of all stakeholders in contemporary educational programs and services is the need of the hour. In essence, when developing educational programs and services all those direct and indirect stakeholders in digital education must be taken into account.

Who should take more responsibility in developing, promoting, and maintaining quality education in the era of the digitalized world? In fact, all those interested and benefiting parties, as shown in Figure 1.2, are and should be held responsible for the initiatives, designing, delivering, and maintaining of quality digital education. As a matter of fact, the providers of the educational programs and services, such as board members, administrators, teachers, and the overall school or university system parties, have the primary responsibility to visualize technological advances and their impacts on the education sector and realizing the emerging educational needs and demands of contemporary learners. Likewise, as the key clients and beneficiaries of the educational programs and services, the stakeholders, like parents, industries, technological firms, communities, government institutions, and other allies, must work together in a collaborative manner to share ideas, experiences, and best practices, in order to ensure the timely availability and delivery of quality education services which are affordable and inclusive. In particular, technological firms and

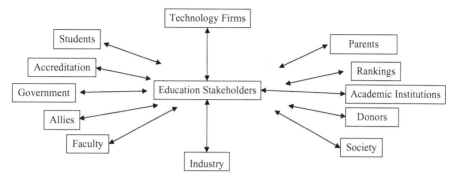

FIGURE 1.2   Stakeholders in digital education

academic institutions must form alliances and partnerships to make sure that educational technological tools are available, user-friendly, and affordable to all learners and teachers. For example, education stakeholders have an interest in the school curriculum designed, or to be designed, for the students; and they are the ones to be influenced by the design of the curriculum. Therefore, all education stakeholders, irrespective of their relative influence on the education system, must be informed about and involved in designing the school curriculum.

1.5   BEYOND THE BENEFITS OF THE DIGITIZATION OF EDUCATION: WHAT ARE THE UNEXPECTED DISADVANTAGES AND CHALLENGES TO BE FACED BY KEY PARTIES?

1.5.1   THE POTENTIAL BENEFITS AND ADVANTAGES OF EDUCATIONAL DIGITALIZATION

The digitalization of education has numerous positive outcomes. As pointed out by the OECD (2016), in addition to the continuity of educational services/programs to learners, in letter and spirit in all circumstances, digital technologies can ensure: (1) innovative pedagogic models (gamification, online laboratories, and real-time assessment); (2) simulations such as remote or virtual online laboratories, providing relatively low-cost flexible access to experiential learning; (3) international collaborations, overcoming barriers of geography and formal classroom hours; (4) real-time formative assessment and skills-based evaluations, allowing teachers to monitor student learning as it happens and adjust their teaching accordingly; and (5) e-learning, open educational resources, and massive open online courses, mainly aimed at autonomous learners.

On the other hand, the implementation of digital learning and educational systems and models offer learners many prominent advantages. First of all, it makes them more engaged and enhances their commitment by motivating them via interesting/ludic/updated/dynamic content and engaging game-based strategies. In addition, digital/online education/learning enables them to plan for their schedules and choose when and where to learn, in view of their availability and preferences. Such flexibility (in time and location) may make them more comfortable by offering them the possibility of progressing at their own rate and creating their own new world of opportunities. More especially, it allows them to individualize and customize their learning programs and courses by level and modality. Meanwhile, as instant and multiple forms of feedbacks and shared files can be afforded to learners on time, it becomes easy for teachers and trainers to drive a customized learning, based on transparency, pacing, sharing, and exchange.

1.5.2 THE DRAWBACKS, REQUIREMENTS, AND CONSTRAINTS OF DIGITALIZED EDUCATION

According to Makosa (2013), digital and online education can cause some difficulties for learners who become stuck to their computer screens in order to read documents and files, to do online courses, to complete required tasks, and so on. Spending many hours on computers may engender insomnia due to the decrease in the level of melatonin, which regulates the biological rhythm of human life. It may also cause hand and spine defects, headaches, and other diseases which can arise from faulty postures.

The light emitted by computer screens may also have a negative impact on learners' eyes and their nervous system, resulting in anxiety, depression, or memory loss.

Some problems can be related to the dependence of learners/trainers on technology, computers, and digital tools. In some situations they cannot perform the required tasks without affecting the interoperability and security of their systems. Indeed, working with technology related tools can engender many risks in many situations, due to the loss of privacy of personal information, online fraud, and/or other cyber security hazards.

In addition, new technological and digital tools (new software versions, applications, changes of hardware) need preparation and development. In fact, they require not only investment (in terms of time for new training programs, and money for new infrastructure needs/affordable broadband at home) from both of the parties, but also adaptability competencies to handle them effectively and rapidly. This can be a challenge for many persons who manifest a certain resistance to change and prefer to remain in their comfort zones. In this regard, some previous studies underlined that a technology-based education has been strongly resisted by teachers, administrators, and parents alike in order to avoid risks and constraints associated with new educational models and paradigms. For instance, Grand-Clement (2017) argued that, above all, transforming educational programs and services will require both operational readiness and strategic direction from academic institutions. More precisely, he stated that management staff as well as faculty/school members will have to generate a flexible organizational culture and system, beyond the comfort zone of every party (learners, trainers, administrative staff). Furthermore, the lack of strategic technology planning and/or financial resources can impede the effectiveness of digital technologies for education systems and models (Grand-Clement, 2017).

1.6 WHAT IS THE FUTURE OF EDUCATION AND DIGITAL EDUCATION?

The future of digital education will depend on the explicit answers to these questions:

1. Are new innovative and alternative learning approaches viable for all potential and current beneficiaries?
2. Are they sufficiently affordable for all potential and current learners and teachers globally?
3. Are they sufficiently globally scaled to meet the expectations and needs of all interested stakeholders, institutions, and organizations?
4. What have politicians and policy makers to do to make such options and alternatives viable for their citizens and national institutions?
5. Are there global norms and standards for designing and monitoring a digital education model?

According to Grand-Clement (2017), there are at least eight other requirements that can switch digital education into successful platforms, models, or systems, in order to guarantee the achievement of expected learning outcomes. Such requirements include the following skills and capabilities: (1) knowledge management capacities (e.g., validation of information, quality assurance of information); (2) self-learning and lifelong learning habits (via the motivation for consistent self-searching and gathering of updated information); (3) discerning the limits and barriers of technological and digital tools (e.g., knowing whether the technology is helpful); (4) communication skills (via written, spoken, and public speaking); (5) literacy (media and digital literacy, reading, and typing); (6) critical thinking and judgment (by criticizing actual products and systems and suggesting new innovative ones); (7) problem solving competencies and personal resilience; and finally (8) team building and the ability to work in groups (face to face or virtual meetings and conversations).

To wrap up all the requirements stated above, Grand-Clement (2017) concluded by revealing that a successful digital education turns around four main factors related to cost, accreditation, access, and specific skills. He mentioned that the cost of digital content has to be driven down via a push. For accreditation, he added that the validity of digital education certifications to the relevant academic institutions should be enhanced in order to increase trust around online learning. By "access", the author refers to social inclusion as a priority. Finally, the focus should be on setting up the necessary infrastructure to map the skills gap and predict what additional skills are needed. It would be important to point out that, to succeed in this digital world, digital and/or digital navigation skills are technical competencies required for enabling persons to use digital technologies.

From a global perspective, a comprehensive national strategic approach to the human development support system in each country can be proposed, as illustrated in Figure 1.3. Indeed, digitalization of educational systems and services is not solely a single organizational task and responsibility. It rather demands a strong synergistic partnership among all key stakeholders directly and indirectly involved in the design, delivery, and outcomes of the educational services/programs.

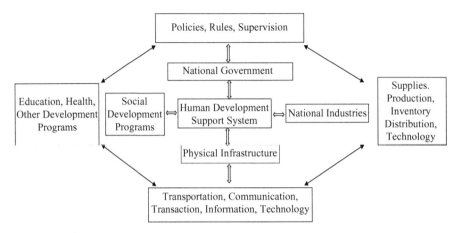

FIGURE 1.3   Human development support system

Those human development support systems represent the educational institutions regardless of their sizes, origins, nature of ownership, and categories or levels. It functions like a spider's web. Meanwhile, all the actors and sub-actors around the human development support system have to work together hand in hand and collaborate in: (1) defining policies and programs for the educational programs and services; (2) diagnosing the needs, demands, and interests of different stakeholders in education; (3) designing, delivering, and evaluating the educational programs and services in accordance with the preferences and interests of those stakeholders; (4) investing in research and innovation in the field; and lastly (5) ensuring the affordability of technological, financial, and human resources for the deserving and needy learners, by supporting the availability of educational programs/services for them.

1.7   CONCLUSION, IMPLICATIONS, AND RECOMMENDATIONS

As we move on to progress on many fronts – economic, commercial, political, technological, financial, infrastructural, and socio-cultural – we are moving towards more flexible, innovative, efficient, affordable, accessible, and shorter educational programs and services in the world. We have become a global village for good or for bad. The current global pandemic, Covid-19, has brought the global human support system to a standstill where the mobility of all human and business forces has been left at the mercy of self or forced quarantining. But it has not stopped us learning and continuing with our educational programs and services. Technological-educational platforms such as Canvas and Blackboard, with the help of Collaborate, Zoom, Google Hangouts, Microsoft Teams, and Skype to name a few, have facilitated digital classes, online classes, and e-learning, regardless of time zones and geographic location. The leadership, students, and faculty are all equally satisfied with the outcome of such education models. Students especially were found to be feeling more engaged with the learning in general and self-learning in particular (Marks, 2019). Digital education is also about creativity, innovation, and making the learning ecosystem dynamic, flexible, and learner centered. So, the digital education era has just begun.

Though the overall good aspects of digitalizing the education system are far superior than the few transitional hindrances, we foresee a few implications associated with it for educational leaders, policy makers, and strategy implementers. For example, educational institutions will have to design and implement education models with educational technologies at the core. This will demand that the academic leadership provides all the required resources and support the transition from a brick-and-mortar to a technology based educational infrastructure and system of operation. This in return will demand training and development programs for the administrators, faculty, and students. In addition, building organizational capabilities and capacities to meet the expectations and needs of all stakeholders in quality education will need reorganization, restrategizing, and refurbishing of the socio-cultural systems of the institutions of higher education. In order to make sure that the digital era is successfully embraced by academia for the benefit of learners and teachers, we present a few recommendations for the academic leadership to pursue:

- Educational technologies should be at the core of the school education model. The school mission, vision, and strategies have to demonstrate the inclusion of technological advances in the education model (pedagogy, curriculum, and school operation).
- Alliances with technological firms should be developed to share ideas with them and get support from them (technologies and technical trainings, for example).
- All the necessary resources (information, money, time, space, people) should be allocated to support and promote the digital based education model.
- The use of technological tools, software, and simulation is to be promoted for teaching and learning purposes. Such a change can help moving from a 100 percent brick-and-mortar based education model to a mixed or hybrid model where certain courses or certain sections of some courses could be taught online/virtually.
- Some of the school functions and activities like conferences, meetings, and other small scale ceremonies might be organized digitally.
- Informational and educational programs should be developed for the different stakeholders involved in education, reinforcing the scope and importance of digital education.

1.8   RESEARCH LIMITATIONS AND FUTURE ORIENTATIONS

The current chapter's analyses and conclusions are based on a review of the existing literature, observations, and the authors' personal experience as professors, administrators, researchers, and trainers in the field of higher education. This research work will help in identifying areas of opportunities and development in the field of digital education; however, systematic and quantitative or qualitative studies will bring about results and conclusions relatively more reliable and valid for the academic leadership and policy makers to make strategic decisions and actions as to whether to digitalize the whole education model or not. Interviews of focus groups and individuals in the field of higher education will be an added value to such studies in the future.

REFERENCES

Barrows, H. S. (1986). A taxonomy of problem-based learning methods. *Medical Education*, 20(6), 481–486.

Beckmann, K. (2018). The importance of digital education. Retrieved from www.thefuture-transformation.com.

Belleau, J. (2015). *Neuropédagogie: cerveau, intelligences, et apprentissages*, Centre de documentation collégiale.

Chauhan, A. (2018). 11 digital education tools for teachers and students. Retrieved from https://elearningindustry.com/digital-education-tools-teachers-students.

Crahay, M. (2015). *Psychologie de l'éducation (Education psychology)*, Editions Presses Universitaires de France (PUF).

Essays, UK. (2018, November). Stakeholders in education. Retrieved from www.ukessays.com/essays/education/stakeholders-in-education.php?vref=1.

Galaise, C. (2001). Approche pédagogique d'apprentissage par problèmes et connaissances conditionnelles en expertise comptable au premier cycle universitaire. Thèse de doctorat inédite, Université du Québec à Montréal.

Grand-Clement, S. (2017). Digital learning: Education and skills in the digital age. Retrieved from www.rand.org/content/dam/rand/pubs/conf_proceedings/CF300.

Makosa, P. (2013). Advantages and disadvantages of digital education. *Biuletyn Edukacji Medialnej*, 2, 21–31.

Marks, M. (2019). The future of effective digital learning and its role in the education system. Retrieved from https://elearningindustry.com/the-future-of-effective-digital-learning-and-its-role-in-the-education-system.

Nahas, G. (1994). *Conceptualisation et langue d'enseignement (Teaching conceptualization and language)*, Doctoral Thesis, René Descartes University. Retrieved from www.georgesnahas.com.

Ngoasong, M. Z. (2018). Digital entrepreneurship in a resource-scarce context: A focus on entrepreneurial digital competencies. *Journal of Small Business and Enterprise Development*, 25(3), 483–500. doi:10.1108/JSBED-01-2017-0014.

OECD (2016). *Innovating education and educating for innovation: The power of digital technologies and skills*, OECD Publishing, Paris.

OECD (2017). Rapport annuel 2016 sur les Principes directeurs de l'OCDE à l'intention des entreprises multinationales (Annual Report on guiding principles for multinationals). Retrieved on June 23, 2020 from: www.oecd.org/fr/daf/inv/mne/2016-Annual-Report-MNE-Guidelines-FR.pdf.

Omar, W. (2017). University presidential forum. Retrieved on June 23, 2020 from: www.moe.gov.my/muat-turun/teks-ucapan-dan-slide/2017/1527-redesign-he-4-0-higher-education-4-0-current-status-and-readiness-in-meeting-the-fourth-industrial-revolution-challenges/file.

Omrane, A. (2019). Oser se découvrir pour s'épanouir- s'épanouir- Guide des méthodes et outils de développement personnel (Dare to discover ourselves to blossom- Guide of personal development methods and tools), EMS Editions.

Omrane, A. (2020). Which are the appropriate skills needed for the entrepreneurial success of startups in the era of digitalization? (pp. 115–235), In Sbesteva, J. (eds.) *Developing Entrepreneurial Competencies for Start-Ups and Small Business*, IGI Global Editions.

Perrenoud, P. H. (1999). *Dix nouvelles compétences pour enseigner. Invitation au voyage (10 new competencies for teaching. Invitation for a journey)*, Paris, ESF Editions.

Piaget, J. (1972). *Essai de logique opératoire*, Editions Dunod.

Roland Berger Consulting (2016). Skill development for industry 4.0. Retrieved from www.globalskillsummit.com/Whitepaper-Summary.pdf/.

Schwab, K. (2016). *The fourth industrial revolution*, Geneva: World Economic Forum.

Shahroom, A. A., & Hussin, N. (2018). Industrial revolution 4.0 and education. *International Journal of Academic Research in Business and Social Sciences*, 8(9), 314–319.

Stoodnt Guest Author (2018). 10 best digital tools for students. Retrieved June 20, 2020, from www.stoodnt.com/blog/10-best-digital-tools-for-students/.

Tremblay, D.-G. (2003). Nouvelles formes de travail, nouvelles modalités d'apprentissage dans l'économie du savoir; que nous apprend le cas du multimédia au Québec? Dans Tremblay, D.-G. (2003). *La nouvelle économie : où ? quoi ? comment ?* Québec: Presses de l'université du Québec, 159–185.

Tremblay, P. (2009). Inclusion scolaire d'élèves présentant des troubles/difficultés d'apprentissage : co-formation entre enseignants de l'enseignement ordinaire et spécialisé dans le cadre d'expériences de co-enseignement (communication orale à la 6e journée d'études des chercheurs belges francophones en éducation, Université libre de Bruxelles, 9 septembre 2009).

United Nations (2020). Latin America and the Caribbean and the COVID-19 pandemic economic and social effects. Retrieved April 3, 2020, from https://repositorio.cepal.org/bitstream/handle/11362/45351/6/S2000263_en.pdf.

Veldman, F. (2004). *Haptonomie, amour et raison (Haptonomy, love and reason)*, PUF Editions.

Zarefard, M., & Cho, S. O. (2018). Entrepreneurs' managerial competencies and innovative start-up intentions in university students: Focus on mediating factors. *International Journal of Entrepreneurship*, 22(2), 141–163.

2 Innovation for Sustainable Development
The Role of Knowledge Management

*Adil Azzahidi, Mustapha Bengrich,
and Amina Omrane*

CONTENTS

2.1   INTRODUCTION

The liberalization and regulations imposed on the world today have positioned companies and innovative entrepreneurs in a critical situation, appealing to them to face the issues related to sustainable development (SD). Indeed, such a new perspective aspires to maintain a balance between preserving the environment with its natural resources and meeting current market expectations with its future needs. A set of declared technological and economic advances do not necessarily lead to the welfare of the society and its social development. It is therefore a question of the scarcity of natural resources, pollution, and greenhouse gas emissions, the depletion of non-renewable energy, and so on.

However, the search for an SD model has become a major objective for economic actors (producers and consumers), especially in industrialized countries. Thus, companies are encouraged to develop socially responsible behaviors in their activities and interactions with their stakeholders. It is an environmental commitment, like all kinds of commitments whose objective is the pursuit of a competitive advantage, as well as a sustainable performance with a socio-environmental vision. Furthermore, improving a company's environmental performance does not necessarily lead to higher costs, but can lead to better economic or financial performance (Porter and Van der Linde, 1995).

It is in fact a new concept of sustainable performance based on eco-innovations (green innovation, environmental innovation, sustainable innovation, etc.) aiming at SD and economic performance reorientation (Cecere et al., 2019; Gardebat and Uzinidis, 2012). These new practices implemented by various institutions (such as companies, public actors, and associations) refer to various types of innovation that make it possible to reduce environmental impacts. Such practices call upon good management, which guarantees competitive advantages in terms of costs and competition (Orsato, 2006), as well as the well-being of humans, by keeping ecological issues at the center of entrepreneurial actions.

Yet, in the context of globalization, innovating, by encouraging initiatives in favor of SD, requires that some companies take into account the integration of knowledge and its positive outcomes for green technology and green innovation management (Abbas and Sagsan, 2019). In addition, an analysis of the different knowledge risks provides companies with better knowledge management (hereafter KM) that can be profitable as an environmental investment, as well as a means for yielding competitive advantages (Durst and Zieba, 2019).

The objective of this chapter is then to examine the possible relationships between SD and innovation. Meanwhile, while the notion of SD has attracted widespread interest among practitioners, policy makers, and researchers (Reynaud, 2010), little research has focused on its linkages with innovation and even less on the role of KM in fostering these linkages. From this perspective, this work focuses on the field of SD, while trying to explore models of value creation, in prior theoretical and empirical researches.

First, the concept of SD and its best practices will be discussed from a value-creation perspective. Then, the focus will be on its essential role as a key vector of innovation, while trying to reconcile the two concepts from an environmental perspective. Finally, we will explore the impacts of KM and its dimensions on the different concepts mentioned above, and present a modeling of the links and associations between SD, innovation, and KM.

2.2 THE NEW GREEN ECONOMY: FROM GROWTH TO SUSTAINABILITY

In recent years, the green economy (GE) has become an issue of concern in various sectors to combat the negative environmental impacts of economic development. Moreover, this new concept has been used as an instrument for transforming value systems, business models, and the working habits of companies. Admittedly, it is a valuable investment that makes it possible to achieve SD, considered both as a leverage for economic growth and as a basic element in the fight against environmental degradation.

2.2.1 The Green Economy: The Emergence of a New Concept

The concept of GE was first used in the 1980s, notably by Pearce, Markandya, and Barbier (1989) in their report *Blueprint for a Green Economy*. The first conceptualization of the term was attempted by Jacobs (1991), who referred to the political ideology of green parties, as well as the academic discipline of environmental economics in his book. According to Barbier, a GE results in "improved human well being and social equity, while significantly reducing environmental risks and ecological scarcities" (Barbier, 2016, p. S1).

However, throughout this period until the early 2000s, the concept of GE was only rarely used, as it disappeared in international development circles (Brown et al., 2014). This is largely due to the emergence of the new concept of SD, which has attracted a great deal of interest, particularly after the Rio Summit in 1992. Indeed, according to United Nations reports, it has been considered an essential factor for SD, while offering economic and environmental solutions.

Institutionally, this concept of GE has been reinforced at the international level by the United Nations Environment Program (UNEP), which first launched the GE Initiative in 2008, and subsequently called for a Global Green New Deal (Barbier, 2009). Indeed, the United Nations Conference on Sustainable Development (Rio+20) in 2012 had identified GE as one of its main areas of focus.

The UNEP defined GE as "one that results in improved human well-being and social equity, while significantly reducing environmental risks and ecological scarcities" (UNEP, 2010, p. 5). Furthermore, its reports are generally considered as a conceptual "benchmark" and refer to four essential interpretations of the term. The first one is related to the reduction of market imperfections through the internationalization of environmental externalities. The second interpretation seeks a broader system for integrating as many environmental challenges as possible with economic ones. The third proposes links between social and economic objectives; and finally, the last interpretation points out the achievement of SD in a new macroeconomic framework.[1]

It is also argued that in a GE:

> Growth in income and employment should be driven by public and private investments that reduce carbon emissions and pollution, enhance energy and resource efficiency, and prevent the loss of biodiversity and ecosystem services. These investments need to be catalysed and supported by targeted public expenditure, policy reforms and regulation changes.

(UNEP, 2011, p. 16)

Finally, this concept of GE, which has evolved over time, has recently been well adapted by major international organizations as a policy response to global financial crises, as well as to the environmental problems experienced by our socio-economic systems (Bina and La Camera, 2011; Death, 2015). However, it has been established as an operational strategy that combines both economic recovery and adequate levels of social protection, while allowing for more sustainable growth in the future (Barbier, 2012; Bowen et al., 2009; Georgeson et al., 2017).

Today, progress on green growth has become increasingly promising in many of the world's countries, with new stimulus measures and "green investments", such as low-carbon energy, renewable energy, energy efficiency, pollution reduction and materials recycling, conservation of natural resources and ecological restoration, and finally environmental compliance, education, training, and public awareness (Barbier, 2016, p. S1).

However, the question that arises today in our current context is: Could such a nascent GE flourish in such a globalized world? In other words: Will the "green" sectors always remain a small niche in a "brown" global economy or not? Of course, whenever the term GE appears, it is often followed by the conditions related to SD. This leads to the conclusion that GE and SD together promote a new wave of sustainable innovation based on R&D and KM, replacing the brown economy.

2.2.2 THE CHALLENGES OF SUSTAINABLE DEVELOPMENT

SD can be regarded as a new concept. It refers to new theories of economic growth and development that consider the environment not as a constraint, but as a limit that requires changes in growth patterns to reach long-term sustainability. Moreover, in recent decades, the combination of economic growth and environmental protection has always been seen as a zero-sum social wealth game (Cohen and Winn, 2007). SD should thus be synonymous with sustainable growth that fits the ecological economy by promoting its sustainability.

Although the term "sustainable development" was first used in the 1980s, it did not gain prominence until the publication of the Brundtland Report of the World Commission on Environment and Development (WCED)[2] entitled "Our Common Future". This was the first overview that looked at the environmental aspects of development from an economic, social, and political perspective.

In the same context, the Commission's report clarified the basis of SD by defining it as follows: "Development which meets the needs of current generations, without compromising the ability of future generations to meet their own needs" (WCED, 1987, p. 37). This definition is still considered to be the most widely used in the world. According to this report, "We are unanimous in our conviction that the security, well-being, and very survival of the planet depend on such changes, now" (WCED, 1987, p. 38).

Thus, according to Pearce et al. (1989), the Brundtland report states that sustainability can be reached only when we achieve alleviation of poverty and deprivation, and when we improve the use of resources permanently. Therefore, the broadening of the concept of development covers not only the economic aspect, but also the social and cultural dimensions, and finally associate economics with ecology in decision making at all scales.

Today, we can say that the term SD has gained an immense popularity around the world as it has attracted the attention of policymakers and public leaders in both developed and developing countries. It recognizes not only the needs and rights of present generations, but also those of the next ones who require a healthy common environment that ensures equity between all humans.

While taking into account the 27 principles proposed at the Earth Summit in Rio De Janeiro in 1992, it appears that the concept of SD was considered a "milestone" for countries wishing to achieve development for their current and future generations in various ways and different contexts (Kori and Gondo, 2012; Mebratu, 1998).

The issue of sustainability is nowadays interpreted from several angles, while acknowledging that it is the current and future levels of production and consumption that must be sustainable, and that the growth of the world population will lead to increased environmental demands. This implies that advocates of SD stress the need to broaden the basis for the consumption of goods and services. They argue that most of these goods and services are presently unsustainable, appealing to us to downsize and change our consumption patterns (Redclift, 2005).

Thus, the different approaches of "sustainability" reflect different patterns of behaviors or commitments. Indeed, defining the needs of some prevents others from meeting their own, while creating a long-term crisis in the sustainability of livelihoods for future generations. In other words, widening our choices may reduce those of others.

According to Goodland and Daly (1996), SD is a cross-cutting issue that involves three essential dimensions (i.e., social, economic, and environmental). Social sustainability refers to poverty reduction. Economic sustainability is assimilated to the long-term sustainability of renewable and non-renewable resources engaged in sustaining production systems and providing long-term benefits for the economy. Finally, environmental sustainability deals with the preservation and maintenance of existing life forms on earth (Goodland, 1995; Kori and Gondo, 2012; Sutton, 2004).

In the same context, it will be important to mention that, historically, the world economy has always been developed around the idea of scarcity. Thus, the role of technology is mainly associated with an increased production based on scarce resources. Indeed, developed and industrialized countries transfer a large part of their toxic waste and polluting technologies to other poor regions, whilst providing themselves with energy, food, and minerals.

However, people's lifestyles reflect the way by which natural resources are devoted to satisfying their needs (Martinez-Alier, 1995). Therefore, it can be deduced that we are exploiting the present at the expense of the future. This leads to social inequalities between generations and may explain why SD has become a subject of debate among environmentalists and economists (Ayres et al., 2001).

Given this situation, many countries have strengthened their national economies through green entrepreneurship strategies. Indeed, green entrepreneurship has now moved from being just a buzzword to becoming an indispensable element in policy-making for the development of a sustainable green economy. In this regard, green innovation, referring to green entrepreneurship or eco-innovation, is pursued as the driving force behind this shift towards long-term sustainability and a "green growth".

2.3 INNOVATION AT THE HEART OF SUSTAINABLE DEVELOPMENT: NEW MODES OF PRODUCTION AND CONSUMPTION

In a changing environment of globalized competition, it appears that the wealth of nations is being created in ways that compromise sustainability, given the difficulty of reconciling "development" and "sustainability". Indeed, the present ecological and environmental pressures are boosting all types of businesses to become entrepreneurial and innovative in a new framework of greening the economy (Kusi-Sarponga et al., 2019). It is so very timely that the negative impacts of economic progress should be minimized to maintain the sustainability of resources in the course of the SD perspective.

In this sense, the challenge is to implement responsible-performance-improvement actions focused on new green business models and green innovation. To this end, particular attention should be paid to eco-entrepreneurs for their dynamism and flexibility in favor of such green innovation.

2.3.1 SUSTAINABILITY AS AN INTEGRAL PART OF ENTREPRENEURSHIP

Over the past two decades, entrepreneurship has become a subject of growing interest for researchers, as it has the potential to influence many areas. It has made a great contribution to economic, social, and cultural development, as well as generated overall positive impacts on all kinds of human activities.

Yet, in recent decades, many studies conducted in the field of the natural environment have confirmed that growth does not work hand in hand with nature conservation. Indeed, the environmental scene has witnessed much damage arising from climate change, high levels of pollution, loss of biodiversity, and soil degradation. Such changes require the involvement of an environmental responsibility in entrepreneurship, or what is known as "ecopreneurship" (Isaak, 1999).

From the same perspective, the combination of environmental objectives and other traditional goals associated with entrepreneurship (economic, psychological, and social) leads us to deduce that sustainable entrepreneurship may be considered as "how opportunities to bring into existence 'future' goods and services are discovered, created, and exploited, by whom, and with what economic, psychological, social, and environmental consequences" (Cohen and Winn, 2007, p. 35).

Since the 1970s, sustainability has always been used as an index of the balance between the economy and ecological support systems. Nevertheless, environmentalists have been blaming economic growth for much of the time, and have demanded limits with a steady-state to address environmental concerns. Indeed, based on the existing approaches of entrepreneurship, definitions closer to the concept of sustainable entrepreneurship can be suggested. On the other hand, Venkataraman's (1997) definition is considered as the main one that has cut across several fields and levels, since it integrates both contemporary orientation and early economic thinking. According to him: "entrepreneurship as a scholarly field seeks to understand how opportunities to bring into existence future goods and services are discovered, created, and exploited, by whom, and with what consequences" (Venkataraman, 1997, p. 120).

Such a definition focuses on the capture and exploitation of opportunities. It attempts to broaden the boundaries of entrepreneurship outwards by moving the entrepreneurial spirit into a larger social context that brings about successful SD.

The concept of SD has been raised recently to address more comprehensively the contribution of entrepreneurial activities to sustainability. In this sense, the relationship between entrepreneurship and SD has been examined in the literature, in light of various research perspectives and by using different terms (e.g., eco-entrepreneurship, ecopreneurship, environmental entrepreneurship, sustainable entrepreneurship, ecological entrepreneurship).

Relatively recently, the concept itself has received an increasing attention since the 1990s, and the majority of authors who have dealt with sustainability and entrepreneurship have often cited the environment as one of their main thrusts. Besides, the majority of them have focused on "green entrepreneurs" (Isaak, 1999). In this regard, green entrepreneurs, or so-called "green" companies, are adopting cleaner and greener business models and production techniques, while seeking competitive costs and marketing advantages. This implies that those entrepreneurs are the most inclined to develop innovation, take into account the wasting of resources, and often seek improvement and efficiency while remaining conscious of environmental issues.

To do this, green entrepreneurs need to find and take advantage of hidden opportunities related to environmental investments to eventually transform them into sources of competitive advantages. Moreover, these opportunities allow green companies to be active in low-potential technological fields, even if they are less innovative, as well as to move easily to green innovation (Cecere et al., 2019). Nevertheless, "being green" costs more; that is why companies wishing to invest in SD initiatives should identify the circumstances that favor the generation of both economic and environmental advantages.

According to Orsato (2006), a differentiation strategy may enable a company to make its ecological investments profitable. Indeed, productivity remains the key purpose for companies that aim at gaining competitiveness, while making better use of resources and waste, obtaining ISO certification, or owning eco-labeled products. To this end, competitive environmental strategies can help managers optimize the economic return on environmental investments and transform them into sources of value creation.

Despite those issues, much less attention has been paid to sustainable entrepreneurship as a concept integrating the environmental and social aspects. According to Cohen and Winn (2007), several current environmental challenges stem from market imperfections (business inefficiency, externalities, faulty pricing mechanisms, information asymmetry), and environmental degradation. This latter provides incentives for entrepreneurial innovation, radical technologies, and innovative business models for their resolution.

Previously, every opportunity or innovation aimed at satisfying consumer needs was liable to damage the planet's natural resource base. Indeed, our goal is to achieve entrepreneurial success through innovations that can reverse or mitigate the existence of unsustainable conditions, by taking advantage of all the opportunities that lead to the development of a model of sustainable entrepreneurship.

2.3.2   HOW TO DRIVE INNOVATIONS TOWARD A GREEN ECONOMY?

Innovation owes its originality to the Austrian School and to Schumpeter (1939)[3] who considered the entrepreneur as the person who creates businesses and jobs, and thus who subsequently participates in the renewal and restructuring of the economic fabric. In the Schumpeterian view, innovation refers to 'doing things differently' in the realm of economic life, and the company is a creative source of new combinations that correspond to different types of innovation" (Schumpeter, 1939, p. 84)[4].

Beyond these different types of innovation, Schumpeter (1934) stressed the importance of technological innovation. Then, every scientific problem solved by the human and technological intervention could be considered as a future threat to sustainability. Indeed, this technological bias in the analysis of the relationship between innovation and SD has been pursued both as a source of problems (pollution, soil degradation, etc.) and as a solution (green, restorative, depolluting, or decontaminating technology) (Djellal and Gallouj, 2009).

Prior research on innovation management has highlighted the particular challenges faced by firms (Christensen, 1997) and has proposed ways to overcome them. However, when a new technological paradigm emerges, it results in a creative destruction (Schumpeter, 1934) of existing skills, by improving the environment for small firms and other industry-related external actors. It enables companies to be more flexible and to seize new opportunities without affecting existing assets (Tushman and Anderson, 1986).

In this sense, the rise of SD in many fields has recently contributed to the multiplication of reflections on the interactions between innovation (especially the technological one) and SD. For Kusi-Sarponga et al. (2019), the sustainability of manufacturing supply chains and the progress of SD relies primarily on sustainable innovation. This has given rise to a new conception, based on the development and implementation of clean technologies that seek to improve the sustainability of companies (Bhupendra and Sangle, 2015).

Although there is much confusion about the social and environmental aspects of SD, most literature on sustainable entrepreneurship takes two key approaches. The first stream places environmental innovation at the heart of the reflection: It is organized around the theme of eco-innovation, which has recently given rise to the term "clean technologies" (Boehnke and Wüstenhagen, 2007; Hart and Milstein, 1999). The second perspective values the concept of social innovation, which refers to product or process innovations with a social purpose (health, education, etc.), or even the process of creating and improving social enterprises (Prahalad and Hammond, 2002).

Nowadays, the majority of industrialized countries should maintain a follow-up of the evolution of their economic stakeholders' behaviors (producers and consumers), considered as a catalyst for the implementation of a SD model (Gardebat and Uzinidis, 2012). Indeed, these parties are encouraged to integrate more responsible social and environmental actions into their activities, as well as into their interactions with all their other partners. Such actions involve initiating a sustainable production and consumption program (Tukker et al., 2010). To this end, entrepreneurial strategies for developing green or eco-innovations remain important for governments and researchers in economics, business, and engineering.

In this regard, a semantic approach has made it possible to identify the plurality of terms designating innovations aiming at SD, as well as the reorientation of economic performance. Indeed, whether it is a "green innovation", "sustainable innovation", "environmental innovation", "eco-innovation" or "ecological innovation" (Schiederig et al., 2012), it is a question of a decrease in environmental impact, referring to all products, services, processes, and methods, that seeks to reduce the negative impacts of the general activity on the environment[5] (Gardebat and Uzinidis, 2012).

Nevertheless, it will be important to mention that sustainable innovation incorporates a social dimension, unlike other concepts such as ecological innovation, eco-innovation, and environmental innovation, which are mainly limited to the environment. As for green innovation, its "green" quality enables its appearance in various sectors and in different components (products, processes, services, methods, business models, etc.).

The "degree of greenness" that these different types of innovations may have can be explained in a way by the level of reduction in environmental problems, such as the management of non-renewable energy, the increase in pollution levels, the preservation of biodiversity, or the development of recycling (Chen, 2008).

According to Lin et al. (2017), innovativeness has a positive impact on green brand loyalty, particularly in the presence of green knowledge. This requires resource allocation, as well as improved environmental knowledge. In this perspective, the adoption of innovation and green technologies by companies promotes the creation of customer confidence and loyalty, which in turn leads to increased profitability and competitive advantages (Albort-Morant et al., 2018).

Yet, today's emerging economies – in the context of a new and stricter form of environmental regulation – not only help to combat pollution and environmental degradation, but undoubtedly generate disruptive eco-innovations; and the idea that sustainable development promotes disruptive innovation (Christensen, 1997) is central to all sustainability debates (Cohen and Winn, 2007; Hockerts, 1999; Wüstenhagen et al., 2008).

We can conclude that sustainable entrepreneurship was conceived as a "breakthrough discipline for innovation" (Fussler, 1996), a "source of creative destruction" (Hart and Milstein, 1999, p. 23), as well as the start toward a "next industrial revolution" (Braungart and McDonough, 1998, p. 82; Senge and Carstedt, 2001, p. 24). Besides, a large number of new markets have been created through environmental innovation.[6]

Indeed, this transition to a GE not only encourages investment in the development of natural capital but also reorients the innovation activities of manufacturing firms toward building up a stock of environmental protection knowledge (Kesidou and Wu, 2020), thus toward integrating them into new technologies that behave differently from traditional ones (Conti et al., 2018). Such a view allows us to emphasize the concept of KM and knowledge-based SD as a new paradigm for global economic growth, prosperity, and sustainability.

2.4   KNOWLEDGE MANAGEMENT: A LEVEL FOR GREEN INNOVATION AND SUSTAINABLE DEVELOPMENT

Over the last three decades, the environment, for organizations around the world, has undergone several social, technological, and environmental changes, and companies must continually innovate to adapt to the new economic context marked by fierce competition, rising costs, and increasingly demanding customers. Moreover, innovation today represents the result of a process based on various interactions and exchange of knowledge between several independent actors.

In this sense, KM is considered to be a key resource for the company which facilitates the efficient exchange of information, the satisfaction of needs at lower costs, and the maintenance of competitive advantages (Mardani et al., 2018). It is one of the main forms of reducing uncertainty in the reform of technical systems (Carrillo and Gaimon, 2004; Nonaka and Takeuchi, 1995), as well as the key to innovation activities (Afuah, 1998; Porter, 1998).

Moreover, for quite some time now, KM has been attracting the interest of the business community as an essential element in the design of strategies and the development of new products and services (Mardani et al., 2018), as a body of authors and researchers, indicate that improved KM leads to improved innovation performance (Arikan, 2009; Belso-Martinez et al., 2011; Casanueva et al., 2013; Koskinen, 2000).

Today, the adoption and dissemination of innovation in the context of SD requires process companies to implement KM approaches for their conservation and mobilization (Dudezert, 2013). This allows an efficient use of resources, as well as a better development of the GE. However, the debate on the emerging knowledge-based economy serves more to strengthen green innovation activities and SD.

2.4.1 Knowledge Management: Foundation and Main Effects

The concept of KM emerged in the mid-1990s, first under the name of "intellectual capital management" before being transformed in the early 2000s. However, managing knowledge within an organization is not always a straightforward process.

KM uses a variety of tools and methods to acquire, develop, capitalize, and share knowledge within an organization. Its dynamic vision allows it to be deployed, adding the role of creating new knowledge and replacing the obsolete one.

According to Barthelme-Trapp and Vincent (2001, p. 5), KM "covers a set of models or methodologies that can implement information processing and communication tools aimed at structuring, enhancing and enabling access by the entire organization to the knowledge developed and put into practice within it".

Furthermore, Bolisani and Bratianu (2018) confirmed that knowledge[7] is an abstract concept that comes from the intangible world and has two dimensions, namely "tacit" and "explicit" (Nonaka and Takeuchi, 1995). Indeed, "explicit" knowledge is easily manifested, becoming formalizable and transferable. It can be codified, verbalized, and presented in written forms such as reports, books, and manuals (Ooi, 2014).

Regarding the implicit part of the so-called "tacit" or hidden knowledge, this is often unwritten and exists only in people's minds (Maravilhas and Martins, 2019). It is difficult to externalize it since its acquisition requires experience, learning, and interaction with several people (Johnson et al., 2019). This is undoubtedly where the role of KM takes on its full meaning, since it allows tacit knowledge to be converted into explicit knowledge, and then be transferred and shared within the company without any obstacles (Yang, 2008).

In the same context, since it is no longer simply a matter of stable and finite knowledge, tacit knowledge requires the capacity for analysis, action, and continuous training. It is an intuition formed and transformed through conjunctures, interactions, exchanges, information gathering, and daily experience. However, this tacit knowledge can never be left totally "frozen" (Nonaka and Konno, 1998).

Yet, given the increasing complexity of the organizations' environment, Durst and Zieba (2019) analyzed the different knowledge risks they face. This discipline, little discussed in the field of KM, requires particular interest from organizations that often seek to use and develop good knowledge for their account. Indeed, they established a map of knowledge risks (human, technological, and operational) that organizations need to handle to avoid certain consequences, such as knowledge attrition, knowledge loss, knowledge leakage, knowledge spillover, lost reputation, and lost sustainability.

By lost sustainability, Durst and Zieba (2019, p. 8) refer to "a situation when a company loses its ecologically-balanced approach towards the operations and does not follow the rules of sustainable development any more". However, this loss can be avoided by a learning approach[8] which allows innovative firms to become learning organizations that continually create, exchange, transform, and use new knowledge (Lundvall, 1995). Indeed, with information systems and the advent of information and communication technologies (ICT) (intranet, groupware, discussion forums, etc.), KM promotes interactivity and cooperative work, while allowing an exchange, transfer, and development of knowledge by linking knowledge and individuals. It is a collaborative space for knowledge sharing within the organization, as well as an evolving database that enables the organization to become more efficient and innovative (Yusr et al., 2017).

In this sense, innovation can be exemplified by different sources of knowledge, including those stemming from science, market needs, links between actors, and finally from technological and social networks. However, these different types of external knowledge require an absorption capacity that allows the enterprise to identify valuable knowledge, assimilate it, transform it, and apply it through innovation or the development of competitive actions (Cohen and Levinthal, 1990).[9]

It appears then that a successful KM has a positive impact on the innovative capacity of enterprises, helping them to adapt to technological changes, market fluctuations, and the rapid commercialization of goods and services (Adama, 2017).

If we take into account clean technologies and green innovation (Bekhet and Latif, 2018) as an ambitious entrepreneurial approach that attempts to make substantial innovations in SD, we can consider that KM is at the heart of innovative companies. These ones may well succeed in stimulating green innovation with more profitability, while creating future markets, far away from classic customer demands. KM will then facilitate for them the achievement of SD and green growth (Abbas and Sagsan, 2019).

2.4.2 What Are the Effective Contributions of Knowledge Management to Green Innovation and Sustainable Development?

With the increased importance given to the role of knowledge in promoting innovation, the relationship between KM and SD is becoming more and more important for organizations. Indeed, they will be not only innovative by changing old ideas, but also able to explore new directions of sustainability and green growth (Breznik, 2018). Furthermore, knowledge is seen as a driving force that contributes to individual, organizational, and national development (Maravilhas and Martins, 2019).

According to The United Nations Global Compact (UNGC, 2018), all companies must respect the environment, adopt the latest technologies that generate the efficient use of resources, and reduce the negative impacts of organizational activities (mainly those industrial ones) on the natural environment. To this end, knowledge sharing and R&D represent the main sources of SD (Habib et al., 2019). Indeed, its main objective is to use these technologies to develop existing products and processes, and thus finally to replicate other more "ecological" and "green" outputs. In this case, organizational performance can be improved not only from an economic perspective, but also from an environmental and social viewpoint (Stanovcic et al., 2015). However, to get a constructive model of sustainability, it is strongly recommended to grow the KM processes that allow the creation and use of cognitive resources in a sustainable way, while taking into consideration the social, environmental, and economic context of the organization (Lim et al., 2017). In other words, in addition to the importance of knowledge as a productive resource within the firm, its acquisition through organizational learning is a key determinant of innovation and performance (Lundvall, 1993, 1995). Indeed, a learning organization that emphasizes the implementation of KM in its overall strategies easily achieves sustainability in all its aspects (Davenport et al., 2018). Besides, KM can also enhance the sustainability of companies through the knowledge absorptive capacity that positively influences their environmental performance (Shahzad et al., 2019).

We can then argue that the knowledge-based economy can be assimilated to a leverage for economic growth and SD as well, considering the efforts sustained to improve and resolve situations of growing resource scarcity and climate disruption. Such improvements allow companies to balance economic benefits with their environmental impact (Siva et al., 2016).

From this perspective, Abbas and Sagsan (2019) examined, in a study undertaken in Pakistan, whether KM processes – including knowledge creation, acquisition, sharing, and application – have an influence on green technology and green innovation. They subsequently studied their outcomes concerning the environmental, social, and economic aspects of sustainability. Indeed, their findings confirm that KM has a significant effect on green innovation and SD activities. On the other hand, they put the emphasis on green learning for successful green innovation, as well as an orientation towards an environmental organizational culture.

We can conclude then that the optimal use of resources might be profitable as an environmental investment if it is a knowledge-based one. Yet, Rezny et al. (2019) detected the failure of the knowledge economy in its attempt to reduce the dependence on increasingly scarce and expensive natural resources (before 2008). This confirms the importance of knowledge as an essential means of maintaining a competitive advantage.

Indeed, according to Conti et al. (2018), the European Union (EU), which had undertaken a study on its innovation systems in the field of renewable energy sources (RESs) between 1985 and 2010, has witnessed a positive evolution. It has strengthened its position as a source of knowledge on RESs compared to the United States, thanks to the integration of its countries' knowledge bases. In the same context, Pencoa et al. (2020) referred to the concept of the "city of knowledge" and its positive outcomes yielding to stimulate urban entrepreneurship in 60 EU cities that are considered as important knowledge centers.

Today, the transition to a GE is an encouraging trend that motivates continuous investment in natural capital and social equity. That is why Kesidou and Wu (2020) have underlined that strict environmental regulation linked to emerging economies helps, not only to combat pollution but also to redirect the innovation activities of firms toward building up a stock of knowledge focused on environmental protection. In the same line of reflection, the recent framework proposed by Zhou, Govindan, and Xie (2020) aim to demonstrate that stimulating SD through green innovation across the supply chain is a complex network activity in which a large number of partners is involved in a fair and reasonable exchange process. Furthermore, knowledge integration and sharing in the sustainable supply chain has a key role to play in achieving green innovation. In turn, green innovation is often associated with the SD of organizations (Abbas and Sagsan, 2019).

To achieve the goals of SD, dynamic organizations adopt multiple strategies oriented towards efficient and improved innovation activities. The purpose of the present work is to explore the relationship between sustainable entrepreneurship, exemplified by the adoption of green technologies and green innovation, and SD, which represents social, economic, and environmental sustainability. By examining some empirical studies already mentioned above, we can state that the integration of KM may have a moderating role in the relationship between sustainable entrepreneurship (based on green innovation and technologies) and SD.

In this regard, the proposal of a model that depicts these relationships will highlight the different hypothetical links between concepts that should be taken into consideration. From this perspective, a future study can opt for the empirical validation of the modeling of these hypothetical links and would provide relevant and promising results (see Figure 2.1).

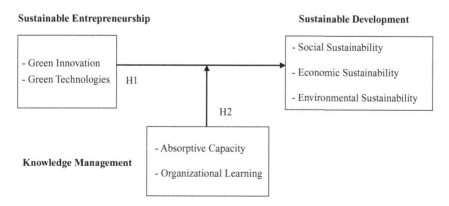

FIGURE 2.1   The proposed conceptual model

2.5 CONCLUSION, LIMITATIONS, AND FURTHER RESEARCH ORIENTATIONS

Following the global trends of SD, stimulating green innovation is the best strategy that might be implemented to slow down negative ecological and environmental impacts. Indeed, the rhetoric on green innovation and SD has been steadily increasing on the agenda of policymakers, entrepreneurs, and business people around the world. Thus, in the years following the socio-environmental crises, the concepts of "sustainability" and "greenness" gained a considerable attention. Nevertheless, their development requires the adoption of a KM orientation, while proposing that the ambition is to build on the knowledge economy as a key element of sustainable entrepreneurship.

Based on an analysis of the literature review, the present research has analyzed the existing interactions between several different concepts, but which all have the same objective related to solving problems associated with situations of increasing resource scarcity and climate disruption. Indeed, in the first section, we saw the emergence of GE in terms of environmental performance and the accompanying challenges related to SD. Then, we examined the roles of "green" innovation and "clean" technologies, considered to be the driving forces behind sustainable entrepreneurship. Finally, we elucidated the concept of KM and emphasized its importance in these interactions, leading us to propose it as a lever of green innovation for the achievement of SD.

A synthetic analysis of the different concepts and interactions between SD, GE, and KM has allowed us to deduce that the promotion of innovation and green technologies within the framework of sustainable entrepreneurship still requires support from governments. Those technologies are conveyed to encourage businesses to produce high-quality goods and services, with fewer natural resources and less environmental disruption. To this end, KM can serve to nurture the innovation process by strengthening research and analysis activities through organizational learning and knowledge absorption capacity. This enables in turn social, economic, and environmental sustainability to be achieved within the framework of a sound SD model based on the transition to a GE.

Consequently, it appears that the attractive force of an economic and a dynamic system can function and remain stable whenever all its mechanisms and phenomena are taken into consideration. Indeed, in the light of all the observations resulting from this analysis, a few lines of research can be suggested toward carrying out an empirical investigation to test the proposed theoretical model in the Moroccan context and elsewhere. Such a study would also gain in relevance by testing the effects of innovation-oriented green business strategies in the fight against global crises (such as the present Covid-19 one).

NOTES

1 Report of the General Secretary, A/CONF.216/PC/2, April 1, 2010, paragraph 44.
2 An independent body established by the United Nations in 1983 to promote the interdependence of the environment and development.
3 He was the first to state that innovations by entrepreneurs are at the heart of economic

development and growth, as well as contributing to the evolution of the economic system. According to him, the entrepreneur is seen as the actor of a "creative destruction", which gives rise to economic imbalance, by putting an end to obsolete products/services to make available others that are rather new on the market.

4 This refers to, new objects of consumption (products/services), new methods of production and transport, new markets and sources of supply, and new types of organizations and structures.

5 That is, reducing greenhouse gas emissions, producing waste, wasting natural resources, preserving biodiversity, as well as enhancing energy security (Gardebat and Uzinidis, 2012, p. 10).

6 According to Fussler, the majority of firms do not adopt sustainable entrepreneurship as a strategy for gaining market share, and they no longer know that this "innovation lethargy" is the one that will hold up in the coming years (Fussler, 1996, p. 9).

7 "Knowledge is quite different from information about its substance and process of acquisition, which is based on learning continuously to produce further new knowledge" (Gallaire, 1987, p. 267).

8 Several forms of learning have been analyzed concerning innovation, such as learning by doing (Arrow, 1962), learning through the use of advanced technologies (Rosenberg, 1982), learning through interaction (Lundvall, 1993), and learning through R&D (Rogers, 1983).

9 It is a set of four capabilities that naturally combine and build on each other to create new knowledge and business results. It measures a firm's ability to acquire, assimilate, transform, and exploit external knowledge (Flatten et al., 2011).

REFERENCES

Abbas, J., & Sagsan, M. (2019). Impact of knowledge management practices on green innovation and corporate sustainable development: A structural analysis. *Journal of Cleaner Production*, 229, 611–620.

Adama, T. Y. M. (2017). Facteurs clés de succès de la gestion des connaissances et capacités d'innovation des entreprises nigériennes (Key factors of success of knowledge management and innovation capacities of Nigerian entreprises). *La revue Gestion et Organisation*, 9(1), 11–24.

Afuah, A. (1998). *Innovation management: Strategies, implementation, and profits.* New York: Oxford University Press.

Albort-Morant, G., Leal-Rodriguez, L., & De Marchi, V. (2018). Absorptive capacity and relationship learning mechanisms as complementary drivers of green innovation performance. *Journal of Knowledge Management*, 22(2), 432–452.

Arikan, A. T. (2009). Interfirm knowledge exchanges and the knowledge creation capability of clusters. *Academy of Management Review*, 34(4), 658–676.

Arrow, K. (1962). The economic implications of learning by doing. *Review of Economic Studies*, 29, 155–173.

Ayres, R., Van den Berrgh, J., & Gowdy, J. (2001). Strong versus weak sustainability: Economics, natural sciences, and consilience. *Environmental Ethics*, 23(2), 155–168. doi:10.5840/enviroethics200123225.

Barbier, E. (2009). Rethinking the economic recovery: A global green new deal. United Nations Environment Programme.

Barbier, E. (2012). The green economy post Rio+20. *Science*, 338(6109), 887–888. Retrieved from https://www.researchgate.net/deref/http%3A%2F%2Fdx.doi.org%2F10.1126%2F science.1227360.

Barbier, E. B. (2016). Building the green economy. *Canadian Public Policy*, 42(1), S1–S9.

Barthelme-Trapp, F., & Vincent, B. (2001). *Analyse comparée de méthodes de Gestion des connaissances pour une approche managériale. Xième Conférence de l'Association Internationale de Management Stratégique, 13–15 juin 2001, Université Laval Québec.*

Bekhet, H. A., & Latif, N. W. (2018). The impact of technological innovation and governance institution quality on Malaysia's sustainable growth: Evidence from a dynamic relationship. *Technology in Society*, 54, 27–40.

Belso-Martinez, J. A., Molina-Morales, F. X., & Mas-Verdu, F. (2011). Clustering and internal resources: Moderation and mediation effects. *Journal of Knowledge Management*, 15(5), 738–758.

Bhupendra, K. V., & Sangle, S. (2015). What drives successful implementation of pollution prevention and cleaner technology strategy? The role of innovative capability. *Journal of Environmental Management*, 155, 184–192.

Bina, O., & La Camera, F. (2011). Promise and shortcomings of a green turn in recent policy responses to the double crisis. *Ecological Economics*, 70(12), 2308–2316. doi:10.1016/j.ecolecon.2011.06.021.

Boehnke, J., & Wüstenhagen, R. (2007). Business models for distributed energy technologies-evidence from German cleantech firms. Academy of Management Annual.

Bolisani, E., & Bratianu, C. (2018). The elusive definition of knowledge. *Emergent Knowledge Strategies*. Strategic Thinking in Knowledge Management, 4, 1–22. Meeting, 1–40.

Bowen, A., Fankhauser, S., Stern, N., & Zenghelis, D. (2009). An Outline of the Case for a "Green" Stimulus. Policy Brief 2009 from the Grantham Research Institute on Climate Change and the Environment and the Centre for Climate Change Economics and Policy.

Braungart, M., & McDonough, W. (1998). The next industrial revolution. *The Atlantic Monthly*, 282(4), 82–92.

Breznik, K. (2018). Knowledge management: From its inception to the innovation linkage. *Procedia – Social and Behavioral Sciences*, 238, 141–148. doi:10.1016/j.sbspro.2018.03.017.

Brown, E., Cloke, J., Gent, D., Johnson, P. H., & Hill, C. (2014). Green growth or ecological commodification: debating the green economy in the global south. *Geografiska Annaler: Series B, Human Geography*, 96(3), 245–259. doi:10.1111/geob.12049.

Carrillo, J. E. & Gaimon, C. (2004). Managing knowledge-based resource capabilities under uncertainty. *Management Science*, 50(11), 1504–1518.

Casanueva, C., Castro, I. & Galán, J. L. (2013). Informational networks and innovation in mature industrial clusters. *Journal of Business Research*, 66, 603–613.

Cecere, G., Rexhäuser, S., & Schulte, P. (2019). From less promising to green? Technological opportunities and their role in (green) ICT innovation. *Economics of Innovation and New Technology*, 28(1), 45–63. doi:10.1080/10438599.2018.1423766.

Chen, Y.-S. (2008). The driver of green innovation and green image-green core competence. *Journal of Business Ethics*, 81(3), 531–543.

Christensen, C. M. (1997). *The innovator's dilemma: When new technologies cause great firms to fail*. Boston, MA: Harvard Business School Press.

Cohen, B., & Winn, M. I. (2007). Market imperfections, opportunity and sustainable entrepreneurship. *Journal of Business Venturing*, 22, 29–49.

Cohen, W. M., & Levinthal, D. A. (1990). Absorptive capacity: A new perspective on learning and innovation. *Administrative Science Quarterly*, 35, 128–152.

Conti, C., Mancusib, M. L., Sanna-Randaccioa, F., Sestinia, R., & Verdolinic, E. (2018). Transition towards a green economy in Europe: Innovation and knowledge integration in the renewable energy sector. *Research Policy*, 47, 1996–2009.

Davenport, M., Delport, M., Blignaut, J. N., Hichert, T., & van der Burgh, G. (2018). Combining theory and wisdom in pragmatic, scenario-based decision support for sustainable development. *Journal of Environnemental Planning Management*, 62(4), 692–716. doi:10.1080/09640568.2018.1428185.

Death, C. (2015). Four discourses of the green economy in the global south. *Third World Quarterly*, 36(12), 2207–2224.

Djellal, F., & Gallouj, F. (2009). Innovation dans les services et entrepreneuriat: Au-delà des conceptions industrialistes et technologistes du développement durable. *Innovations*, *1*(29), 59–86.

Dudezert, A. (2013). *La connaissance dans les entreprises*. La Découverte, Collection Repères: Geneva.

Durst, M., & Zieba, M. (2019). Mapping knowledge risks: Towards a better understanding of knowledge management. *Knowledge Management Research & Practice*, 17(1), 1–13. doi:10.1080/14778238.2018.1538603.

Flatten, T. C., Engelen, A., Zahra, S. A., & Brettel, M. (2011). A measure of absorptive capacity: Scale development and validation. *European Management Journal*, 29, 98–116.

Fussler, C. (1996). *Driving eco-innovation, a breakthrough discipline for innovation and sustainability*. London: Pitman.

Gallaire, H. (1987). *La représentation des connaissances. La recherche en intelligence artificielle*. Seuil: Paris.

Gardebat, J., & Uzinidis, D. (2012). Innovation verte et durabilité des territoires. *Marché et organisations*, 2(16), 9–20.

Georgeson, L., Maslin, M., & Poessinouw, M. (2017). The global green economy: a review of concepts, definitions, measurement methodologies and their interactions. *Geo: Geography and Environment*, 4(1), 1–23.

Goodland, R. (1995). The concept of environmental sustainability. *Annual Review of Ecology and Systematics*, 26(1), 1–24. doi:10.1146/annurev.es.26.110195.00024.

Goodland, R., & Daly, H. (1996). Environmental sustainability: Universal and non-negotiable. *Ecological Applications*, 6(4), 1002–1017. doi:10.2307/2269583.

Habib, M., Abbas, J., & Noman, R. (2019). Are human capital, intellectual property rights, and research and development expenditures really important for total factor productivity? An empirical analysis. *International Journal of Social Economics*, 46(6), 756–774. doi:10.1108/IJSE-09-2018-0472.

Hart, S. L., & Milstein, M. (1999). Global sustainability and the creative destruction of industries. *Sloan Management Review*, 41(1), 23–33.

Hockerts, K. (1999). The sustainability radar – a tool for the innovation of sustainable products and services. *Greener Management International*, 25, 29–49.

Isaak, R. (1999). *Green logic: Ecopreneurship, theory and ethics*. West Hartford, CT: Kumarian.

Jacobs, M. (1991). *The Green Economy: Environment, Sustainable Development, and the Politics of the Future*. London: Pluto Press.

Johnson, T. L., Fletcher, S. R., Baker, W., & Charles, R. L. (2019). How and why we need to capture tacit knowledge in manufacturing: Case studies of visual inspection. *Applied Ergonomics*, 74, 1–9.

Kesidou, E., & Wu, L. (2020). Stringency of environmental regulation and eco-innovation: Evidence from the eleventh Five-Year Plan and green patents. *Economics Letters*, 190, 1–5.

Kori, E., & Gondo, T. (2012). *Environmental sustainability: Reality, fantasy or fallacy? 2nd International Conference on Environment and BioScience, IPCBEE, 44(22), IACSIT Press*, Singapore. doi:10.7763/IPCBEE.

Koskinen, K. U. (2000). Tacit knowledge as a promoter of project success. *European Journal of Purchasing and Supply Management*, 6, 41–47.

Kusi-Sarponga, S., Gupta, H., & Joseph Sarkis, J. (2019). A supply chain sustainability innovation framework and evaluation methodology. *International Journal of Production*, 57(7), 1990–2008. doi:10.1080/00207543.2018.1518607.

Lim, M. K., Tseng, M. L., Tan, K. H., & Bui, T. D. (2017). Knowledge management in sustainable supply chain management: Improving performance through an interpretive structural modeling approach. *Journal of Cleaner Production*, 162(20), 806–816.

Lin, J., Lobo, A., & et Leckie, C. (2017). The influence of green brand innovativeness and value perception on brand loyalty: The moderating role of green knowledge. *Journal of Strategic Marketing*, 27(1), 81–95. doi:10.1080/0965254X.2017.1384044.

Lundvall, G. A. (1993). Explaining interfirm cooperation: The limits of transaction cost approach. In G. Grabher (Ed.), *The embedded firm: On the socioeconomics of industrial networks* (pp. 52–64). London: Routledge.

Lundvall, G. A. (Ed.), (1995). *National systems of innovation*. London: Pinter Publishers.

Maravilhas, S., & Martins, J. (2019). Strategic knowledge management a digital environment: Tacit and explicit knowledge in Fab Labs. *Journal of Business Research*, 94, 353–359. doi:10.1016/j.jbusres.2018.01.061.

Mardani, A., Nikoosokhan, S., Moradi, M., & Doustar, M. (2018). The relationship between knowledge management and innovation performance. *The Journal of High Technology Management Research*, 29(1), 12–26.

Martinez-Alier, J. (1995). Political ecology, distributional conflicts and economic incommensurability. *New Left Review*, 9(3), 295–323.

Mebratu, D. (1998). Sustainability and sustainable development: Historical and conceptual review. *Environmental Impact Assessment Review*, 18(6), 493–520.

Nonaka, I., & Konno, N. (1998). The concept of Ba: Building a foundation for knowledge creation. *California Management Review*, 40(3), 40–54.

Nonaka, I., & Takeuchi, H. (1995). *The knowledge creating company*. New York: Oxford University Press.

Ooi, K. B. (2014). TQM: a facilitator to enhance knowledge management? A structural analysis. *Expert Systems with Applications*, 41(11), 5167–5179.

Orsato, R. J. (2006). Competitive environmental strategies: When does it pay to be green? *California Management Review*, 48(2), 127–143.

Pearce, D. W., Markandya, A., & Barbier, E. B. (1989). *Blueprint for a green economy*. London, UK: Earthscan.

Pencoa, L., Ivaldib, E., Bruzzia, C., & Mussoa, E. (2020). Knowledge-based urban environments and entrepreneurship: Inside EU cities. *Cities*, 96, 1–17.

Porter, M. (1998). *On competition*. Boston: Harvard Business Review.

Porter, M. E., & Van der Linde, C. (1995). Green and competitive: Ending the stalemate. *Harvard Business Review*, 73, 124–134.

Prahalad, C. K., & Hammond, A. (2002). Serving the world's poor, profitably. *Harvard Business Review*, 80(9), 48–57.

Redclift, M. (2005). Sustainable development (1987–2005): An oxymoron comes of age. *Sustainable Development*, 13(4), 212–227. doi:10.1002/sd.281

Reynaud, E. (2010). *Le management stratégique durable,* in Reynaud E., Stratégies d'entreprises en développement durable. L'Harmattan.

Rezny, L., White, J. B., & Maresova, P. (2019). The knowledge economy: Key to sustainable development? *Structural Change and Economic Dynamics*, 51, 291–300.

Rogers, E. (1983). *Diffusion of innovations*. New York: The Free Press.

Rosenberg, N. (1982). *Inside the black box: Technology and economics*. Cambridge University of Press: New York.

Schiederig, T., Tietze, F., & Herstatt, C. (2012). Green innovation in technology and innovation management: An exploratory literature review. *R&D Management*, 42(2), 180–192.

Schumpeter, J. (1934). *The theory of economic development*. Cambridge, MA: Harvard University Press.

Schumpeter, J. (1939). *Business cycles*, vol. 1. New York: Mc Graw-Hill.

Senge, P., & Carstedt, G. (2001). Innovating our way to the next industrial revolution. *Sloan Management Review*, 42(2), 24–39.

Shahzad, M., Ying, Q., Ur Rehman, S., Zafar, A., Ding, X., & Abbas, J. (2019). Impact of knowledge absorptive capacity on corporate sustainability with mediating role of CSR: Analysis from the Asian context. *Journal of Environmental Planning Management*, 63(2), 148–174. doi:10.1080/09640568.2019.1575799.

Siva, V., Gremyr, I., Bergquist, B., Garvare, R., Zobel, T., & Isaksson, R. (2016). The support of quality management to sustainable development: A literature review. *Journal of Cleaner Production*, 138(2), 148–157.

Stanovcic, T., Pekovic, S., & Bouziri, A. (2015). The effect of knowledge management on environmental innovation: The empirical evidence from France. *Baltic Journal Management*, 10, 413–431.

Sutton, P. (2004). A perspective on environmental sustainability. Paper on the Victorian Commissioner for Environmental Sustainability, 1–32.

Tukker, A., Cohen, M. J., Hubacek, K., & Mont, O. (2010). Sustainable consumption and production. *Journal of Industrial Ecology*, 14(1), 1–3.

Tushman, M. I., & Anderson, P. (1986). Technological discontinuities and organizational environments. *Administrative Science Quarterly*, 31(3), 439–465.

UNGC (2018). *United Nations global compact*. New York: United Nations.

United Nations Environment Programme (UNEP) (2010). *Green economy: Developing countries success stories*.

United Nations Environment Programme (UNEP) (2011). *Towards a green economy: Pathways to sustainable development and poverty eradication*. Nairobi: UNEP.

Venkataraman, S. (1997). The distinctive domain of entrepreneurship research: An editor's perspective. In J. Katz & R. Brockhaus (Eds.), *Advances in entrepreneurship, firm emergence and growth* (vol. 3, pp. 119–138). Greenwich, CT: JAI Press.

WCED (World Commission on Environment and Development) (1987). *Our Common Future: Report of the World Commission on the Environment and Development* [Bruntlandt Report]. General Assembly, United Nations, Forty-second Session, Supplement No. 25, A/42/25. Also published as *Our Common Future*. 1987. Oxford and New York: Oxford University Press.

Wüstenhagen, R., Hamschmidt, J., Sharma, S., & Starik, M. (2008*). Sustainable Innovation and Entrepreneurship*, Cheltenham, UK; Northampton, MA, USA: Edward Elgar.

Yang, J. (2008). Managing knowledge for quality assurance: An empirical study. *International Journal of Quality & Reliability Management*, 25(2), 109–124.

Yusr, M. M., Mokhtar, S. S. M., Othman, A. R., & Sulaiman, Y. (2017). Does interaction between TQM practices and knowledge management processes enhance the innovation performance? *International Journal of Quality & Reliability Management*, 34, 955–974.

Zhou, M., Govindan, K., & Xie, X. (2020). How fairness perceptions, embeddedness, and knowledge sharing drive green innovation in sustainable supply chains: An equity theory and network perspective to achieve sustainable development goals. *Journal of Cleaner Production*, 260, 1–14.

3 How to Improve Environmental Performance via Green Innovation-Oriented Practices and Managerial Environmental Concerns?

Lahcene Makhloufi, Meirun Tang, Amina Omrane, and Mohamad Zamhari Bin Tahir

CONTENTS

3.1 INTRODUCTION

Due to rapid industrialization and its awful effects on humanity, local authorities in many countries have faced severe environmental concerns mainly caused by the poor efficiency of resource use and pollution. More especially in China, environmental degradation has become a problematic issue that has contributed to the development of Chinese initiatives toward greening ventures and green trade as a whole (Haibo, Ayamba, Agyemang, Afriyie, & Anaba, 2019). From this perspective, eco-innovation has been pursued as one of the major driving forces for caring about environmental sustainability. Such specific form of innovation refers mainly to eco-friendly products, green entrepreneurship, green industry, and/or innovation based on ecological concerns. To eco-innovate, the Chinese authorities called upon developing green innovation practices, regarded as a vital lever for advancing industry, customer satisfaction, and business growth in line with responding to environmental concerns (ecological and socioeconomic ones) (Chen, Yi, Zhang, & Li, 2018). Indeed, according to Feng and Chen (2018), performance of green innovation practices can be obviously increased through: deploying clean technologies to filter hazardous waste, greening production systems to reduce CO_2 emissions, and affording a green workplace environment to encourage employees' performance. Such practices will in turn lead to sustain eco-friendly products that suit customers' environmental awareness and advance green business growth, taking into account environmental concerns (Feng & Chen, 2018).

In the same way, green entrepreneurship aims not only at producing green products that meet customers' environmental awareness, but also at balancing business growth and ecological concerns. Such an equilibrium between economic and ethical objectives can be achieved by applying and supporting green innovation practices (Gibbs & O'Neill, 2014). Consequently, firms' managers and entrepreneurs are called upon to renew and restructure their businesses through employing green innovation strategies in order to address economic and ecological sustainability challenges. Green entrepreneurship orientation (GEO) fosters employees' creativity by providing them with a green and safe workplace environment, which in turn leads to a collective learning and an interpersonal effective communication among workers. Such an environment will motivate them to develop high quality and eco-friendly products, keeping in line with minimizing hazardous waste and responding to ecological concerns (Chen & Chang, 2013). For these reasons, entrepreneurs, who are seriously committed to developing and exploiting green practices based on seizing new opportunities, will use their know-how to improve their green innovation performance (GIP) in order to grow their environmental performance (Shirokova & Ivvonen, 2016).

From another perspective, due to increased public environmental awareness, GEO faces intense pressure to fulfill customer needs based on eco-friendly products and to advance owners' businesses. In addition, governmental and other stakeholders' pressures has a significant effect on entrepreneurs' orientation. Hence, GIP plays a strategic role in balancing external pressures and entrepreneurs' goals. It also influences in turn green practices to respond efficiently to environmental threats by reducing environmental degradation and by focusing on socioeconomic welfare (Cao & Chen,

2019). Therefore, investigating the relationship between green practices, GEO, and environmental performance is both theoretically and practically significant for entrepreneurs and academicians.

In prior studies related to green innovation, the core focus was set on how to achieve socioeconomic and ecological benefits for the three building blocks of sustainability (Chen et al., 2018). Moreover, the majority of them examined the role of green entrepreneurship in developing GIP (Gast, Gundolf, & Cesinger, 2017; Kirkwood & Walton, 2014). However, understanding and demonstrating how green practices support and compel entrepreneurs to take steps toward ecological sustainability has been theoretically and practically overlooked in this previous research. To fit such a shortfall, the present study relies mainly on dynamic capability and entrepreneurship theories to test to what extent green innovation capabilities and managerial concerns foster environmental performance. In other words, this research aims at supporting the proposition that achieving the entrepreneur's key goals will lead ultimately to maintain environmental sustainability (Tang, Walsh, Lerner, Fitza, & Li, 2018).

This study proposes also to test the critical role of green absorptive capacity (GAC) and collaboration among employees based on environmental concerns, on improving the performance of green innovation practices, and on supporting entrepreneurs' actions toward greening their businesses. It further assesses the critical role of managerial concerns in multiplying green entrepreneurial actions and their environmental sustainability. Meanwhile, it explores whether top managers are sufficiently aware of environmental concerns that lead to environmental performance.

Due to the inconclusive results and the lack of studies linking managerial environmental concerns and green entrepreneurship toward achieving environmental performance (Tang et al., 2018), the current research tests the moderating influence of managerial environmental concerns. Its purpose is also to enrich the GIP and GEO literature with an evidence-based empirical investigation of the relationship between GIP, GEO, and environmental sustainability performance (ESP), along with the potential moderating effect of environmental regulations.

In doing so, the present empirical investigation contributes to the body of knowledge on sustainable entrepreneurship by providing significant evidence on the missing links between GIP, GEO, and sustainable development performance. The core focus of GEO is to create eco-friendly products and services that match economic and ecological benefits through the implementation of green innovation practices, which in turn leads to resolving environmental problems. In this regard, the main objective of sustainable entrepreneurship concerns the extent to which entrepreneurs will innovatively be able to explore the opportunities and exploit green innovation capabilities so as to yield economic benefits and enrich social equity in line with environmental conditions.

In summary, sustainable entrepreneurship represents the process by which entrepreneurs can achieve a balance between human needs development (HND) and environmental sustainability requirements (ESRs). Such a critical achievement can be obtained only through the implementation of green innovation practices at all levels (the individual, organizational, and national ones.). To conduct the empirical investigation, a deductive approach was applied with a cross-sectional survey targeting the top management (CEO and managers) of petrochemical firms located in Guangdong province in China. The study was carried out on a full sample of 586 petrochemical companies based in Guangdong province in China. Therefore, this research is designed to provide a framework of the main environmental issues in China, especially those which are related to the industry sector, the primary driver of air and water ground pollution, CO_2 emission, ecological and agriculture problems, along with other community welfare problems. For this purpose, a partial least squares structural equation modeling (PLS-SEM) data analysis was carried out as the best appropriate statistical tool. The detailed results of the structural tested model will be considered and discussed in the subsequent sections. The last section concludes by emphasizing the relevance of the findings for academics, entrepreneurs, and policy makers, as well as the research limitations and further future orientations.

3.2 LITERATURE REVIEW AND HYPOTHESES DEVELOPMENT

3.2.1 GREEN ABSORPTIVE CAPACITY AND ITS EFFECTS ON GREEN INNOVATION PERFORMANCE

The dynamic capability theory viewed absorptive capacity (AC) as a critical driver for organizational knowledge management, supporting firms' owners and enabling them to yield, transmit, and integrate the obtained knowledge into the existing knowledge (Alves, Salvini, Bansi, Neto, & Galina, 2016). As a result, developed managerial and business processes will in turn empower managers and enable them to create unique intangible resources that lead to sustainable competitive advantages (Alves, Salvini, Bansi, Neto, & Galina, 2016). In other words, the dynamic capability theory underlined the importance of AC in terms of its capabilities to upgrade organizational learning, R&D, advanced business production systems, employees' skills and knowledge, workplace environment, and management systems, contributing in turn to fostering green innovation practices into firms. From this perspective, Jin et al. (2019) argued that green innovation practices may help entrepreneurs to increase the efficiency of their resource use by reducing hazardous waste, impeding further CO_2 emissions, and thereby leading to healthy eco-systems.

AC is then acknowledged as the strategic source that supports green innovation activities. It implies that it might foster firms' internal capabilities and improve business growth and environmental performance. Keeping in line with this orientation, several research works undertaken by Albort-Morant, Leal-Rodríguez, and De Marchi (2018), Cohen and Levinthal (1990), Gluch, Gustafsson, and Thuvander (2009), and Pacheco, Alves, and Liboni (2018) pointed out that AC has a substantial impact on different environmental and organizational aspects, supporting GIP and GEO in maintaining environmental performance. Similarly, the present study considers AC as an organizational capability that enables entrepreneurs to seize new opportunities and internally develop the exploitation of new competencies in order to cope with green innovation activities.

Therefore, we posit that AC refers to the capability gained by firms, empowering them to adapt to a highly fluctuated environment (presenting internal and external pressures) by fostering their internal skills. It also contributes to ensure the continuity of green businesses in line with caring for environmental concerns.

The lack of studies dedicated to exploring the links between AC and GIP, as well as their consequences, justify our recourse to the current study in order to understand in depth to what extent AC may enhance green innovation activities, by combining the yielded and integrated new knowledge with green innovation plans, programs, training, and production system techniques. Consequently, this research suggests examining the likely effects of the GAC of firms, and more specifically of new ventures, on their GIP, enabling entrepreneurs to reach their strategic goals and contribute to environmental sustainability. The first research hypothesis can then be proposed as follows:

H1: The new venture's GAC has a significant positive impact on its GIP.

3.2.2 ENVIRONMENTAL COOPERATION AND ITS EFFECTS ON GREEN INNOVATION PERFORMANCE

Several previous studies addressed the effects of environmental cooperation (EC) on businesses' financial and non-financial performance (Younis, Sundarakani, & Vel, 2016), and on green supply-chain management to improve firms' performance (Perotti, Zorzini, Cagno, & Micheli, 2012). For instance, Diabat, Khodaverdi, and Olfat (2013) examined the influence of suppliers' and customers' cooperation and green supply chains on firms' performance. Nevertheless, few studies have attempted to explore the underlying mechanism of cooperation as an organizational capability that fosters GIP. Moreover, investigating the relationships between cooperation based on environmental concerns, GIP, GEO, and ESP has been neglected, leading to an emerging critical theoretical and practical gap.

The present work aims at filling this shortfall by answering the following question: To what extent does cooperation, based on environmental problems between different partners, enable entrepreneurs to improve their green innovation (GI)?

In order to bring up elements of responses to this question, we referred to Kong et al. (2016) who addressed the effects of EC on green innovation products and processes. It was also of interest to consider the writings of Albort-Morant et al. (2018). In fact, these scholars argued that inter-relationships with stakeholders play a significant role in strengthening green practices through the transmission of knowledge into products and managerial systems.

On the basis of these few assumptions, we posit that EC, known as the cooperation between an organization and its external partners (such as suppliers, stakeholders, customers, authorities, and laboratories under the umbrella of the entrepreneurial eco-system), can be viewed as a driving force of superior environmental performance. EC is hence considered as one of the key solutions that allows entrepreneurs to maintain the survival of their new ventures, fighting also against the scarcity of resources and the restriction of their use in a highly turbulent environment. According to Weber and Heidenreich (2018), EC is about establishing external relationships with experts and laboratories, and expanding internal knowledge by upgrading inter-organizational learning. By doing so, EC supports the innovation capabilities of the firm in order to create new strategic resources (via know-how, learning, and routines). Such new resources will be so ambiguous and original as to be duplicated by competitors. In this regard, the performance of green innovation will be improved, leading to superior advanced production systems, resource efficiency, and eco-friendly products (Weber & Heidenreich, 2018). Environmental degradation will also be minimized, generating the development of sustainable environmental performance (Kong, Feng, & Ye, 2016).

Thus, this chapter is dedicated to investigate the influence of EC on new ventures' GIP, in order to achieve a superior environmental performance. The corresponding research hypothesis can be formulated as follows:

H2: EC has a significant positive impact on new ventures' GIP.

3.2.3 The Relationships Between Green Entrepreneurial Orientation, Green Innovation Performance, and Green Environmental Performance

Generally, GIP is pursued as one of the best solutions taken into account by policy makers and citizens in many countries in order to tackle environmental degradation problems. Indeed, in this era of industrial revolution 4.0, economic development is based on green growth, which is required to suit the enormous pressures of societal, governmental, and environmental restrictions (Haibo et al., 2019). Consequently, firms compulsorily apply clean technologies, green production systems, and/or green raw materials in order to increase their resources' efficiency and productivity and thereby avoid further hazardous waste (Melander, 2017).

These environmental restrictions and pressures encourage entrepreneurs and managers to combine environmental strategies into their overall business strategy, so that they can ensure the green business survival of their ventures in line with caring for environmental concerns. In this regard, greening processes and products allow firms to capture and extend their market share and customer loyalty, which in turn yield considerable benefits related to environmental performance sustainability. In keeping with this orientation, Tariq, Badir, Tariq, and Bhutta (2017) affirmed that, by deploying clean technologies, firms could control energy savings, reduce CO_2 emissions, clean up the workplace environment, develop a green mindset for employees, and manage hazardous waste. By doing so, they will respond to customers' environmental awareness through producing eco-friendly products based on customer loyalty, extending market share, and then caring about societal and environmental concerns (Tariq, Badir, Tariq, & Bhutta, 2017).

Nowadays, entrepreneurs face not only strong and fierce competition, but also pressure from customers and authorities, along with volatile societal expectations. That is why the survival of new ventures becomes difficult. In this perspective, GIP has been regarded as a strategic orientation, which can enable entrepreneurs to overcome the environmental constraints and limitations through applying green technologies that ensure green business growth (Albort-Morant, Leal-Millán, & Cepeda-Carrión, 2016). This implies that GIP empowers entrepreneurs so that they can perform actions and implement strategies toward the development of green business processes and products. GIP helps entrepreneurs to proactively respond to organizational and environmental concerns by achieving a balance between economic benefits and ecological concerns (Chen et al., 2018).

Previous studies (i.e., Cao & Chen, 2019; Chen et al., 2018; Lin, Zeng, Ma, Qi, & Tam, 2014; Lin & Chen, 2018; Qi, Shen, Zeng, & Jorge, 2010) have explored the main factors that can influence green entrepreneurship, including green innovation. However, some studies undertaken mainly by Demirel, Li, Rentocchini, & Tamvada (2019) and Lin & Chen (2018), have stressed the relationship between green innovation and green entrepreneurship. More specifically, Tang et al. (2018) stated that the failure of entrepreneurship to achieve superior performance is due to their reluctance to apply green innovation practices along with customer and societal environmental awareness. Huang et al. (2016) added that it is also caused by high costs generated by applying green technologies. Other researchers (i.e., Albort-Morant et al., 2018; Kong et al., 2016) have mentioned that firms may enjoy considerable earnings while applying green technologies in the creation of eco-friendly products. They also revealed that governmental facilities enable entrepreneurs to engage in actions toward improving green business growth and environmental performance (EP).

Entrepreneurs are then called upon to explore those core activities that yield environmental and organizational issues and allow them to exploit their green capabilities in order to face these environmental concerns. According to Saunila, Ukko, & Rantala (2018), this can be accomplished by deploying clean technologies, green production systems, and controlling hazardous waste, which in turn leads to mastering environmental degradation challenges. From this perspective, green entrepreneurship interferes as a proactive process based on the rationalization of the use of resources, resulting in the exploitation of big opportunities and business growth improvement. For example, it was argued that entrepreneurs, who apply green chemistry in order to reduce hazardous waste and CO_2 emissions, succeed in minimizing environmental degradation caused by the considerable use of petrochemical products (Matus, Xiao, & Zimmerman, 2012).

According to Chuang and Yang (2014), green engineering and chemistry help firms to provide a clean and green workplace environment, which motivates employees and increases their labor performance and satisfaction, allowing managers to control harmful emissions and toxicology, and thereby environmental degradation issues.

Consistent with this reflection, the dynamic capability theory proposes that green entrepreneurship seeks to create green value-added while considering environmental concerns through the deployment of green engineering at the process and product level which meets ISO 14000 requirements, thus responding to human environmental concerns.

The following set of hypotheses can be asserted:

H3: The new venture GIP has a significant positive impact on its GEO.
H4: The new venture GIP has a significant positive impact on its EP.
H5: Managerial environmental concerns moderate the relationship between GEO and EP.

From another perspective prior studies have highlighted the significant effects of GIP on green entrepreneurship (Lin & Chen, 2018) and environmental performance

(Albort-Morant et al., 2018). Top management is the main center of decision making related to green innovation practices that allow responding to increased environmental concerns, either societal, governmental, or ecological ones. Notably, it has been proved that the efforts sustained by managers and their commitment toward environmental degradation issues can be explained by their awareness of GIP as a strategic solution for advancing businesses, ensuring the survival of the industry through deploying a green organizational culture and mindset of employees (El-Kassar & Singh, 2019).

In addition, managers of production and environmental units play a significant role in applying green chemistry and green engineering as required for green innovation production processes. According to Bhupendra and Sangle (2015), this green engagement enables them to reduce CO_2 emissions and hazardous waste, along with the pollution of groundwater, and then to improve environmental performance. Hence, the current study posits that managerial environmental concerns contribute to the successful implementation of green practices that ensure green business growth and maintain environmental performance (Xue, Boadu, & Xie, 2019). Previous researches (i.e., Ar, 2012; Saudi, Obsatar Sinaga, & Zainudin, 2019; Tang et al., 2018) underlined that MEC has a significant association with environmental innovation strategies.

The literature on entrepreneurship and green innovation has shown a lack of studies addressing the relationships between MECs, GIP, and sustainable environmental performance (SEP). Two research works have been found which seem to have examined the links between MECs, GEO, and EP (Saudi et al., 2019; Tang et al., 2018). In such studies, MECs refer to top management's thoughts that foster green innovation capabilities and green production systems and processes in order to overcome environmental constraints. Hence, GEO is highly influenced by the extent to which MECs pay more attention to green activities (e.g., green employees behavior, green training, green workplaces, green engineering, green chemistry, green strategies) to green the overall corporate strategy of the venture/firm. In other words, MECs put the emphasis on responding to customer environmental awareness along with other contextual pressures (exerted by authorities, suppliers, society, and stakeholders). MECs then support and maintain the new venture's environmental performance.

For more precision, three empirical studies, conducted respectively by Tang et al. (2018), Saudi et al. (2019), and Xue et al. (2019), pointed out that MECs moderate the relationship between green products' innovation and firm performance or EP. The lack of research dedicated to the moderating effect of MECs on the association between GEO and EP is also notable, along with inconclusive past results. Therefore, the current study aims to address the moderating effects of MECs on the relationship between GEO and EP. Top managers are called upon to interpret the ecological concerns into environmental innovation strategies in order to ensure the economic sustainability and EP.

Thus, the two following testable hypotheses can be proposed:

H6: MECs have a significant positive influence on the new venture's EP.

H7: MECs positively moderate the relationship between GEO and the new venture's EP.

3.3 RESEARCH METHODOLOGY

3.3.1 SAMPLING, TARGETED INTERVIEWEES, AND COLLECTION PROCEDURE

The current research seeks to address the effects of managerial concerns and innovation-oriented practices on EP. The theoretical model, presented in Figure 3.1, illustrates the different relationships between the different concepts of our framework, taking into account the different testable hypotheses proposed above in the literature review.

Therefore, selecting the appropriate respondents is known to be an important step for providing accurate and valuable data to test the specific relationships between all the corresponding variables of the research model.

To carry out the empirical investigation, the present study was conducted on top managers (e.g., managers of the environment, marketing, and production departments) and entrepreneurs of large Chinese petrochemical firms located in Guangdong (China). To be more precise, the focus was only on Guangdong province, where 586 large petrochemical firms exist. The first sample size of this study included 330 entrepreneurs and top managers working in the technology and science departments. However, out of 330 distributed questionnaires (in a dual language, along with a cover letter explaining the importance of the study and its main ethical/scientific goals), only 234 completed surveys were gathered with complete data, representing a response rate of 71%.

As stated by Krejcie and Morgan (1970) and Israel (2003), quantitative studies can be characterized by low response rates, caused mainly by uncompleted and missing responses given by interviewees.

In order to examine the content validity and reliability of items of the different constructs used in the current research, we invited four academic experts from the School of Management of the University of Guizhou, China, and two managers working in petrochemical firms.

The final draft of the questionnaire was then made after taking into account the feedback and comments emanating from those professionals and academic experts. Thereafter, 18 managers operating in petrochemical firms were also asked to participate in a pilot study. The obtained results indicated an adequate internal consistency and an acceptable composite reliability of all the constructs, ranging from 0.782 to

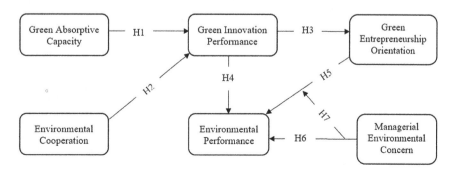

FIGURE 3.1 Theoretical framework

0.876. The final version of the survey was established before being distributed to the respondents. The constructs were measured using a seven-point Likert type scale, ranging from "1" representing "strongly disagree" to "7" representing "strongly agree".

3.3.2 MEASURES OF VARIABLES

This study reviewed and analyzed the past literature to cover the issue of environmental entrepreneurs and environmental sustainability. Thus it took past research and adopted the measurements there so as to be suitable in the context of Chinese petrochemical firms in order to measure the constructs of the research model.

In order to measure GAC, we referred to the works of Pacheco et al. (2018) and obtained a ten-item scale appropriate to our study. However, items measuring MECs were adopted from the research undertaken by Eiadat, Kelly, Roche, and Eyadat (2008) and Xue et al. (2019), leading to a four-item scale. For the EC, we adopted the items from the writings of Vachon and Klassen (2006) and Younis et al. (2016) to

TABLE 3.1
Respondents' profiles and companies' backgrounds

Respondents' profiles	Frequency	Percentage
Education		
Degree	97	41.4
Master's	87	37.1
PhD	23	9.82
Others	27	11.5
Years of experience		
1 to 5 years	42	17.98
5 to 10 years	116	49.57
More than 10 years	76	32.47
Firm's age		
1 to 5 years	23	9.82
6 to 10 years	53	22.64
More than 10 years	158	67.52
Ownership structure		
Private firms	135	57.69
Collective and state-owned firms	65	27.78
Foreign firms	36	15.38
Position		
Production manager	42	17.95
Executive director	46	19.65
Engineer	27	11.53
Director of the environment	37	15.81
R&D director	19	8.11
CEO	48	20.51
Other	13	5.55
Number of employees		
Less than 100	49	20.94
100 to 250	69	29.48
251 to 500	82	35.04
More than 500	34	14.52

obtain an eight-item scale. To operationalize GIP, we also employed the scales developed by Chen, Lai, and Wen (2006) and Pacheco et al. (2018), yielding a ten-indicator scale. However, to assess GEO, items were extracted from the proposed scales of Covin and Miller (2014) and Jiang et al. (2018).

For this research used the six more precision, items proposed by those scholars to operationalize the GEO. Finally, items chosen for measuring ESP were adopted from the studies of El-Kassar and Singh (2019) and Jiang et al. (2018).

3.3.3 FIRMS' CHARACTERISTICS AND RESPONDENTS' PROFILES

Upper-managerial executives and CEOs of petrochemical firms represent the targeted population and the interviewees of the current study. Their respective profiles are described in Table 3.1.

As shown in Table 3.1, the respondents are mostly well educated: 41.4% of them hold a bachelor's degree, whereas 47% of them obtained postgraduate degrees and have gained considerable experience in the industry sector. Such percentages support the proposition that intellectual capital plays a critical role by spreading a green organizational culture in line with the corporate strategy of the venture. In addition, the descriptive analysis supported the evidence that the correspondent firms have built a consolidated experience in the petrochemical industry, reflecting the major role of this field in improving a Chinese economic growth.

3.4 DATA ANALYSIS

The present study was based on the partial least squares (PLS) analysis to assess three main things: (1) the convergent and the discriminant validity (by the measurement model) (Henseler, Ringle, & Sarstedt, 2015); (2) the path coefficients and the effect size of all the latent constructs; and (3) the R^2 value and the predictive relevance Q^2 of the model (by using the blindfolding procedure) (as supported by Hair Jr. et al., 2014).

This research employed an independent samples t-test to check whether gathered data were free of non-response bias or not. It was found that the Levene test for equality of variance was greater than 0.05 for all the variables, demonstrating that this study is free of non-response bias. Hence, the significance level requirement was attained (Pallant, 2011).

In addition, the multi-collinearity test was applied to determine whether data were free of common method variance (CMV) or not. Results showed that the values of the variance inflation factors (VIFs) were below 3.3, corroborating that the study is free of CMV (Kock, 2015).

3.4.1 MEASUREMENT/OUTER MODEL (CONSTRUCT VALIDITY AND RELIABILITY) ASSESSMENT

The outer model of this current study was assessed by measuring the factor loadings of each construct, its composite reliability (CR), as well as its average variance extraction (AVE). As presented in Table 3.2, for each construct, factor loadings (corresponding to the different items), which have values greater than 0.707, were

TABLE 3.2
Measurement model results

Constructs	Items	Loadings	Cronbach's alpha	CR	AVE
Environmental cooperation	EC1	0.75	0.93	0.94	0.67
	EC 2	0.79			
	EC 3	0.84			
	EC 4	0.90			
	EC 5	0.71			
	EC 6	–			
	EC 7	0.86			
	EC 8	0.80			
Green absorptive capacity	GAC1	0.72	0.92	0.94	0.70
	GAC 2	–			
	GAC 3	–			
	GAC 4	0.75			
	GAC 5	0.82			
	GAC 6	0.92			
	GAC 7	–			
	GAC 8	0.90			
	GAC 9	0.87			
	GAC 10	0.82			
Managerial environment concern	MEC 1	0.84	0.87	0.90	0.70
	MEC 2	0.82			
	MEC 3	0.83			
	MEC 4	0.85			
Green innovation performance	GIP 1	0.64	0.93	0.94	0.64
	GIP 2	0.84			
	GIP 3	0.83			
	GIP 4	0.85			
	GIP 5	–			
	GIP 6	0.87			
	GIP 7	0.82			
	GIP 8	–			
	GIP 9	0.76			
	GIP 10	0.73			
Green entrepreneurial orientation	GEO 1	0.82	0.85	0.89	0.63
	GEO 2	0.82			
	GEO 3	0.80			
	GEO 4	0.79			
	GEO 5	0.73			
Environmental performance	EP 1	0.86	0.85	0.91	0.65
	EP 2	0.88			
	EP 3	0.87			
	EP 4	0.75			
	EP 5	0.77			
	EP 6	–			

retained. However, those ones showing values less than 0.707 were excluded, in accordance with Hair Jr. et al.'s (2014) statements. The CR related to all the factors is higher than 0.7, the threshold value proposed by Chin (1998). Moreover, all the obtained values of the AVE are higher than 0.5, the threshold value suggested by Hair Jr. et al. (2017).

TABLE 3.3
Fornell–Larcker criterion and the heterotrait–monotrait ratio

	EC	EP	GAC	GEO	GIP	MEC
EC	.82					
EP	.73(.74)	.80				
GAC	.78(.68)	.767(.76)	.83			
GEO	.704(.81)	.684(.79)	.609(.82)	.79		
GIP	.667(.70)	.604(.746)	.637(.68)	.712(.80)	.84	
MEC	.423(.47)	.367(.41)	.434(.47)	.39(.43)	.323(.32)	.78

The discriminant validity is another compulsory test related to the measurement model, which refers to the extent to which the constructs differ from one another (Zhu et al., 2006; cited by Omrane, 2015). For the PLS approach, the discriminant validity requires three key criteria, namely (1) the Fornell and Larcker criterion, (2) cross-loadings, and (3) the heterotrait–monotrait (HTMT) ratio (Hair Jr. et al., 2017). The Fornell and Larcker criterion is generally assessed by conducting a comparison of the absolute value of the correlation of inter-constructs of the whole research model with the square root of the AVEs for the same factors. As presented in Table 3.3, related to the correlation matrix, the absolute values of the inter-construct correlation (off-diagonal elements) were less weak than the square root of the AVEs of the diagonal elements. The first criterion is then well respected. The criterion of the cross-loadings was examined by checking whether factor loading for each construct exceed the cross-loadings for all other constructs. Similarly, the obtained results make it possible to validate the second criterion.

The last important test, called the HTMT ratio, corresponding to the third criterion and recommended by Henseler et al. (2015), suggests that the discriminant validity may not be proved by considering only cross-loadings and the Fornell and Larcker criterion. Hence, the HTMT ratio was measured for the current research. As stated in Table 3.3, all values presented in parentheses are less weak than/or equal to 0.85, the threshold value required for this HTMT criterion (Kline et al., 2012). Moreover, HTMT inference results revealed that the confidence interval did not show any value of 1 for the latent variables of this study (Henseler et al., 2015). Consequently, all the criteria of the discriminant validity are respected for the present research.

3.4.2 STRUCTURAL MODEL AND PLS STATISTICS

Results of the structural model indicated whether the seven proposed hypotheses were confirmed. Table 3.4 and Figure 3.2 show that both GAC and EC have a positive impact on GIP as the path coefficients' values were significant ($B = 0.149$, t = 3.19, p < 0.001 and $B = 0.479$, t = 10.96, p < 0.001). Hypotheses H1 and H2 are then supported.

In addition, findings revealed that GIP influences positively both GEO and EP, as the path coefficients' values were significant ($B = 0.123$, t = 5.89, p < 0.001 and SPE $B = 0.233$, t = 4.86, p < 0.001). Similarly, H3 and H4 are also confirmed.

TABLE 3.4
Results of structural model analysis

Hypotheses	Tested relationships	Beta	t-value 1-tailed	P-value	f^2	Decision
H1	GAC -> GIP	0.149	3.191	0.000	0.231	Supported
H2	EC -> GIP	0.479	10.968	0.000	0.309	Supported
H3	GIP -> GEO	0.123	5.892	0.000	0.160	Supported
H4	GIP -> ESP	0.233	4.867	0.000	0.263	Supported
H5	GEO -> ESP	0.418	8.532	0.000	0.374	Supported
H6	MEC -> ESP	0.191	3.932	0.000	0.318	Supported
H7	GEO*MEC -> ESP	0.179	1.891	0.000	0.279	Supported

Notes: EC: environmental cooperation; GAC: green absorptive capacity; MEC: managerial environmental concern; GIP: green innovation performance; GEO, green entrepreneurship orientation; EP: environmental performance.

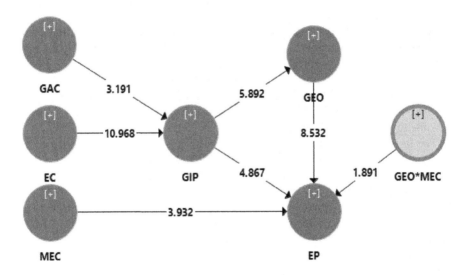

FIGURE 3.2   Structural model results

On the other hand, the results presented in Table 3.4 show that GEO has also a positive and significant effect on EP, given the fact that the corresponding path coefficients' values were significant ($B = 0.418$, t = 8.53, p < 0.001). Thus, H5 was confirmed. The effect of MECs on EP was also positive and statistically significant ($B = 0.191$, t = 3.93, p < 0.001), leading us to confirm hypothesis H6.

Finally, the moderating effect of MEC on the relationship between GEO and EP was statistically significant (as $B = 0.179$, t = 1.89). For this reason, the hypothesis H7 is supported.

3.4.3 EFFECT SIZE TESTING

For the PLS approach, the robustness of a structural model is checked by assessing the difference between the square multiple correlations, via the effect size related to every single endogenous construct on exogenous ones (Hair Jr. et al., 2014). As illustrated in Table 3.4, the effect size of all constructs is moderate for all interactions (between variables), except for GEO which has a substantial effect size on EP (Cohen, 1988). Furthermore, GAC and EC show a strong contribution, explaining the high-value variance of the coefficient of determination–R^2 (26%) respectively for GIP, GEO (24%), and EP (41%). This implies that the tested relationships are reliable and justified.

3.4.4 MODERATING EFFECT TESTING

In order to test the proposed hypothesis H7, the product indicator approach (Henseler & Fassott, 2010) was applied via PLS software (version 3.0). The moderating effect of MECs on the relationship between GEO and EP was assessed by examining the interaction effect 'MEC*GEO' on the one hand, and EP on the other hand. (Please see Figure 3.3).

Table 3.4, as well as Figure 3.2, and Figure 3.3 show that the interaction effect (MEC*GEO) has a t-value of 1.89, indicating a positive relationship between MEC and GEO ($B = 0.179$). Such results provide several lines of evidence related to the

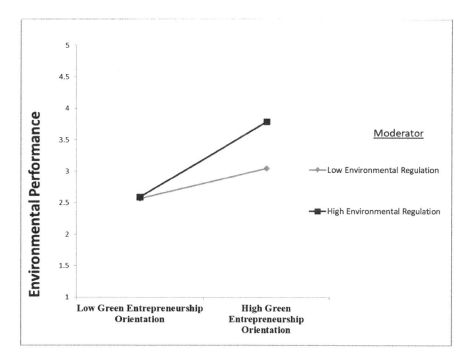

FIGURE 3.3 The moderating effect of MECs on the relationship between GEO and EP

TABLE 3.5
Variance explained and predictive relevance

Constructs	R^2	Q^2
Green innovation performance	0.26	0.34
Green entrepreneurship orientation	0.24	0.25
Sustainable environmental performance	0.41	0.43

hypothesis H7, suggesting that entrepreneurs' orientation about entrepreneurs' environmental concerns moderates the relationship between their GEO and their EP. Figure 3.2 confirms that this moderating effect is both positive and strong.

To assess the predictive relevance Q^2 of the structural model, PLS (version3.0) proponents (e.g., Geisser, 1975) suggest recourse to blindfolding procedures. According to Geisser (1975), values which are greater than "0" indicate that the model is relevant. In addition, the more the Q^2 value is near to "1", the higher the model relevance and strength is (Chin, 1998). As shown in Table 3.5, Q^2 values of GIP, GEO, and EP are respectively 0.34, 0.25, and 0.43, proving that the model is relevant.

3.5 DISCUSSION OF RESULTS

This study mainly addresses the critical role of GIP on achieving superior EP through GEO. Indeed, we posited that entrepreneurs, who are highly influenced by GIP outcomes, will in turn lead to control the balance between economic benefits and ecological constraints. We also considered that examining these three fundamental building blocks of sustainability may allow us to understand in depth the extent to which green innovation practices act as an enabler of GEO and an intermediary factor between GEO and EP. Entrepreneurs' proactive decision making is profoundly impacted by the results of a green corporate strategy focused on responding to the ecological and socioeconomic concerns about environmental degradation. Therefore, emphasizing and testing the relationship between these three key levers of sustainability in a single model is very relevant. It provides significant evidence about the central role of GIP as it may empower entrepreneurs to explore business opportunities and exploit green capabilities in order to extend their new ventures' market share and green value-added.

From another perspective, the dynamic capability theory viewed GAC as a critical driver, which helps entrepreneurs to adapt their ventures to the highly fluctuated environment and to ensure a superior business performance. Meanwhile, GAC acts as an intermediary factor that enforces green innovation practices to achieve green entrepreneurship goals (Jiang et al., 2018). In addition, EC enhances new ventures' internal green capabilities through fostering cooperation with several external partners in order to support the overall green strategy (Hájek & Stejskal, 2018). The sustainable entrepreneurship theory argues that entrepreneurs seeking ecological concerns may explore the market failure opportunities thanks to their AC and can exploit them by their internal capabilities.

This study demonstrates then the essential role of AC in improving GIP and enhancing environmental sustainability in order to capture market opportunities.

The results of this research support all the testable relationships between GIP, GEO, and EP, along with the moderating effect of MEC. They are consistent with previous empirical studies (i.e., Guo, Wang, & Chen, 2020; Guo, Wang, & Xie, 2018; Y.-H. Lin & Chen, 2018), which have explored the specific determinants of GIP, such as GEO and EC.

However, the present study brings up novelty by testing the effects of GIP, as an enabler and an intermediary variable between GEO and EP. It confirms the hypotheses suggesting that entrepreneurs are seeking to develop their market shares based on environmental and ecological sustainability. Hence, the current work extends the existing body of knowledge by providing empirical evidence about the need to foster new ventures' GIP by capturing and exploiting market failure opportunities, in line with responding to ecological sustainability concerns.

On the other hand, the significant impact of GAC on GIP indicates that entrepreneurs' AC plays a strategic role in fostering new ventures' green practices through the integration of new knowledge with the actual one, yielding a collective organizational learning. Such a dynamic process contributes to reconfiguring business processes and production systems, a clean environment, and then eventually to improving ecological sustainability. Green entrepreneurship is thus improved via green practices, which in turn enable firms to care about ecological concerns.

The findings reveal also the crucial role of GAC in supporting entrepreneurs' actions toward exploring market failure opportunities (the failure of responding to ecological concerns) and translating them into internal resources to fit market gaps. In fact, the dynamic capability theory advances that GAC is a main driving force for business growth, whereas the sustainable entrepreneurship theory underlines that GIP enables entrepreneurs to capture opportunities based on environmental and ecological aspects via green practices.

Furthermore, this study support that EC has a substantial effect on GIP, which means that collaboration with external partners based on environmental innovation is highly associated with green practices and organizational learning. It implies that petrochemical firms are paying much attention to collaboration with external partners such as university laboratories, data interchange, consultants, and government facilities (e.g., R&D funding, tax preferences), which in turn leads them to restructure, reconfigure, and reengineer the business structure and production systems, as well as marketing tools. The responsiveness of new ventures to environmental, societal, and governmental conditions will be enhanced, ensuring survival and advancement of businesses.

This study provides also a significant evidence for top managers and policymakers to achieve and support collaboration within and between different partners. It proves that inter-organizational relationships contribute to improving the green business growth of several industries, which in turn may create business networks, supporting economic and ecological sustainability.

As pointed out by Hájek and Stejskal (2018), it is recommended that entrepreneurs establish a close network with other partners in order to improve the performance of the green practices they engage in, leading them to adjust to customer environmental awareness, and governmental, social, and ecological pressures. Likewise, Higgins and Yarahmadi (2014) argued that cooperation based on environmental innovation (EC) fosters GIP, and generates a superior EP. The same scholars added that EC could emanate from governmental structures (tax preferences and funds, subsidies investment in clean technologies), private and public laboratories (extending the knowledge-base through R&D, patents ownership), inter-organizations (suppliers are required to supply green raw materials), and/or customers (eco-friendly products).

Additionally, results show that MECs have a positive impact on GIP, indicating that managers and executives of petrochemical firms are well educated and responsible about environmental concerns related to their production systems. They make efforts and time for deploying green technologies and controlling production systems across all business units. Their social responsibility may in turn lead to improving the workplace environment, reducing hazardous waste, and thereby impeding further CO_2 emissions in order to respond to ecological concerns and customer eco-friendly expectations. Consequently, enabling entrepreneurs to invest more in clean technologies to implement green production systems and to enhance resource efficiencies and reduce production costs will ultimately yield many competitive advantages.

Results obtained in this study corroborate the prior recent researches undertaken by Saudi et al. (2019) and Tang et al. (2018), by testing the moderating effect of MECs on the relationship between GEO and EP. MECs are one of the primary sources for formulating a corporate strategy based on developing: (1) a green workplace environment, (2) green training for employees, and (3) a green organizational culture to establish a green atmosphere, which would influence the mindset of the workforce toward ecological concerns.

This study is the first that explores and tests the moderating effect of MECs on the relationship between GEO and EP, indicating that green entrepreneurs' actions toward exploring market failure opportunities are enhanced via the support of MECs about ecological sustainability. This means that the more entrepreneurs are engaged in green plans and activities, the greater they will influence their new ventures' EP, by investing in those emerging opportunities. They may not only extend their market share, but also advance their business growth in line with ecological problems.

3.6 THEORETICAL AND PRACTICAL RESEARCH IMPLICATIONS

The sustainable entrepreneurship theory advanced is that entrepreneurs can succeed in enhancing their EP based on GIP, by responding to ecological concerns and exploiting market failure opportunities.

The dynamic capability theory views GAC and GEO as the key strategic solutions for fostering GIP. GIP capabilities also empower entrepreneurs to capture opportunities that suit ecological and organizational concerns, improving in turn their EP.

Thus, it appears that this study extends the sustainable entrepreneurship theory by explaining the need for better green entrepreneurial capabilities. Indeed, by integrating those two theories, this research explores the main factors that lead to EP based on GIP. This suggests that petrochemical entrepreneurs should respond to ecological concerns to avoid any unexpected market failure due to increased external pressures and competition. Their GAC enables them to adapt to the fluctuating environment and to absorb external pressures, by switching organizational combined knowledge into effective green business practices.

The results of this study suggest also that managers and entrepreneurs engage in actions to absorb and convert the explored threats into opportunities through updating, restricting, and reengineering business structure and production systems to suit environmental, societal, governmental, and ecological constraints. They are called upon to focus on environmental aspects and to strike a balance between the exploitation of emerging opportunities (market failure ones) based on environmental aspects and the adoption of other organizational and environmental pressures (e.g., customers, suppliers, governments, stakeholders). Consequently, applying the two theories stated above would allow us to explain the importance of GAC, EC, and GIP in enabling entrepreneurs to capture market failure opportunities and exploit them through their green practice capabilities (toward ecological sustainability issues). Indeed, entrepreneurs are well able to explore and gain benefit from market failure opportunities if they can manage, assimilate, and integrate their newly obtained knowledge with existing knowledge to formulate a new strategy (e.g., mobilizing resources, spreading a green culture, training) that can match environmental concerns. Entrepreneurs could also create the know-how, strategic resources, and practices which become inimitable resources (for competitors).

Furthermore, the findings of the current study prove that cooperation with external partners is beneficial for managers/entrepreneurs as it empowers them to understand and discover external threats and opportunities through inter-organizational learning. EC sensitizes entrepreneurs toward customer preferences, suppliers' information, governmental regulations and facilities, laboratories, advanced production systems, and so on., leading them to formulate good plans and actions, and to prepare for any unexpected business changes. The study suggests thus that managerial concerns about environmental degradation significantly play a considerable role in balancing up entrepreneurs' goals and ecological sustainability.

To conclude, it will be important to underline that the present study is the first empirical research that addresses the moderating impact of MECs on the relationship between GEO and EP. It provides some evidence that MECs strengthen green entrepreneurs' initiatives to capture and exploit green opportunities, helping them to ensure business growth in line with ecological sustainability.

3.7 RESEARCH LIMITATIONS AND FURTHER FUTURE ORIENTATIONS

This study has some limitations, which must be highlighted for subsequent research.

First, there are five major provinces with petrochemical firms in China. All of them are affected by pollution caused by this industry. However, this research targeted large petrochemical firms existing in one single province: Guangdong. We suggest that future research replicates the same investigation for the other four provinces, so that findings can be generalized. Second, this research is focused on the branch of industry that causes serious ecological problems, while other industries may also have a substantial impact on environmental sustainability.

Third, this study applied a cross-sectional survey to examine the proposed hypotheses, which means that we were unable to observe the issue over time to judge their reliability. By the way, examining this model through a longitudinal survey could provide strong evidence concerning the progress of GIP and EP. It may enable researchers to confirm the associations between different constructs of the model.

Fourth, this research proposes the testing and exploring of the effects of GIP (by considering its triggers) and MECs on EP. It is then recommended that future studies take into consideration other individual, organizational, and environmental factors to explore their influence on the predictive relevance of the present investigation.

Finally, the proposed model can also be tested in other different contexts in order to assess the effects of cultural and regulatory discrepancies.

REFERENCES

Albort-Morant, G., Leal-Millán, A., & Cepeda-Carrión, G. (2016). The antecedents of green innovation performance: A model of learning and capabilities. *Journal of Business Research*, 69(11), 4912–4917.

Albort-Morant, G., Leal-Rodríguez, A. L., & De Marchi, V. (2018). Absorptive capacity and relationship learning mechanisms as complementary drivers of green innovation performance. *Journal of Knowledge Management*, 22(2), 432–452.

Alves, M. F. R., Salvini, J. T. S., Bansi, A. C., Neto, E. G., & Galina, S. V. R. (2016). Does the size matter for dynamics capabilities? A study on absorptive capacity. *Journal of Technology Management & Innovation*, 11(3), 84–93.

Ar, I. M. (2012). The impact of green product innovation on firm performance and competitive capability: The moderating role of managerial environmental concern. *Procedia-Social and Behavioral Sciences*, 62, 854–864.

Bhupendra, K. V., & Sangle, S. (2015). What drives successful implementation of pollution prevention and cleaner technology strategy? The role of innovative capability. *Journal of Environmental Management*, 155, 184–192.

Cao, H., & Chen, Z. (2019). The driving effect of internal and external environment on green innovation strategy: The moderating role of top management's environmental awareness. *Nankai Business Review International*, 10(3), 342–361.

Chen, X., Yi, N., Zhang, L., & Li, D. (2018). Does institutional pressure foster corporate green innovation? Evidence from China's top 100 companies. *Journal of Cleaner Production*, 188, 304–311.

Chen, Y.-S., & Chang, C.-H. (2013). The determinants of green product development performance: Green dynamic capabilities, green transformational leadership, and green creativity. *Journal of Business Ethics*, 116(1), 107–119.

Chen, Y.-S., Lai, S.-B., & Wen, C.-T. (2006). The influence of green innovation performance on corporate advantage in Taiwan. *Journal of Business Ethics*, 67(4), 331–339.

Chin, W. W. (1998). The partial least squares approach to structural equation modeling. *Modern Methods for Business Research*, 295(2), 295–336.

Cohen, J. (1988). *Statistical power analysis for the behavioral sciences*. 2nd ed.: Hillsdale, NJ: Erlbaum.

Cohen, W. M., & Levinthal, D. A. (1990). Absorptive capacity: A new perspective on learning and innovation. *Administrative Science Quarterly*, 35, 128–152.

Covin, J. G., & Miller, D. (2014). International entrepreneurial orientation: Conceptual considerations, research themes, measurement issues, and future research directions. *Entrepreneurship Theory and Practice*, 38(1), 11–44.

Demirel, P., Li, Q. C., Rentocchini, F., & Tamvada, J. P. (2019). Born to be green: New insights into the economics and management of green entrepreneurship. *Small Business Economics*, 52(4), 759–771.

Diabat, A., Khodaverdi, R., & Olfat, L. (2013). An exploration of green supply chain practices and performances in an automotive industry. *The International Journal of Advanced Manufacturing Technology*, 68(1–4), 949–961.

Eiadat, Y., Kelly, A., Roche, F., & Eyadat, H. (2008). Green and competitive? An empirical test of the mediating role of environmental innovation strategy. *Journal of World Business*, 43(2), 131–145.

El-Kassar, A.-N., & Singh, S. K. (2019). Green innovation and organizational performance: The influence of big data and the moderating role of management commitment and HR practices. *Technological Forecasting and Social Change*, 144, 483–498.

Feng, Z., & Chen, W. (2018). Environmental regulation, green innovation, and industrial green development: An empirical analysis based on the Spatial Durbin model. *Sustainability*, 10(1), 223.

Gast, J., Gundolf, K., & Cesinger, B. (2017). Doing business in a green way: A systematic review of the ecological sustainability entrepreneurship literature and future research directions. *Journal of Cleaner Production*, 147, 44–56.

Geisser, S. (1975). A new approach to the fundamental problem of applied statistics. *Sankhyā: The Indian Journal of Statistics, Series B*, 385–397.

Gibbs, D., & O'Neill, K. (2014). Rethinking sociotechnical transitions and green entrepreneurship: The potential for transformative change in the green building sector. *Environment and Planning A*, 46(5), 1088–1107.

Gluch, P., Gustafsson, M., & Thuvander, L. (2009). An absorptive capacity model for green innovation and performance in the construction industry. *Construction Management and Economics*, 27(5), 451–464.

Guo, Y., Wang, L., & Chen, Y. (2020). Green entrepreneurial orientation and green innovation: The mediating effect of supply chain learning. *SAGE Open*, 10(1), 1–13.

Guo, Y., Wang, L., & Xie, Y. (2018). Green Innovation, Green Entrepreneurial Orientation and Supply Chain Learning: Evidence from Manufacturing Firms in China. Manufacturing Firms in China. MDPI. Preprints 2018, 2018050232. Switzerland.

Haibo, C., Ayamba, E. C., Agyemang, A. O., Afriyie, S. O., & Anaba, A. O. (2019). Economic development and environmental sustainability: The case of foreign direct investment effect on environmental pollution in China. *Environmental Science and Pollution Research*, 26(7), 7228–7242.

Hair Jr., J. F., Sarstedt, M., Ringle, C. M., & Gudergan, S. P. (2017). *Advanced issues in partial least squares structural equation modeling*. SAGE Publications.

Hair Jr., J. F., Sarstedt, M. Hopkins, Kuppelwieser, V. (2014). Partial least squares structural equation modeling (PLS-SEM): An emerging tool in business research. *European Business Review*, 26(2), 106–121.

Hájek, P., & Stejskal, J. (2018). R&D cooperation and knowledge spillover effects for sustainable business innovation in the chemical industry. *Sustainability*, 10(4), 1064.

Henseler, J., & Fassott, G. (2010). Testing moderating effects in PLS path models: An illustration of available procedures. *Handbook of partial least squares* (pp. 713–735). Berlin: Springer.

Henseler, J., Ringle, C. M., & Sarstedt, M. (2015). A new criterion for assessing discriminant validity in variance-based structural equation modeling. *Journal of the Academy of Marketing Science*, 43(1), 115–135.

Higgins, P. G., & Yarahmadi, M. (2014). *Cooperation as a driver of development and diffusion of environmental innovation*. Paper presented at the *IFIP International Conference on Advances in Production Management Systems*. (pp. 374–381). Berlin, Heidelberg: Springer.

Huang, X.-X., Hu, Z.-P., Liu, C.-S., Yu, D.-J., & Yu, L.-F. (2016). The relationships between regulatory and customer pressure, green organizational responses, and green innovation performance. *Journal of Cleaner Production*, 112, 3423–3433.

Israel, G. D. (2003). *Determining Sample Size.* Program Evaluation and Organizational Development, IFAS, University of Florida. PEOD-5: October.

Jiang, W., Chai, H., Shao, J., & Feng, T. (2018). Green entrepreneurial orientation for enhancing firm performance: A dynamic capability perspective. *Journal of Cleaner Production*, 198, 1311–1323.

Jin, W., Zhang, H.-Q., Liu, S.-S., & Zhang, H.-B. (2019). Technological innovation, environmental regulation, and green total factor efficiency of industrial water resources. *Journal of Cleaner Production*, 211, 61–69.

Kirkwood, J., & Walton, S. (2014). How green is green? Ecopreneurs balancing environmental concerns and business goals. *Australasian Journal of Environmental Management*, 21(1), 37–51.

Kline, E., Wilson, C., Ereshefsky, S., Tsuji, T., Schiffman, J., Pitts, S., & Reeves, G. (2012). Convergent and discriminant validity of attenuated psychosis screening tools. *Schizophrenia Research*, 134(1), 49–53.

Kock, N. (2015). Common method bias in PLS-SEM: A full collinearity assessment approach. *International Journal of e-Collaboration (IJeC)*, 11(4), 1–10.

Kong, T., Feng, T., & Ye, C. (2016). Advanced manufacturing technologies and green innovation: The role of internal environmental collaboration. *Sustainability*, 8(10), 1056.

Krejcie, R. V., & Morgan, D. W. (1970). Determining sample size for research activities. *Educational and Psychological Measurement*, 30(3), 607–610.

Lin, H., Zeng, S., Ma, H., Qi, G., & Tam, V. W. (2014). Can political capital drive corporate green innovation? Lessons from China. *Journal of Cleaner Production*, 64, 63–72.

Lin, Y.-H., & Chen, H.-C. (2018). Critical factors for enhanceing green service innovation. *Journal of Hospitality and Tourism Technology*, 9(2), 188–203.

Matus, K. J., Xiao, X., & Zimmerman, J. B. (2012). Green chemistry and green engineering in China: Drivers, policies and barriers to innovation. *Journal of Cleaner Production*, 32, 193–203.

Melander, L. (2017). Achieving sustainable development by collaborating in green product innovation. *Business Strategy and the Environment*, 26(8), 1095–1109.

Omrane, A. (2015). Entrepreneurs' social capital and access to external resources: The effects of social skills. *International Journal of Entrepreneurship and Small Business*, 24(3), 357–382.

Pacheco, L. M., Alves, M. F. R., & Liboni, L. B. (2018). Green absorptive capacity: A mediation-moderation model of knowledge for innovation. *Business Strategy and the Environment*, 27(8), 1502–1513.

Pallant, J. (2011). *SPSS Survival manual: A step by step guide to data analysis using SPSS.* New South Wales: Allen & Unwin.

Perotti, S., Zorzini, M., Cagno, E., & Micheli, G. J. (2012). Green supply chain practices and company performance: The case of 3PLs in Italy. *International Journal of Physical Distribution and Logistics Management*, 42(7), 640–672.

Qi, G., Shen, L. Y., Zeng, S., & Jorge, O. J. (2010). The drivers for contractors' green innovation: An industry perspective. *Journal of Cleaner Production*, 18(14), 1358–1365.

Saudi, M. H. M., Obsatar Sinaga, G., & Zainudin, Z. (2019). The effect of green innovation in influencing sustainable performance: Moderating role of managerial environmental concern. *International Journal of Supply Chain Management*, 8(1), 303.

Saunila, M., Ukko, J., & Rantala, T. (2018). Sustainability as a driver of green innovation investment and exploitation. *Journal of Cleaner Production*, 179, 631–641.

Shirokova, G., & Ivvonen, L. (2016). Performance of Russian SMEs during the economic crisis: The role of strategic entrepreneurship. Working Paper. Saint Petersburg University, Russia.

Tang, M., Walsh, G., Lerner, D., Fitza, M. A., & Li, Q. (2018). Green innovation, managerial concern and firm performance: An empirical study. *Business Strategy and the Environment*, 27(1), 39–51.

Tariq, A., Badir, Y. F., Tariq, W., & Bhutta, U. S. (2017). Drivers and consequences of green product and process innovation: A systematic review, conceptual framework, and future outlook. *Technology in Society*, 51, 8–23.

Vachon, S., & Klassen, R. D. (2006). Extending green practices across the supply chain: The impact of upstream and downstream integration. *International Journal of Operations & Production Management*, 26(7), 795–821.

Weber, B., & Heidenreich, S. (2018). When and with whom to cooperate? Investigating effects of cooperation stage and type on innovation capabilities and success. *Long Range Planning*, 51(2), 334–350.

Xue, M., Boadu, F., & Xie, Y. (2019). The Penetration of Green Innovation on Firm Performance: Effects of Absorptive Capacity and Managerial Environmental Concern. *Sustainability*, 11(9), 2455.

Younis, H., Sundarakani, B., & Vel, P. (2016). The impact of implementing green supply chain management practices on corporate performance. *Competitiveness Review*. 26(3), 216–245.

Zhu, K., Kraemer, K. L., & Xu, S. (2006). The process of innovation assimilation by firms in different countries: a technology diffusion perspective on e-business. *Management Science*, 52(10), 1557–1576.

4 The Promotion of Social Entrepreneurship in Morocco in the Service of Sustainable Development

Mohamed Talbi, Khalid Sadiqi, and Nabil Ouarsafi

CONTENTS

4.1 INTRODUCTION

The economic, social, and environmental development of countries is the ultimate goal sought by decision makers and politicians. This objective is demonstrated in particular by the fight against poverty through the creation of jobs and income generating activities, the reduction of inequalities and exclusion, the accumulation of wealth, the achievement of growth, the protection of the planet and the improvement of the living conditions of the population, and so on (Singh & Reji, 2020). However, the concretization of this objective requires the mobilization of many resources which must be encouraged to adopt sustainable development approaches as well as to strengthen their institutional capacities, by creating solutions and actions likely to meet the envisaged challenges.

From this perspective, social entrepreneurship is a major and strategic issue for meeting economic, social, and environmental challenges that needs to be integrated by a large number of countries in their public development policies; it now represents a movement toward inclusive growth and a privileged means of providing potential responses to the challenges posed by sustainable development (Davis, 2002; Brouard, 2007; Omrane & Fayolle, 2010; Stevens, Moray and Bruneel, 2015 cited in Hidalgo, Rialp, & Urbano, 2020; Layadi, Rouggani, & Bouayad, 2019).

Social entrepreneurship as an entrepreneurial dynamic is increasingly successful in emerging and developed countries. Some statistics, at this level, show that one out of every five enterprises created in developed countries is a social enterprise, and more than 10% of the generated wealth by these countries comes from social enterprises (Injaz Al Maghrib, 2016). A particular importance has been attached to this type of business, as several case studies show that this type of business plays a fundamental role in sustainable development. In this sense Singh and Reji (2020) point out that the potential of the social entrepreneurship approach in creating a sustainable social impact not only inspires hope among social entrepreneurs, but also among professionals in various fields and sectors (non-profit, for-profit, and public), yet traditional and old-fashioned methods of solving deep-rooted societal problems do not offer sustainable solutions. In this context, entrepreneurship is increasingly recognized as a tool to foster sustainable development (ibid.). Strengthening countries' efforts to develop and support this sector is increasingly crucial.

The different conceptions related to this concept, or the importance and advantages of this field in terms of sustainable development, challenged us to ask questions about the relationship between this axis and sustainable development and whether such a development of social entrepreneurship would contribute to sustainable development. It is from this perspective that this chapter was written. We wish to shed light on the questions: What are the links that can exist between social entrepreneurship and sustainable development? And what measures should be undertaken in Morocco to promote social entrepreneurship as a contribution to sustainable development?

Aware of this challenge, Morocco has identified entrepreneurship and social innovation as key areas for the development of the economy and society. This path of development is in line with the royal vision of H.M. King Mohammed VI, insofar as it places social entrepreneurship at the heart of Moroccan society's project, both at the national and continental levels. The king stressed that Morocco's investment is in the entrepreneurial culture, adding that "entrepreneurship and innovation are twin values that are both stepping stones to freedom, social mobility and prosperity, if the business environment allows it and the general conditions are met".[1] At the same time, Morocco is particularly concerned to make sustainable development a priority in its development policy. Indeed, it has laid the foundations for sustainable development in the country through several political, institutional, legal, and socio-economic reforms. This process has been strengthened by the adoption of the National Charter for the Environment and Sustainable Development, the implementation of the National Initiative for Human Development (NIHD), and the launch of the National Strategy for Sustainable Development (NSSD).

We first proceed to a global reading of the phenomenon of social entrepreneurship through a synthetic review of the literature. Focusing in particular on its origins, its emergence, its history, and its different conceptions, a synthesis of definitions from the literature and an attempt to define it will be made. Then we will address the second key concept of this chapter, which is sustainable development in Morocco, by clarifying it and highlighting the strategies adopted by the Moroccan state, in order to then identify the links that may exist between social entrepreneurship and sustainable development. Finally, we will set up a framework of action for the promotion of social entrepreneurship, by carrying out a theoretical examination of the various measures and provisions likely to develop it in a sustainable way in Morocco.

4.2 SOCIAL ENTREPRENEURSHIP: A GLOBAL READING

4.2.1 The Phenomenon of Social Entrepreneurship: Origins and Historical Overview

Social entrepreneurship is a young concept that emerged in the 1990s on both sides of the Atlantic (CODES, 2007). The work of Steyart and Hjorth, 2006, cited in Bennani (2019), points out that research on the development of social entrepreneurship has, only in recent years, been conducted by researchers and experts who were generally outside the field of entrepreneurship. Similarly, many researchers confirm that this theme is generally accepted in practice and that social entrepreneurship is not a new phenomenon, though its language may be new and its formulation relatively recent (Dees, 1998; Barendsen & Gardner, 2004; Janssen, Bacq, & Brouard, 2012).

A review of the literature on this subject shows that the origins and conceptions of this research theme have been dealt with under two approaches (Le Velly, 2014): An American approach that places more emphasis on the figure of the entrepreneur, and a European approach that focuses more on certain specificities of social enterprise (Asli & Slitine, 2013).

As for the Anglo-Saxon literature, the concepts of social entrepreneurship emerged in the United States in the late 1970s, first in the literature on social change, before spreading in the 1980s into media and management literature (Anne-Claire and Thierry, 2011), as the diversity of concepts used at that time served to describe entrepreneurial behavior with social objectives. Similarly, the concept of social enterprise appeared in the American landscape from the 1990s, with the launch in 1993 of the "Social Enterprise Initiative" by Harvard Business School, followed by other major universities (such as Columbia and Yale) and various foundations that set up training and support programs for social entrepreneurs and social enterprises (CODES, 2007; Asli & Slitine, 2013; Ilbert & Ciheam, 2014; Bennani and Ghita, 2019).

While in French-speaking literature, this field of research only emerged in the 1990s (Anne-Claire and Thierry, 2011), mainly under the influence of Italian scholars, who in 1991 introduced the specific status of "social cooperative", the aim was to respond to needs not or only poorly met by public services (CODES, 2007; Asli & Slitine, 2013; Bennani and Ghita, 2019). This research topic took on particular importance in the 2000s (Anne-Claire and Thierry, 2011).

In fact, when digging into the literature on the origins and history of social entrepreneurship, historical success stories and revolutionary examples are cited to advantage by many authors, including those founders considered to be the true predecessors in this field, including a social enterprise called the "Grameen Bank" initiated in 1976 by the inventor Muhammad Yunus, to the extent that the Grameen Microcredit Bank has changed the lives of millions of people by providing financial services to the poor, especially women (Mair & Marti, 2006; Ndour, 2017). It has helped these women to restore the profitability of their activities and to fight poverty (Yunus, 2011). In 2003, this bank had lent $4 billion to 2.8 million Bangladeshi villagers, 95% of whom were women (Bornstein, Taudière, & Clarinard, 2005; Ndour, 2017). This allowed these beneficiaries to "better feed their families, build houses covered with corrugated iron to shelter from the monsoon, send their children to school and save for their old age" (Bornstein et al., 2005; Ndour, 2017). A second success story often cited in the academic landscape is that of the Ashoka Foundation, created by its founder Bill Drayton in 1980, to provide start-up funds for socially minded entrepreneurs (Mair & Marti, 2006). Indeed, it was one of the first organizations explicitly designed to finance social entrepreneurs, whose objective was to provide these entrepreneurs with the means to develop themselves and provide financial resources and a professional network within which they could disseminate their ideas and solutions (Barendsen & Gardner, 2004; Guzmán Vásquez & Trujillo Dávila, 2008).

From the perspective of an analysis of the historical evolution of social entrepreneurship in Morocco, and on the basis of a synthetic reading of the literature, we have tried to provide a historical overview of this theme from two angles. First of all, the focus will be on the social and solidarity economy (SSE) sector, which is the real axis for discussing the subject of social entrepreneurship, since the initiatives in this field of research originated in this sector, very often adopting an associative or cooperative legal form (Diani, n.d.). Similarly, the birth and development of social entrepreneurship is closely linked to the growth of the social economy sector (OECD, 2001). Second, emphasis will be placed on the traditional characteristics and conceptions of Moroccan society, which constitute an essential component for discussing the origins and emergence of social entrepreneurship in that country. Indeed, the SSE sector, which has become established in the Moroccan economic and social landscape over the centuries, has undergone political and social changes in the kingdom. This sector aims at providing answers to society's problems in terms of environmental protection and the fight against exclusion, health, equal opportunities, and so on, notably through social innovation, solidarity savings, fair trade, and microcredit. This branch of the economy includes several structures such as cooperatives, associations, mutual societies, companies, and foundations (MTF-GROUP "LE MATIN", 2018).

Similarly, the modern concept of the social economy was mentioned for the first time at a conference organized in Rabat in 1987 by the department in charge of planning for a session of the international cooperative university (Adrdour et al., 2016). In fact, the context of the emergence of this concept coincides with the adoption of the structural adjustment program as early as 1990, which was necessary to restore overall macroeconomic equilibrium (Adrdour et al., 2016). During this period, there was a gradual disengagement of the state in several economic or social areas and a reduction in the availability of jobs and public services. From the same perspective, the economic and social development plans put in place over the period 1988–1992 have made the SSE an alternative in terms of job creation and resource mobilization.

In the same vein, since cooperatives are considered key structures of the SSE, they contribute significantly to the economic and social development of countries. Thus, since independence in 1956, the cooperative economic model has been a strategic choice for Morocco, with a view to ensuring national mobilization for the modernization and development of traditional sectors, particularly agriculture (Ahrouch, 2011), Similarly, Morocco recognized the importance of cooperatives and associations in 1958, when laws on associations were promulgated as a framework for national providence, mutual aid, and solidarity, and in 1963 (CESE (le Conseil Economique, Sociale et Environnemental), 2015; MTF-GROUP "LE MATIN", 2018). These mutuals are particularly active in the areas of social security, provident insurance, and solidarity. They began to take a structured and organized form in the early 2000s (CESE, 2015). While the historical field of the intervention of the associative sector was mainly concerned with the reduction of inequalities of income and access to basic needs (literacy, microcredit, integration of the handicapped and street children, integration of women, etc.), it has been extended to the socio-economic sphere, with local development and the reduction of infrastructure deficits (electrification, drinking water supply, opening up by the construction of roads, tracks, bridges), which had hitherto been the responsibility of the state.

In the same period, in 1963 to be precise, an administrative structure in charge of supporting cooperatives in the areas of training, information, and legal support was set up under the name of "Development Cooperation Office" (Ahrouch, 2011). This office was restructured in 1975 to become a public enterprise with a legal and financial personality as well as administrative autonomy (Ait Haddout & Jaouad, 2003; Ahrouch, 2011).

From the above, and by highlighting the main initiatives and strategies taken by the Moroccan state, it can be argued that the SSE sector has, since the 1990s, occupied a considerable place in the minds of economic and social development programmers. These were strengthened in 2005 with the advent of the NIHD, based on a participatory approach that places SSE enterprises at the center of the human development process. Thus, the main objective of this mechanism was to participate in human development by addressing the challenges of cooperatives, associations, and mutuals (De Miras, 2007; Achour, 2018). Moreover, the implementation of this initiative opened up a huge construction site and was a great opportunity for the country's development through the construction of supports and structures adapted to the SSE (Khatibi, 1998, cited in Achour, 2018).

In addition to initiatives and programs to support the sector, Morocco has set up a multitude of structures whose remit is to accompany and support all the actors in the ecosystem. They include in particular: "The Office for the Development of Cooperation", a body responsible for implementing government policy in the field of cooperatives; "The Social Development Agency", an institution that is dedicated to poverty reduction and the promotion of social development in Morocco; and "Maroc Taswiq", which positions itself as a collective of small and medium producers within the framework of the "Green Morocco Plan". We can cite also: "Entraide nationale", whose mission is to provide all forms of aid and assistance to the population and to contribute to family and social promotion; "National Promotion", whose vocation is to coordinate and implement the achievement of full employment of rural populations in order to foster the development of the national territory; and "The Agency for Agricultural Development", whose objective is to support solidarity-based agriculture through the promotion and implementation of viable projects pertaining to farmers (CESE, 2015).

Particular interest has been given to this theme over time, both by public authorities and researchers from the academic world. The enthusiasm for this theme is explained by the importance of this sector in inclusive and sustainable development.

However, Moroccan society's interest in the solidarity economy is not only related to the economic situation. It can also be explained by a traditional and a cultural dimension, in addition to the demands arising from the social and economic changes brought about by globalization (Luttwak E-N, 1999; Cited in Adrdour et al., 2016). In fact, Moroccan society was characterized early in its history by the practices and know-how of solidarity, collective work, and economic mutualization, rooted in its culture, even if its practices are not known as "social entrepreneurship" (Layadi et al., 2019). However, specific terminology was used at that time, distinguishing these modes of collective participation according to regions and activities, ranging from the Touiza, the Agadir, and the Agoug, to the Chard, the Ouziaa, and the Khattaras (CESE, 2015). Similarly, these modes of solidarity were predominant in the different regions of the kingdom, particularly in rural areas (MTF-GROUP "LE MATIN", 2018).

4.2.2 SOCIAL ENTREPRENEURSHIP: CLARIFYING THE CONCEPT OF SOCIAL ENTREPRENEURSHIP

In order to better understand the phenomenon of social entrepreneurship and to clarify this research theme, a definition of the concepts is necessary. In this sense, Brouard (2007) confirms that the definition of a concept is the first step in order to fully understand what it is about. However, in this chapter, we do not cite an exhaustive list of all existing definitions in the entrepreneurial literature. We will rather precede an overview of social entrepreneurship by presenting several definitions of this concept. By emphasizing the commonalities and characteristics between them, we attempt to provide a definition of this field of research in the Moroccan context.

Social entrepreneurship is a theme that has aroused growing interest among practitioners and researchers alike. This interest is justified by two main factors. On the one hand, by the importance that the latter can play in social and economic development. On the other hand, by the growing number of studies devoted to this subject, as well as symposia and seminars organized by academic circles and by the public as well as private sector (OECD, 2001).

In general, social entrepreneurship is inspired by traditional entrepreneurship and provides new solutions by tackling issues such as microfinance, humanitarian, social, and/or environmental problems (Al Maghrib, 2016).

In this regard, it is important to note that, since the early 1980s, many definitions of social entrepreneurship and social entrepreneurs have emerged (Chauffaut, Lensing-Hebben, & Noya, 2013). However, to date there is no uniform vocabulary and acceptance of the concept (OECD/Union européenne, 2013). Moreover, several concepts such as social entrepreneurship, social entrepreneur, and social entrepreneurial organization, or even social enterprise, have often been indiscriminately used in the literature to express the same idea (Brouard & Larivet, 2010; Bacq & Janssen, 2011; Janssen et al., 2012). This proliferation of definitions of the concept can be explained by the absence of a unifying paradigm (Janssen et al., 2012). From the same perspective, many authors confirm that there is not a unanimously recognized, universal, and clear definition of this field (Hemingway, 2005 cited in Khassal, 2018; OECD, 2001; Peredo & McLean, 2006; Anne-Claire and Thierry, 2011). Indeed, some authors employ the term to refer exclusively to the demand of market logic for social utility purposes. Some others, such as the international association Ashoka, use it to qualify the social vocation and initiatives taken by emblematic entrepreneurs of systemic change, while other scholars still highlight the highly innovative nature of the projects carried out by those entrepreneurs.

4.2.3 SOCIAL ENTREPRENEURSHIP: SYNTHESIS OF DEFINITIONS IN THE LITERATURE

For the sake of analysis, it seemed necessary to begin by defining the concept of social entrepreneurship through a review of the literature. The various founding definitions in the literature on the subject are as follows.

TABLE 4.1
Synthesis of definitions for the concept of social entrepreneurship

Authors	Synthesis of definitions
(Boschee, 1995, p. 4)	Social entrepreneurship is defined as "the action of non-profit leaders who pay attention to market forces, without losing sight of their mission to balance moral imperatives and profit motives, and whose conjunction is the heart and soul of the movement".
(Dees, 1998, p. 4)	"Social entrepreneurs act as agents of change in the social sector through their mission to create and maintain social value, and through the recognition and relentless pursuit of new opportunities. To serve that mission, as part of a continuous process of innovation, adaptation and learning, they act boldly without being limited, a priori, by available resources and by demonstrating a strong sense of commitment and responsibility to their mission and its social impacts".
(OECD, 1999, p. 11; OECD, 2001)	"Any private activity of general interest which is organized on the basis of an entrepreneurial approach and whose main purpose is not the maximization of profits but the satisfaction of certain economic and social objectives, as well as the ability to develop, through the production of goods and services, innovative solutions to the problems of exclusion and unemployment".
(Alvord, Brown, & Letts, 2004, p. 262)	"Social entrepreneurship enables the creation of innovative solutions for immediate social problems and the mobilization of ideas, capacities, resources and social arrangements necessary of sustainable social transformations".
(Mair & Marti, 2006, p. 37)	"We define social entrepreneurship broadly, as a process involving the innovative use and combination of resources to seize opportunities to catalyze social change and/or respond to social needs".
(Austin, Stevenson, & Wei–Skillern, J., 2006, p. 371)	"Social entrepreneurship is an innovative activity that creates social value and can take place within or across the associative, commercial or public sectors".
(Peredo & McLean, 2006, p. 64)	"Social entrepreneurship occurs when a person or group has the objective(s) of creating social value ... demonstrates an ability to recognize and take advantage of opportunities ... is innovative ... accepts an above-average degree of risk ... and is particularly resourceful ... in the pursuit of their social project".
(Guzmán Vásquez & Trujillo Dávila, 2008, p. 10)	"Social entrepreneurship is a specific type of entrepreneurship that seeks solutions to social problems through the construction, evaluation and realization of opportunities to generate sustainable social value with different types of organizations".
(Zahra, Gedajlovic, Neubaum, & Shulman, 2009, p. 519)	"Social entrepreneurship encompasses the activities and processes undertaken to discover, define and exploit opportunities to increase social wealth by creating new businesses or managing existing organizations in innovative ways".
(Bornstein & David, 2012)	"Social entrepreneurship is a process whereby citizens build or transform institutions to advance solutions to social problems, such as poverty, disease, illiteracy, environmental destruction, human rights abuses and corruption, in order to make life better for all".
(Chauffaut et al., 2013, pp. 16–17)	"Social entrepreneurship can be defined as entrepreneurship that aims at providing innovative solutions to unresolved social issues and problems. Therefore, it often goes hand in hand with social innovation processes that aim at improving people's lives by supporting social change ... Social entrepreneurship seeks to solve social problems rather than exploit market opportunities for profit, even though it operates within the market and is subject to its constraints".
(Mendoza, Alva, & Ramos, 2020, p. 5)	"Social entrepreneurship is a process by which people develop or transform institutions, systems or governments to provide solutions to social and environmental problems".

4.2.4 SOCIAL ENTREPRENEURSHIP: DEFINITIONS AND SCOPE

By considering the definitions cited in Table 4.1, it appears that, although each author defines social entrepreneurship in his or her own way, there are some common points between them: Those that revolve around the process and activities that have a social impact; where the social mission is the main concern; a search for sustainable solutions, solving social problems, or meeting social and environmental needs (access to care, energy and housing, education, long-term unemployment, green growth, etc.); the creation of social value.

In other words, there are commonalities and characteristics that can be described as fundamental for defining social entrepreneurship. For example, a common element in various definitions is related to finding solutions to social problems. The second facet is the social component, which is essential for the definition of social entrepreneurship. Moreover, without the social mission, social entrepreneurship cannot exist. In fact, social entrepreneurship and social innovation are part of the solution, since both are specifically aimed at providing innovative solutions to unresolved social problems, thereby placing social value creation at the heart of their mission to improve the lives and well-being of individuals and communities.

In short, social entrepreneurship can be defined as a set of innovative entrepreneurial actions, implemented by an individual or a group of individuals from the public or private sector, and identifying the opportunities that arise on the market. It aims at solving the social, environmental, and humanitarian problems of the community, as well as changing the mentalities and behaviors of its members.

4.3 SUSTAINABLE DEVELOPMENT IN MOROCCO

After presenting a global reading of the phenomenon of social entrepreneurship through a synthesis of the literature, we will now present the second key concept of this chapter, namely sustainable development in Morocco, in order to identify the links that may exist between social entrepreneurship and sustainable development. In this regard, we present some definitions of the concept and then approach the different strategies for sustainable development in Morocco, before discussing the links between social entrepreneurship and sustainable development on the basis of the work that has been done at this level.

4.3.1 DEFINITION OF SUSTAINABLE DEVELOPMENT

As with social entrepreneurship, the concept of sustainable development is considered by the majority of actors in the field to be extremely broad in its scope. It is also of crucial importance to researchers and decision makers, since this subject remains topical and gives rise to much debate and reflection. Moreover, this subject is of concern to both Western and developing countries, which are also beginning to attach considerable importance to it.

The concept of sustainable development has been the subject of various definitions. For example, according to the report of the World Commission on the Environment (Brundtland, 1987), the concept is defined as "a form of development that meets the needs of the present without compromising the ability of future generations to meet their own needs". It is of course a question of organizing the world's societies in such a way as to preserve the resources of the environment, while meeting our needs. In other words, the underlying idea is that it is possible to devise a policy and strategy to ensure the continuity over time of economic and social development, while respecting the environment and without compromising the natural resources indispensable to human activity (Mancebo, 2010).

Based on this definition by Brundtland (1987), sustainable development seeks to reconcile a dynamic economy, a high level of education, health protection, social and territorial cohesion, and environmental protection in a peaceful and secure world while respecting cultural diversity.

For the OECD, sustainable development as a concept "encompasses concerns for equity and social cohesion, as well as the need to address threats to the common goods of humankind". Thus, the search for a balance between environmental protection, economic development, and social well-being is fundamental.

Sustainable development is mainly and simultaneously based on three dimensions: Economic, social, and environmental (Tilley & Young, 2009; De Hoe & Janssen, 2014). In other words, it is a combination of economic, environmental, and social aspects, the aim of which is to improve the living conditions of humans while respecting the limits of ecosystems.

From a sustainable development perspective, it is essential to reconcile these three basic elements, which are interdependent and all indispensable for the well-being of individuals and societies: Economic growth, social inclusion, and environmental protection.

4.3.2 SUSTAINABLE DEVELOPMENT STRATEGY IN MOROCCO

Since 1992, during the Rio Summit, the Kingdom of Morocco presented a "Vision of the Fundamentals for the Construction of a New Model of Society". Since then, Morocco has committed itself to meeting the major and imperative challenges of sustainable development by making this policy a genuine social project and a new model of development. This commitment has been translated into successive reforms aimed at establishing a solid foundation for economic development and improving social conditions, while accelerating the pace of environmental achievements through both preventive and corrective measures (MEMDD, 2017).

The evolution of Morocco's trajectory over the last 20 years, in terms of building a social project based on sustainable development, has gone through three major phases: A first phase (1992–2000) of economic and social upgrading; a second phase (2000–2011) of the establishment of fundamental levers and an acceleration of development); and a third phase (2011) of constitutional reform and operationalization of sustainable development. This last phase reflects the progress of the institutionalization of sustainable development in Morocco.

The year 2011 marks a turning point in the realization of the company's project, when the New Constitution of Morocco integrated sustainable development into its text which thus strengthens governance (MEMDD, 2017). The advisory commission also delivered its report on advanced regionalization. It seeks an "integrated and sustainable development in economic, social, cultural and agro-environmental terms".

Finally, in March 2014, the Framework Law No. 99-12 on the National Charter for the Environment and Sustainable Development was published with a view to adopting the NSSD. The elaboration of the NSSD in Morocco thus marks the culmination of a strong commitment to the realization of sustainable development.

This strategy outlines a common project for all of the nation's actors, both public and private, to support them in their efforts to achieve sustainability, each in its own field around strategic choices and indicators that have been the subject of a broad consensus.

Before presenting the principles, the major issues, the strategic axes as well as the challenges of the NSSD adopted by Morocco, it will be judicious to first define the sustainable development strategy.

4.3.3 WHAT IS A SUSTAINABLE DEVELOPMENT STRATEGY?

According to MEMDD (2017), the sustainable development strategy "is an ongoing process based on commonly accepted issues and objectives to be achieved. It is a new way of understanding development by pooling the efforts and contributions of each stakeholder". Accordingly, it promotes a vision of progress that reflects the history and core values of the country and the direction in which it is moving. It also examines the options available, by translating this broad vision into a number of specific short and long-term objectives at the national and local levels through different actors and cross-sectorial policy integration. Finally, a sustainable development strategy has the capacity to monitor current social, economic and environmental conditions, and foreseeable future trends.

For the OECD (2002), a sustainable development strategy «is not a "grand design" or a set of plans, but rather a set of instruments and ways of working that address the challenges of sustainable development in a coherent and dynamic way».

4.3.4 THE PRINCIPLES OF THE NATIONAL STRATEGY FOR SUSTAINABLE DEVELOPMENT IN MOROCCO

According to MEMDD (2017), the national sustainable development strategy must necessarily respond to a number of guiding principles set by the government to ensure consistency in its design. As a result, the adopted principles to develop this strategy in Morocco are shown in Figure 4.1.

-01- INTERNATIONAL CONFORMITY	The strategy is in line with international best practices, and takes up at a minimum the challenges to which the Kingdom is committed in terms of sustainable development, namely the fight against climate change, the fight against desertification and the protection of biodiversity.
-02- IN CONFORMITY WITH THE PRINCIPLES OF THE FRAMEWORK LAW	The strategy is in line with the principles of the Framework Law 99-12 on the Environment and Sustainable Development Charter, namely: integration, territoriality, solidarity, precaution, prevention, responsibility and participation.
-03- COMMITMENT	The national sustainable development strategy is conceived as an ongoing process of engaging different stakeholders to achieve common goals that contribute to addressing key sustainable development issues.
-04- OPERATIONAL	The strategy is intended to be operational by building on the strategies, plans and programs currently being implemented. The strategy is in no way a break with the development choices made by the Kingdom. The strategy is based on concrete measures with monitoring and/or results indicators.

FIGURE 4.1 The four main principles that guided the elaboration of the NSSD in Morocco

Source: Final report project[2] of the NSSD of the Kingdom of Morocco 2030.

4.3.5   THE VISION OF THE NATIONAL STRATEGY FOR SUSTAINABLE DEVELOPMENT IN MOROCCO

The Strategy aims at implementing the foundations of a green and inclusive economy in Morocco by 2030 in order to achieve the intergenerational goal of sustainable development. The proposed vision is the result of an in-depth diagnosis, and is based on the integration of the four fundamental pillars of sustainable development; namely economic, social, environmental, and cultural. The economic pillar is undoubtedly the driving force behind the vision because without a healthy and efficient economy no sustainable development is possible. Figure 4.2 encompasses the four fundamental pillars of the NSSD.

4.3.6   THE CHALLENGES TO BE MET BY THE NATIONAL SUSTAINABLE DEVELOPMENT STRATEGY

In order to pursue the evolution of this great social project and build a new model of development, the NSSD has to take up several challenges. These are mainly related to climate change, conservation and management of natural resources, public health, risk prevention and management, poverty, transport and sustainable mobility, sustainable production and consumption, governance, social inclusion, demography and immigration, education and training, and research and development.

According to our understanding, and with reference to the theoretical foundations of social entrepreneurship, which we discussed in the first section, it appears that

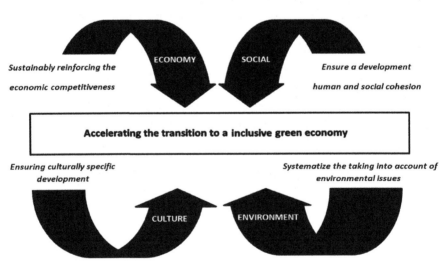

FIGURE 4.2   Vision of the NSSD to ensure a transition to a green and inclusive economy

Source: Final report project of the NSSD of the Kingdom of Morocco 2030.

social entrepreneurship can be a promising approach to sustainable development and still be an instrument of crucial importance in responding to one of the challenges raised by the latter.

In this section, we have outlined and discussed the various elements constituting the NSSD in Morocco, which must consolidate its regional bases as a national economy managed by regional growth engines that are complementary in their constitution and open to innovation regarding job creation, through the promotion of social entrepreneurship and the pooling of efforts as well as the contribution of each stakeholder. This is the only measure of success of any entrepreneurial strategy. This leads us back to the question posed in this research: What is the link between social entrepreneurship and sustainable development? In answering this question, we will build on previous work done.

4.4 LINKS BETWEEN SOCIAL ENTREPRENEURSHIP AND SUSTAINABLE DEVELOPMENT

The literature emphasizes that social entrepreneurship is a major issue in meeting the challenges of sustainable development. It sees social entrepreneurship as a preferred means of providing potential responses to the challenges posed by sustainable development (Davis, 2002; Brouard, 2007; Stevens, Moray and Bruneel, 2015, cited in Hidalgo et al., 2020; Omrane & Fayolle, 2010; Layadi et al., 2019).

To be able to undertake such action in this sense, research becomes effective in explaining the relationship and link between these two variables. A number of studics have been carried out on this subject concerning, among other things, the contribution and relationship of social entrepreneurship to sustainable development.

At the conceptual level, some authors such as De Hoe and Janssen (2014) point out that social entrepreneurship and sustainable development entrepreneurship are two distinct concepts that should not be confused with each other or with other concepts with which they have similarities, such as, for example, corporate social responsibility. They also specify that sustainable development entrepreneurship with an environmental goal is in addition to the social and economic goals it has in common with social entrepreneurship. Thus, sustainable development entrepreneurs take into account social, environmental, and economic objectives at the same time.

For Tilley and Young (2009), the sustainability entrepreneur is the only "entrepreneurial" way to achieve sustainable development. Thus, economic, social, and environmental entrepreneurs can partially contribute to the achievement of sustainability goals; however, they alone cannot meet all the challenges of sustainable development. According to these authors, being a "sustainable development entrepreneur" implies taking into account three components: Economic, social, and environmental. As for the relationship between social entrepreneurship and sustainable development, Tilley and Young (2009) emphasize that sustainable development entrepreneurship and social entrepreneurship are quite distinct: the former includes environmental and social considerations in addition to economic concerns, while the latter deals only with social and economic issues.

Similarly, some experiments and research have highlighted the contributions of social entrepreneurship to sustainable development. In the view of Muhammad Yunus, the pioneer of social entrepreneurship and founder of the Grameen Bank, "Development only makes sense if it is sustainable", and social entrepreneurship, in his view, provides an effective solution to some of our most difficult problems. He strongly encourages OECD countries to explore and embark on this path toward sustainable development (OECD, 2015). Indeed, the social enterprises he created in Bangladesh, be they sole proprietorships or large joint ventures with multinationals, have provided evidence that social enterprise can ensure development. This social entrepreneur defends his point of view through the achievements of the social enterprises he founded. This is the case of a joint company, Grameen Danone Foods Ltd. It is a privately owned company, created by the Grameen Group in 2006, with one of the world-leading food producers, the Danone Group, in a rural area of Bangladesh. Its vocation is to solve certain social problems by producing a yoghurt enriched with micronutrients that could help reduce child malnutrition in disadvantaged areas in Bangladesh. As a result, children love this healthy and delicious product. While neither Danone nor Grameen is reaping any dividends from this joint venture, the board of directors, management, and staff are devoting their creativity and dynamism to its social impact, bearing in mind the need to cover costs and ensure the company's expansion (OECD, 2015).

In the research of Loudiyi and Louhmadi (2016) on the contribution of social entrepreneurship and particularly on cooperative entrepreneurship and its participation in the fulfillment of sustainable development requirements, it has been shown that cooperatives take into consideration the dimensions of sustainable development. More especially, from an economic dimension, they contribute to the eradication of the informal sector and the partial reduction of unemployment as well as the fight against poverty and social exclusion, and the improvement of human living conditions. However, this study also demonstrates that social entrepreneurship as an entrepreneurial dynamic is not sufficiently structured to evolve in the region in which the investigation was conducted. This is due to the multivaried analysis which revealed that the two components of sustainable development, namely the social component and the environmental component, do not contribute positively to the evolution of social entrepreneurship and specifically of traditional craft cooperatives.

Other research work in this same field, an examination of the role of social entrepreneurship in strengthening sustainable development in the Lebanese context, showed that social entrepreneurship positively contributes to the objectives of sustainable development. The results of the research revealed that the effects of social entrepreneurship on education, health, and economic sectors are quite positive. Thus, it seems that social entrepreneurship plays an important role in reducing poverty, ensuring employment, economic prosperity, environmental justice, pollution-free ecology, good governance, reducing terrorism, ensuring peace, controlling corruption, and so on, and provides opportunities to achieve sustainable development goals in Lebanon (Diab, 2019).

The literature, therefore, provides important answers to our research question regarding the contribution of social entrepreneurship to sustainable development. Based on the various research findings cited, we will tackle below some avenues for reflection and a framework for action likely to promote sustainable social entrepreneurship in Morocco.

4.5 FRAMEWORK OF ACTIONS FOR THE DEVELOPMENT AND PROMOTION OF SUSTAINABLE SOCIAL ENTREPRENEURSHIP IN MOROCCO

Having presented the theoretical framework of the two notions, namely social entrepreneurship and sustainable development, we can, therefore, address the measures likely to foster sustainable entrepreneurship in Morocco.

The readings of various academic contributions has allowed us to formulate certain avenues of reflection while identifying some measures likely to contribute to the development of sustainable social entrepreneurship in Morocco. This prior literature includes research reports of some organizations operating in the field of social entrepreneurship, those interested in this theme – such as the Moroccan Centre for Innovation and Social Entrepreneurship (MCISE), the General Confederation of Enterprises of Morocco (GCEM), Group "LE MATIN", the British Council, and the Euro-Mediterranean Forum of Economics Institutes (EMFEI). We also relied on readings of press coverage dealing with themes related to our research question, which allowed us to formulate certain avenues of reflection while identifying some measures likely to contribute to the development of sustainable social entrepreneurship in Morocco.

It should be noted that several factors influence the development of social entrepreneurship, more precisely: The absence of an ecosystem of social innovation, the lack of clear legislation governing social entrepreneurial activities, difficulties in accessing finance, cumbersome administrative procedures for setting up businesses, problems with determining the appropriate clientele, land ownership obstacles,[3] and the lack of human resource know-how (reflected in the lack of skills in management, planning, project design and evaluation, market development, training, etc.) (Tsakas and Moukaddem, 2019; Tallie Hausser, Tsakas, & Moukaddem, 2019).

To cope with these factors hindering the development of social entrepreneurship, it is fundamental to set up support mechanisms and introduce measures and tools to overcome them. We propose, indeed, lines of thought and a framework of actions on the undertaken measures or to be undertaken in Morocco likely to promote a social entrepreneurship to meet the challenges raised by sustainable development.

4.5.1 THE FINANCIAL SYSTEM

As we have pointed out, the lack of funding for social businesses to carry out their missions is one of the main barriers that hinder a social project from developing. This problem relating to financing is at the heart of Morocco's concerns; in this context, it has launched a multitude of products and support programs to facilitate the financing of entrepreneurship. Indeed, there are different types of sources of financing.

4.5.1.1 Collaborative Finance: "Crowdfunding"

Collaborative financing, known internationally as "crowdfunding", "is a process of raising funds to finance a project or business through multiple investors and via an

Internet platform. It, consequently, increases the possibilities of access to capital for entrepreneurs and social organizations" (MTF-GROUP "LE MATIN", 2018).

This is a time-bound communication campaign to engage the community and raise money for a well-defined goal. The principle of this funding mechanism is simple. The project holder launches a campaign on crowdfunding electronic platforms, where he presents his project and the use that will be made of the amount he seeks to collect.

This funding mechanism uses crowds to obtain funding via platforms (Burton et al., 2015 cited in Rouxhet et al., 2017). It is the means of financing favored by innovators and associations, and is also increasingly attractive to start-ups and project holders who are unable or unwilling to finance their projects via traditional formulas (banks, honor loans, etc.). Moreover, crowdfunding has various advantages. It allows the saver to choose the project or structure that will receive his or her money, as well as allowing these project holders to closely monitor the progress made and promote their image as contributors to a project with social impact (Tsakas and Moukaddem, 2019).

In Morocco, this new means of financing is the subject of law project No. 15-18, which aims to provide a framework for this activity and was presented to Parliament in 2018 for promulgation and approval. Thus, this bill is one of the measures aimed at promoting entrepreneurship and innovation and supporting the emergence of social, cultural, and creative projects.[4] According to this, these activities take three forms of financing, namely: Loans, donations, and capital investment.

These new forms of financing are beginning to gain momentum, enabling project leaders to raise funds for development, as some platforms[5] have recently been created in Morocco. Hence, this innovative mechanism can be a lever for the promotion of social entrepreneurship in Morocco.

4.5.1.2 Islamic Finance

With the financing problems faced by social entrepreneurs, which is reflected in their difficulties in accessing the solutions offered by the conventional financial system, Islamic finance can be considered as an innovative alternative to traditional financing. Amzil (2013) emphasizes that the principles of Islamic finance make it a purely entrepreneurial one and that the financial tools offered by Islamic banks would be perfectly adapted to the expectations and needs of social entrepreneurs.

Aware, therefore, of the importance of this alternative system in the social and economic development of countries, Morocco undertook in 2016 actions to develop participatory finance, including the preparation of legislative, regulatory, and fiscal instruments necessary for the launch of Islamic banks called "participatory". So, in 2017, the Moroccan authorities gave the green light to Islamic banks for an effective start of their activities and to be able to issue requests for approval to Bank Al-Maghrib, after having had their requests endorsed by the "Supreme Council of Ulemas. To this end, five requests for the authorization of the creation of participatory banks have been accepted, namely by Umnia Bank, Bank Assafa, Bank Al Yousr, Al Akhdar Bank, and Bank Al-Tamweel wa Al-Inma. Similarly, three other financial institutions of the conventional Moroccan system have chosen to invest in the Islamic finance landscape but this time not through the creation of new banks or subsidiaries but through participative windows. These are Société Générale, BMCI Group, and Crédit du Maroc which have respectively created Dar Al Amane, Nejmah, and Arreda. This will enable these banks to offer services such as deposit accounts, current accounts, withdrawal and payment cards, as well as various participatory products, such as "Murabaha real estate", "Murabaha equipment", "Murabaha autos", "Ijara", and "Investment contracts".

In addition to the solutions proposed by participatory banks, other measures inspired by Islamic finance can be added to the tools that can support social entrepreneurship; these are the Islamic financing methods based on donations and charity (*ihsan*) such as "La Zakat", "Le Waqf", and "Le Credit Without Interest (Qard Hasan)" (Amzil, 2013).

Through these measures, Morocco also leaves the way open for innovation, allowing new modes of financing to emerge, unconventional projects to spring up, and market targets that do not adhere to the principles of conventional finance to be integrated (Tsakas and Moukaddem, 2019).

4.5.1.3 Microfinance

Microcredit is one of the other sources available to social entrepreneurs to finance their business creation or development projects. In this context, microfinance institutions can play a key role in the development of social entrepreneurship as well as in promoting financial inclusion. They offer financing solutions to small and micro-enterprises and low-income populations that generally have difficulty in meeting the guarantee obligations imposed by the credit agents of the traditional banking system. According to the white paper on microcredit in Morocco, this type of financing "consists of providing short-term loans to very low-income people who do not have access to the services offered by conventional financial institutions, to help them launch their activities or develop their business". The principle of microfinance therefore suggests giving the poor access to financial products, which will subsequently constitute a recipe for social and economic development. Also it presents various advantages, in terms of the cost of credit which is less expensive compared to the tariffs of the financial products proposed by the traditional credit agents, in the same way it targets low-income groups that have never had recourse to formal banking services. In addition, it ensures not only the granting of credit but also the accompaniment of these beneficiaries in particular by training actions. Accordingly, microcredit is considered a crucial instrument for ensuring development, the fight against poverty, and financial inclusion. In addition to the latter, it is a source of financial, economic, and social innovation that encourages entrepreneurial initiatives such as commercial, craft, and agricultural micro-activities.

At this level, it is useful to recall an example of success in this field, namely the experience of the most famous social entrepreneurship in the world, founded by the Bangladeshi economist and entrepreneur Mohamad Yunus, who is the pioneer of microfinance in Bangladesh, and known for having founded in 1976 the first microcredit institution, the Grameen Bank, nicknamed the "banker of the poor", to which he dedicated the Nobel Peace Prize awarded to him in 2006. As a consequence, the small credit granted by this institution to groups of destitute women proved to be a financial innovation capable of changing the lives of the poor, which revolutionized the microfinance industry (Lelart, 2007; Coupez, Lutzel, & De Reille, 2009).

The entrepreneurial literature on the subject of microfinance affirms the exclusively social mission of this sector, which has become one of the key issues in contemporary social affairs research and practice (Shams, Vrontis, Belyaeva, Thrassou, & Christofi, 2018), and which provides as many answers in terms of its contribution to supporting the informal sector, the fight against poverty, and the development of social entrepreneurship.

In Morocco, microfinance, as an emanation of the Moroccan civil society supported by the government, has become an important economic sector of the kingdom and maintains an important potential in the short and long term. It has grown considerably over the years and is a real lever for development, so much so that the Moroccan microfinance sector is very diversified with 13 microcredit associations (MCAs; four large MCAs with national scope, three MCAs with regional coverage, and five MCAs that are local associations; Tsakas and Moukaddem, 2019).

It should be noted that efforts are made and measures are implemented and undertaken to point out the strategic prospects of this sector in Morocco. The national microfinance sector is then a key field in the fight against poverty through job creation and income-generating activities. It is also efficient, sustainable, and integrated in the kingdom's public policies.

4.5.2 TRAINING MEASURES

Another important tool for the promotion of social entrepreneurship is the development of human capital through quality training and pedagogy that takes into account the specificities of the world of social entrepreneurship. In this sense, Brouard, Larivet, and Sakka (2012) underlined that training is at the heart of human capital development and entrepreneurial culture. As stated by Fortin (2002), the school's contribution is crucial through the training it offers. It remains the preferred means of discovering entrepreneurial potential, supporting and updating it, and therefore of developing an entrepreneurial culture. To prosper, there is a need for better training by informing the various stakeholders in the ecosystem, such as governments, local elected officials, entrepreneurs, educators, and the general population. This is why it is necessary to introduce entrepreneurial and managerial training in the education and training systems from the earliest stages. Such training measures will promote the development of entrepreneurial attitudes and skills through awareness-raising actions based on self-employment as a professional option. They will also provide the essential and specifically needed skills to foster entrepreneurial initiative and business creation (CESE et al., 2018).

The training of social entrepreneurs needs also to be upgraded (Omrane & Fayolle, 2010). It is therefore a priority to invest in training, since quality education is fundamental for boosting productivity and economic dynamism, facilitating innovation and adaptation to change, as well as fostering the capacity to create jobs (CESE et al., 2018).

In this context, it is necessary to set up support programmers and measures, in particular training in entrepreneurship, in order to encourage the population to become involved in an entrepreneurial dynamic.

It should be noted that any support programmer requires human resources that are both competent and qualified, mastering social issues and capable of implementing

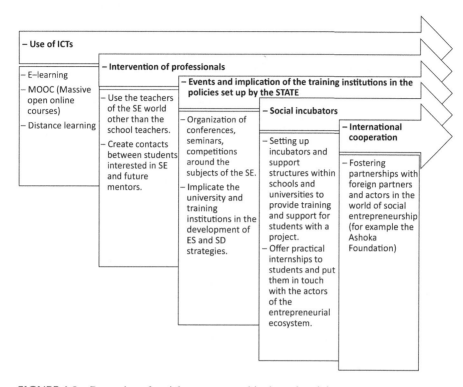

FIGURE 4.3 Promotion of social entrepreneurship through training measures

the actions undertaken, thus meeting the expectations of project leaders and social entrepreneurs. However, it is not enough simply to draw up development policies and set up training programmers, but there must also be rigorous monitoring and permanent control to ensure that the desired objectives are properly implemented and achieved.

Yet, it appears that certain gaps exist in the education and training system that are reflected in the discrepancies between the theoretical education provided to students and the practices of working life. These teachings are now far from their expectations and therefore hamper their progress in business creation or development projects. So, in order to limit the errors of observed referrals, practical trainings should be offered, provided by professionals from the business and entrepreneurial world, with a pedagogical content that is more responsive and ever closer to the reality of social business.

The reflections on the measures to be undertaken in the area of training can be presented as shown in Figure 4.3.

4.5.3 PROMOTION OF SOCIAL ENTREPRENEURSHIP

Without an effective strategy to promote the social entrepreneurship sector, the actions undertaken cannot reach the desired objectives. Such a promotion will consist, on the one hand, in raising awareness and mobilizing the community on the importance of innovation and social entrepreneurship in terms of sustainable development in Morocco; and, on the other hand, in promoting initiatives while playing on the attractiveness of the social entrepreneurship sector.

Certainly, many events and demonstrations have been organized in recent years on the subject of social entrepreneurship, but they may not be visible to target a large number of population groups. A communication strategy will then be a key measure to further promote the sector, while ensuring that communication channels are diversified when disseminating information to target a large number of audiences.

Various tools can be envisaged in this sense, such as the development of websites, software, open platforms and places dedicated to promoting recording, exchange, and the creation of ideas between social idea builders by facilitating physical meetings and online meetings, if possible in creative places dedicated to such exchanges (MTF-GROUP "LE MATIN", 2018).

It is also possible to strengthen the promotion process through the creation of public and private structures, such as MCISE and the innovation cities set up within universities. Besides, the role that can be played by different actors existing in the entrepreneurial ecosystem is considerable. For instance, we can cite regional investment centers, the GCEM, the chambers of commerce, and universities, by means of support, information, and awareness-raising actions. With the same logic, the establishment of such a specific institutional mechanism, with structures in charge of promotion and a statistical system specific to the sector, could make it possible to monitor the main indicators and assess the real impact of social entrepreneurship on sustainable development, social promotion, and job creation.

In addition, the organization of national conferences on the theme of social entrepreneurship with a view to making it a periodic meeting to discuss the advances and limits of the field of development can also contribute to its promotion.

Moreover, the development of a special magazine on social entrepreneurship could also be used to raise awareness of the field. This would constitute a reference medium of crucial importance for capitalizing knowledge in the field, through the dissemination of all events related to this field, success stories in the field of social entrepreneurship, the support mechanisms put in place to assist project leaders, as well as all the news related to this subject.

4.5.4 Legal Framework

In this plan, the questions that arise are, on the one hand, whether the current legal framework in Morocco is adapted to social entrepreneurship; in other words, whether Morocco has already a legal form specific to social business enabling it to supervise this entity. And on the other hand, one wonders whether the current institutional and legal context would be an obstacle to the growth and development of social entrepreneurship in Morocco.

It should be mentioned that the choice of a legal status appropriate to the project is one of the steps to be taken by project promoters and entrepreneurs who wish to set up businesses. It is a decisive choice because each legal form brings certain advantages and subjects business creators to certain obligations, including the legal and fiscal aspects.

As we have already pointed out in this chapter, the development of social entrepreneurship in Morocco remains at an embryonic state. Various studies carried out on this subject stressed the absence of regulations that clearly define the constituent elements of social enterprises, by specifying its rights and obligations. That is why this type of structure is not well known and developed in Morocco. Hence, the need for such a scheme and the development of a legal framework to recognize and promote social businesses will be beneficial for society as a whole.

Despite the absence of clear regulations and a statute dedicated to social business, it exists in other forms, such as cooperatives, associations, or enterprises, which are part of a "social logic" and not a purely economic logic.

Besides, the social entrepreneurship sector in Morocco shows great dynamism, and the potentialities and opportunities to be exploited in this field are many and numerous, but require a favorable legal framework and tax incentives that make this sector more attractive so as to attract a large number of social entrepreneurs to invest in it.

The reflection on this aspect of regulation can indeed strengthen the development and promotion of this sector in Morocco. From this perspective, a draft law on the status of social business is already under discussion; such a project should involve all actors in the entrepreneurial ecosystem. Besides, a set of recommendations is indeed desirable by as many stakeholders, including the adoption of a law on social business officially defining this structure in Morocco and specifying its rights and obligations, while not forgetting the granting of aid and subsidies to this type of enterprise. The creation of a Council for Innovation and Social Entrepreneurship (CIES) is also recommended (MTF-GROUP "LE MATIN", 2018). Its mission is to provide support to social businesses and to strengthen the actions of intervention bodies, to carry out sectorial and territorial studies, to promote public/private partnerships, and to ensure the coordination of all the actors in the entrepreneurial ecosystem.

4.5.5 SUPPORT FOR SOCIAL INNOVATION

Social innovation is an essential aspect to be taken into account in addressing social challenges. It makes it possible to anticipate and produce appropriate responses to social needs that are not or only poorly met under current market and public policy conditions (MTF-GROUP "LE MATIN", 2018). In this sense, Mulgan, Tucker, Ali, and Sanders (2007) consider social innovation as a set of new ideas based on innovative activities and services that are motivated by the objective of meeting a social need.

According to Brouard et al. (2012), social innovation needs concrete recognition and support beyond prizes and competitions. At this level, specific innovation support mechanisms are needed to further encourage social entrepreneurship initiatives and encourage social entrepreneurs to innovate strongly.

The promotion of social entrepreneurship in Morocco must therefore make social innovation a driving force of the national innovation strategy in order to maximize its impact. By developing a culture of social innovation, organizing a national debate on the subject, and expanding knowledge, research, reflection, and transmission of knowledge and good practices on social innovation, social entrepreneurship will be developed.[6] It should also integrate this pillar into the ecosystem of social entrepreneurship and sustainable development entrepreneurship, by encouraging innovative entrepreneurial initiatives that can broaden the community of social entrepreneurs and enhance their attractiveness, as well as by stimulating cooperation, coordination, and synergy between its members. In addition, reflection may also focus on mobilizing public and private funding for the launch and scaling up of innovative initiatives.

In this context, Morocco, aware of the importance of this area and the role it can play in development, has paid particular attention to social entrepreneurship. As it is at the heart of the state's strategic orientation, it is usually reflected in the royal speech:

> the questions related to youth are inseparable from the issues of growth, investment and jobs. That is why, to remedy the precarious situation of young people, it is necessary to innovate through concrete initiatives and projects that can free their energies, provide them with employment and a stable income. Only then will they be able to actively contribute to the country's development.[7]

Sustaining social innovation can then be an indisputable measure and a key source of creativity through which social entrepreneurship and sustainable entrepreneurship can be promoted to address economic, social, and environmental challenges.

4.5.6 REINFORCEMENT OF THE ECOSYSTEMS OF ENTREPRENEURIAL SUPPORT

It should be noted that many project holders, wishing to launch or develop such an activity in the field of social entrepreneurship, may face many difficulties and suffer from various obstacles that hinder the success of their adventure. These difficulties lie in particular in the fact that these entrepreneurs do not necessarily have the means and skills (technical, administrative, management, etc.), nor the sufficient financial capital, to carry out and sustain their projects. Such a support and incubation structure is necessary to meet the expectations of these social entrepreneurs.

Numerous research studies in the field of entrepreneurship show that entrepreneurial support is a determining factor in the development and sustainability of business projects. In this sense, some authors confirm that "businesses that have benefited from support during their creation are more sustainable than other businesses and that, moreover, this support has a positive effect on the development and profitability of young businesses" (Leger-Jarniou, 2008).

While many support structures for entrepreneurship in business exist in Morocco, those related to social entrepreneurship are, to our knowledge, less numerous. Otherwise, they are insufficiently visible to project holders and social entrepreneurs. A reflection in this sense should be conducted as a communication strategy aimed at making these actors more visible.

It would then be opportune to develop tools and support measures adapted to the social entrepreneurship sector in Morocco, taking into consideration the specificities and cultural context of the country as well as the challenges of sustainable development. It would also be necessary to create support structures for social entrepreneurship and social incubators, capable of providing technical support and training to social project holders, promoting the emergence, consolidation, and development of structures with a social vocation, as well as contributing to the mobilization of territories.

4.6 CONCLUSION

To conclude, it is important to underline that Morocco accords a major interest both to the social entrepreneurship sector and to sustainable development. This is reflected in the integration of these two axes in the strategic orientations and public policies of the development of countries. The country is indeed aware of the importance that these two dimensions can represent, in terms of economic, social, and environmental development. This is why its alternatives are key components of the new development model.

It is therefore obvious that this form of entrepreneurship represents a real lever for sustainable development and a real movement toward the promotion of financial inclusion in Morocco. It also represents an essential means of promoting and advancing social change through business.

However, it is necessary not only to draw up development policies and undertake support and promotion actions, but also to ensure the proper implementation of the measures and actions undertaken to provide sustainable, innovative, and appropriate entrepreneurial solutions to each social challenge in Morocco.

In response to the main question of this research, namely whether support measures for social entrepreneurship in Morocco contribute to sustainable development, the present research revealed three main findings. On the one hand, several previous studies confirmed the link between social entrepreneurship and sustainable development (Tilley & Young, 2009; De Hoe & Janssen, 2014). On the other hand, other prior writings carried out in this sense show that social entrepreneurship is a major lever to promote one of the challenges of sustainable development (OECD, 2015; Loudiyi & Louhmadi, 2016). Finally, the results of this research demonstrate that the tools and support measures likely to promote social entrepreneurship are also numerous and multiple. They can be conducted on several aspects such as the financing of social entrepreneurship, training, promotion, entrepreneurial support, and the development of the legal framework. Some of these measures have been already implemented. However, efforts remain to be deployed in order to achieve the objectives of inclusion and the social impact of social entrepreneurship and thus meet the challenges of sustainable development.

The present research can then constitute a preliminary step and a theoretical basis for other subsequent research that will further deepen the reflection on the contribution of social entrepreneurship to sustainable development in Morocco, by measuring the social impact of the various tools supporting social entrepreneurship already identified in this chapter.

As the approach adopted was a theoretical one, it will be also useful to conduct an exploratory study for the actual evaluation of the social impact of the different support measures put in place by the state. In this framework, such a study will constitute a research track of great benefit for researchers, policy makers, and all actors in the entrepreneurial ecosystem.

For a sustainable future, the measures taken by developing countries like Morocco should therefore take into account the organizational environment, which is characterized by turbulence, complexity, and contingency. In fact, such a global environment affects all actions undertaken in terms of development. This is the case today for the Covid-19 pandemic crisis, whose repercussions will be harmful for the entire economy of the countries. It would then be useful to set up "an informational watch" to deal with unforeseen events via the adoption of a sustainable entrepreneurship approach, as a real path to overcome this crisis.

NOTES

1 Excerpt from the Message of H.M. King Mohammed VI to the participants in the Fifth Global Entrepreneurship Summit, held on November 20, 2014 in Marrakech.
2 The report can be downloaded from the website: http://environnement.gov.ma/PDFs/publication/Rapport_Strat%C3%A9gie_Nationale_DD_juin2017_Mai%202017_Web.pdf.
3 According to the results of a national survey of non-profit institutions, published in 2011 by the Office of the High Commissioner for Planning (HCP): More than half of the associations do not own premises and almost 30% are housed free of charge, only 11% are tenants and only 9% own their premises.
4 Following the presentation note of the law project No. 15-18, relating to collaborative financing, of March 2, 2018, of the Ministry of Economy and Finance; www.sgg.gov.ma/portals/0/AvantProjet/164/Avp_loi_15.18_Fr.PDF.
5 Flowingo.ma is the first Moroccan platform for participatory financing which was launched in 2016.
6 As per the note of the event "Le workshop" organized on June 24–25, 2019, by the French development agency and Maroc Impact, downloadable at: https://files.cgem.ma/upload/324501874.pdf.
7 Speech of His Majesty King Mohammed VI on October 13, 2017, on the occasion of the opening of the first session of the second legislative year of the tenth legislature.

REFERENCES

Achour, H.Y. (2018). *L'émergence d'une économie sociale et solidaire au Maroc à la lumière des théories du troisième secteur (The emergence of a social and solidarity economy in Morocco in the light of third sector theories)*, HEC School of Management of the University of Liège.
Adrdour, M., Oudada, M., & Elouardi, A. (2016). L'émergence d'une économie solidaire au Maroc. le cas des Chiffonniers de la ville d'Ait Melloul (The emergence of a solidarity economy in Morocco. Case of the Chiffonniers of the city of Ait Melloul). Paper presented at the XVth Annual International Conference of Territorial Intelligence: Social and Solidarity Economy in the territories: Initiatives, Challenges and Prospects, Charleroi-Liège, Belgium. https://halshs.archives-ouvertes.fr/halshs-01717927.
Ahrouch, S. (2011). Les coopératives au Maroc: enjeux et évolutions (Cooperatives in Morocco: stakes and evolutions). *International Journal of the Social Economy, (322)*, 23–26. https://doi.org/10.7202/1020728ar.
Ait Haddout, A. & Jaouad, M. J. K. D. (2003). *L'économie sociale au Maroc: approches méthodologiques et acteurs en présence (The social economy in Morocco: methodological approaches and actors involved)*.
Injaz Al Maghrib. (2016, Janvier). *Guide de l'entrepreneuriat social, un nouveau business model, Un programme développé pour les villes de Fès, Meknès, Kénitra et Salé dans*

le cadre du partenariat avec le MEP (A programme developed for the cities of Fez, Meknes, Kénitra and Salé in the framework of the partnership with the MEP).

Alvord, S. H., Brown, L. D., & Letts, C. W. (2004). Social entrepreneurship and societal transformation: An exploratory study. *The Journal of Applied Behavioral Science*, 40(3), 260–282. https://doi.org/10.1177/0021886304266847.

Amzil, L. (2013). L'entrepreneuriat social et la finance islamique: Des motivations partagées (Social entrepreneurship and Islamic finance: Shared motivations). *Journal of Management and Economics*, Vol. 1, no. 2, pp. 75–82. 2014.

Anne-Claire, P. & Thierry, S. (2011). *Cas en entrepreneuriat social (Social entrepreneurship cases)*, Collection Case Studies, Editions EMS.

Asli, A., & Slitine, A. E. I. (2013). L'entrepreneuriat social au Maroc, Perception et pistes de développement (Social Entrepreneurship in Morocco: Perception and tracks of development). *Moroccan Journal of Management and Marketing Research*, 8, 235–249.

Austin, J., Stevenson, H., & Wei–Skillern, J. (2006). Social and commercial entrepreneurship: Same, different, or both? *Entrepreneurship Theory and Practice*, 30(1), 1–22., https://doi.org/10.1111/j.1540-6520.2006.00107.x

Bacq, S., & Janssen, F. (2011). The multiple faces of social entrepreneurship: A review of definitional issues based on geographical and thematic criteria. *Entrepreneurship and Regional Development*, 23(5–6), 373–403.

Barendsen, L., & Gardner, H. (2004). Is the social entrepreneur a new type of leader? *Leader to Leader*, (34), 43.

Bennani, G. (2019). Entrepreneuriat social : une problématique de concept (Social entrepreneurship: a problematic of concept). *Moroccan Journal of Prospective in Management Sciences*, (2).

Bornstein, D & David, S. (2012). *Emprendedores sociales: Lo que todos necesitan saber (Social entrepreneurs: What everyone needs to know)*. Edition Debate, Madrid.

Bornstein, D., Taudière, I., & Clarinard, R. (2005). *Comment changer le monde: Les entrepreneurs sociaux et le pouvoir des idées nouvelles (How to change the world: Social entrepreneurs and the power of new ideas)*: Editor-La découverte.

Boschee, J. (1995). Some nonprofits are not only thinking about the unthinkable, they're doing it running a profit. *Across the Board, the Conference Board Magazine* (Vol. 32, No. 3, pp. 20–25).

Brouard, F. (2007). Social entrepreneurship: Towards a better understanding of the concept (L'entrepreneuriat social, mieux connaitre le concept). Available at SSRN 1326958.

Brouard, F., & Larivet, S. (2010). Essay of clarifications and definitions of the related concepts of social enterprise, social entrepreneur and social entrepreneurship. *Handbook of research on social entrepreneurship*, 29–56.

Brouard, F., Larivet, S. & Sakka, O. (2012). Défis et actions pour développer l'entrepreneuriat social (Challenges and actions to developing social entrepreneurship).*The Journal of Management Sciences*, 255–256(3), 17–22. doi:10.3917/rsg.255.0017.

Brundtland, G. (1987). Our Common Future—Call for Action. *Environmental Conservation*, 14(4), 291–294. Doi:10.1017/S0376892900016805.

CESE (le Conseil Economique, Sociale et Environnemental). (2015). *Report on "Social economy: a lever for an inclusive growth".*

CESE: Economic and Social Council of Spain, Economic and Social Council (OKE) of Greece, Economic and Social Council of Jordan et al. (2018). *Final Euromed report euromed on social economy and entrepreneurship in the Euro-mediterranean region.*

Chauffaut, D., Lensing-Hebben, C., & Noya, A. (2013). *L'entrepreneuriat social en France. Réflexions et bonnes pratiques (Social Entrepreneurship in France. Evaluation and good practices) Paris: La Documentation française, Reports and documents from the Centre d'Analyse Stratégique*, (56).

CODES. (2007). *Pour une approche partagée de l'entrepreneuriat social et de son développement (For a shared approach to social entrepreneurship and its development)*, The Codès Note No. 1.

Coupez, B., Lutzel, E., & De Reille, X. (2009). *Paradoxes et défis de la microfinance: Une industrie qui résiste à la crise, une nouvelle classe d'actifs (Paradoxes and challenges of microfinance: A crisis-resistant industry, a new asset class). Report on Money in the World*.

Davis, S. M. (2002). Social entrepreneurship: Towards an entrepreneurial culture for social and economic development. Available at SSRN 978868.

De Hoe, R. & Janssen, F. (2014). L'entrepreneuriat social et l'entrepreneuriat durable sont-ils liés? (Are social entrepreneurship and sustainable entrepreneurship related?). In: *Raymond Guillouzo, Entrepreneurship, sustainable development and territory's: Contextualized Approaches*, http://hdl.handle.net/2078.1/153396.

De Miras, C. (2007). Initiative nationale pour le développement humain et économie solidaire au Maroc pour un accès élargi à l'eau et à l'assainissement (National initiative for human development and solidarity economy in Morocco for a wider access to water and sanitation). *Tiers Monde Journal*, (2), 357–377.

Dees, J. G. (1998). *The meaning of "social entrepreneurship"*. Comments and suggestions contributed from the Social Entrepreneurship Founders Working Group. Durham, NC: Center for the Advancement of Social Entrepreneurship, Fuqua School of Business, Duke University.

Diab, M. B. (2019, May). Towards Social Entrepreneurship and Sustainable Development in Lebanon. *In Proceedings of the International Conference on Business Excellence* (Vol. 13, No. 1, pp. 56–70). Sciendo.

Diani, A. (n.d.). *L'Entrepreneuriat social au Maroc: État des lieux d'un secteur en quête de reconnaissance (Social entrepreneurship in Morocco: An overview of a sector seeking Recognition)*. https://forumess2017.sciencesconf.org/data/pages/DIANI_C7.pdf.

Fortin, P. A., (2002). *La culture entrepreneuriale, un antidote à la pauvreté (The entrepreneurial culture, an antidote to poverty)*. Éditions Transcontinental.

Guzmán Vásquez, A., & Trujillo Dávila, M. A. (2008). Emprendimiento social—Revisión de literatura (Social entrepreneurship—Literature review). *Management Studies*, 24(109), 105–126. https://doi.org/10.1016/S0123-5923(08)70055-X.

Hidalgo, L. F., Rialp, J., & Urbano, D. (2020). Are there really differences between social and commercial entrepreneurship in developing countries? An institutional approach. In *Handbook of research on smart territories and entrepreneurial ecosystems for social innovation and sustainable growth* (pp. 306–325). IGI Global.

Ilbert, H. & Ciheam, I. (2014). *Contribution à l'étude des enjeux et défis de l'entrepreneuriat social dans les pays en développement: cas du Maroc (Contribution to the study of the stakes and challenges of social entrepreneurship in developing countries: the case of Morocco)*. Memory search.

Janssen, F., Bacq, S. & Brouard, F. (2012). L'entrepreneuriat social: Un thème pour la recherche passée, présente et future (Social entrepreneurship: A theme for past, present and future research). *International Journal SME*, 25 (3–4), 17–44. https://doi.org/10.7202/1018416ar.

Khassal, H. (2018). *Entrepreneuriat social: Une approche dynamique par le business model (Social entrepreneurship: A dynamic approach through the business model)*, PhD Thesis.

Layadi, H., Rouggani. K., & Bouayad. A.N. (2019). L'Entrepreneuriat social au Maroc: Définitions, enjeux et réalité (Social entrepreneurship in Morocco: Definitions, challenges and reality). *Moroccan Journal of Foresight in Management Sciences*, no. 2.

Le Velly, R. (2014). Entrepreneuriat social (Social Entrepreneurship). In Chauvin, P-M. et al. *Sociological dictionary of entrepreneurship*, Presses de Sciences Po (PFNSP), 191–203.

Leger-Jarniou, C. (2008). *Accompagnement des créateurs d'entreprise: regard critique et propositions. (Support for creators of companies: critical view and proposals)*. *Market and Organizations*, (1), 73–97. URL: https://www.cairn.info/revue-marche-et-organisations-2008-1-page-73.htm.

Lelart, M. (2007). Le père du microcrédit honoré par le prix Nobel de la paix (The father of microcredit honored by the Nobel Peace Prize). *Journal of Political Economy*, 117(2), 197–208. URL: https://www.cairn.info/revue-d-economie-politique-2007-2-page-197.htm

Loudiyi S. & Louhmadi A. (2016). La contribution de l'entrepreneuriat social dans l'accomplissement des exigences du développement durable. *Cas des coopératives artisanales de Tanger-Fahs Anjra (the contribution of social entrepreneurship in meeting the requirements of sustainable development. The case of artisanal cooperatives in Tangier-Fahs Anjra).*

Mair, J. & Marti, I. (2006). Social entrepreneurship research: A source of explanation, prediction, and delight. *Journal of World Business*, 41(1), 36–44.

Mancebo, F. (2010). Conclusion générale. In *Le développement durable (Sustainable development)* (pp. 295–302). Paris: Armand Colin.

MEMDD. (2017, October). *Stratégie nationale de développement durable 2030 (National Sustainable Development Strategy 2030)*. Executive Summary-Kingdom of Morocco.

Mendoza, E. A. S., Alva, E. E. G., & Ramos, O. C. B. (2020). Habilidades competitivas para el emprendimiento social en el Estado de MÉXICO (Competitive abilities for social entrepreneurship in the State of MEXICO). *Journal of Sustainable Development, Business, Entrepreneurship and Education*, (04).

MTF-GROUP "LE MATIN". (2018, Avril). « *CAP-2022, Tous ensemble pour la promotion de l'entrepreneuriat social au Maroc (All together for the promotion of social entrepreneurship in Morocco).*

Mulgan, G., Tucker, S., Ali, R., & Sanders, B. (2007). *Social innovation: What it is, why it matters and how it can be accelerated. URL:* http://eureka.sbs.ox.ac.uk/761/.

Ndour, M. (2017). *L'évolution du business model de l'entreprise sociale, le cas des entreprises des TIC: une étude comparative de cas France/Sénégal (The evolution of the social enterprise business model, the case of ICT enterprises: a comparative case study France/Senegal)*. PhD thesis.

OECD. (1999). *Les entreprises sociales (Social enterprises)*, OECD, Paris.

OECD. (2001). *Les entreprises sociales*, OECD Publishing, Paris, https://doi.org/10.1787/9789264282339-fr.

OECD (2015), *Renforcer la contribution de l'entrepreneuriat social au développement durable, dans development Co-operation Report 2014: Mobilising resources for sustainable development*, Éditions OECD, Paris. DOI: https://doi.org/10.1787/dcr-2014-20-fr.

OECD/Union européenne. (2013). Synthèse sur l'entrepreneuriat social - L'activité entrepreneuriale en Europe (Synthesis on social entrepreneurship - Entrepreneurial activity in Europe), Luxembourg.

Omrane, A. and Fayolle, A. (2010). *L'entrepreneuriat social et le développement durable: quels modèles d'affaires dans le champ social? (Social entrepreneurship and sustainable development: which business models in the social field?)*, 19th AIMS conference, 01–04 June, Luxembourg, France.

Peredo, A. M., & McLean, M. (2006). Social entrepreneurship: A critical review of the concept. *Journal of World Business*, 41(1), 56–65.

Rouxhet, M. (2017). Le crowdfunding comme outil de financement des coopératives: analyse des motivations des coopératives (Crowdfunding as a tool for financing cooperatives: analysis of the motivations of cooperatives). Master's thesis in management sciences, University of Liège.

Shams, R., Vrontis, D., Belyaeva, Z., Thrassou, A., & Christofi, M. (2018). Historical perspectives on social business enterprises: Looking backward to move forward.

Journal of Social Entrepreneurship, 9(3), 288–293. doi:10.1080/19420676.2018 .1494778.

Singh, A., & Reji, E. (Ed.). (2020). *Social entrepreneurship and sustainable development.* London: Routledge India, https://doi.org/10.4324/9781003042396.

Tallie Hausser, T., Tsakas, C., & Moukaddem, K. (2019). *Developing social entrepreneurship and social innovation in the Mediterranean and Middle East, in collaboration with Mediterranean Institute.*

Tilley, F. & Young, W. (2009). Sustainability entrepreneurs. *Greener Management International*, (55).

Tsakas, C., & Moukaddem, K. (2019). Comment l'entreprenariat social peut-il contri-buer à développer le secteur privé et appuyer la croissance et l'emploi dans les PM? (How can social entrepreneurship help develop the private sector and support growth and employment in the MP?) In Patricia Augier, Constantin Tsakas, Sami Mouley, Karine Moukaddem, & Jocelyn Ventura (Eds.), *Le secteur privé dans les pays Méditerranées principaux dysfonctionnements et opportunités de l'entrepreneuriat social- FEMISE Rapport euromed* (pp. 173).

Yunus, M. (2011). *Pour une économie plus humaine (For a more human economy).* JC Lattès.

Zahra, S. A., Gedajlovic, E., Neubaum, D. O., & Shulman, J. M. (2009). A typology of social entrepreneurs: Motives, search processes and ethical challenges. *Journal of Business Venturing*, 24(5), 519–532.

5 Chemistry for Energy Conversion and Fossil Free Sustainable Enterprise

Isak Rajjak Shaikh

CONTENTS

5.1 INTRODUCTION

As humans, we know that the fossil fuel based energy system costs us a lot. This is not only because of the volatile fuel prices that drive the economy or cause economic instability, but it is also due to the fact that coal, oil, and gas increase human vulnerability. Coal mining and coal-fired power plants face a wide range of occupational health issues, whereas the oil and gas industry, be it onshore or offshore, mostly lives with the serious hazards of gas leakage and catastrophic oil spillages impacting marine life (The Ocean Portal Team, Smithsonian Institution, 2018).

Fossil fuels account for 80% of global energy consumption and nearly 75% of greenhouse gas (GHG) emissions (IPCC, 2014). Indirect GHG emissions from oil and gas operations today are equivalent to around 5200 million tonnes of carbon dioxide (CO_2) (IEA Flagship Report, 2018). World Health Organization (WHO) data shows that air pollution caused by fossil fuel burning takes 4.2 million lives per year.

And more than 80% of the officially categorized urban population on the planet is exposed to air quality levels that exceed WHO limits (WHO, 2016).

As rapid industrialization and the continuing population boom drive the need for energy, the present-day energy system is increasingly seen as inflicting an adverse impact on the recycling of matter and the flow of energy in an ecosystem which consists of a community of organisms in a specific locality together with their abundant biodiversity and physico-chemical regulatory systems, including air, soil, freshwater, oceans, climate, sunlight, and the atmosphere (Zachos et al., 2001; Parmesan, 2006). For some island nations and coastal or riverine areas with "eco-anxiety" or "climate tragedy", there is a profound existential crisis compounded with the realization that this world is much more fragile than once thought, owing to the several negative environmental impacts of fossil fuel burning, global warming, pollution, the associated multiple adverse health effects – especially in children (Frederica, 2017) – unexpected natural calamities and weather change, landscape deterioration, and above all, the scientific uncertainty surrounding it (Butt et al., 2013). So one can only imagine in the future how alarming climate change and/or these safety and environmental health impacts and their consequences could be for human life, life-supporting systems, and the ecological balance.

Though it appears that the cause of climate change is the chemical conversion of carbonaceous materials into thermal and derived forms of energy, the transition from fossil energy systems to more sustainable energy supply systems is impossible without chemical energy conversion. Innovation in green or sustainable chemistry and engineering will be key to transitioning to a cleaner energy system. This chapter seeks to elucidate the pivotal role of chemistry in particular to deliver a set of novel energy conversion methodologies with inventive steps and the possibility of commercialization with high standards of scalability and sustainability. The chapter summarizes the types of chemistry-based renewable energy options noted in the literature and then evaluates entrepreneurial opportunities to address them in a state-of-the-art and in a what-if manner.

5.2 DRIVING FORCES IN FUTURISTIC ENERGY ENTERPRISE

5.2.1 SUSTAINABILITY: ANOTHER LOOK AT RESOURCES AND RENEWABILITY

The issue of climate change is caused by the neglect of a kinetic term describing "the time constant of equilibrium between input of energy carriers and output of waste" (Berntsen et al., 2006; Archer & Brovkin, 2008; Van Hise, 2008). The term "sustainability" refers to the *endurance of systems and processes* and describes *the kinetics and the balance between the input and output.*

Concerted international efforts are underway to practice sustainability or the transition to sustainable production and consumption patterns (UNESCO, 1997). A new business concept "eco-efficiency" (Stephan Schmidheiny with WBCSD, 1992), as the way the private sector is advised to use it today, has emerged for implementing "Agenda-21" (UNCED report, 1992) with regards to sustainable development (Brundtland Commission report, 1987; Johannesburg Declaration, 2002). The United Nations, with a set of 17 interconnected sustainable development goals (SDGs) and

169 associated targets, defines global priorities and aspirations for the year 2030 in order to "achieve a better and more sustainable future for all" (UN General Assembly, 2015). Environmental policy making and the development of a knowledge-based society and "green" economy are highly desirable (UNEP, 2011) through entrepreneurial sustainable development (UN, 2017; Johnson & Schaltegger, 2019). We need values, compliance, and contributions from individuals, governmental institutions, and private players in regards to energy and resource efficiency, occupational safety, and environmental protection, while practicing entrepreneurship and corporate social responsibility (Dekra Sustainability Magazine, 2017–18). The "doughnut model" devised by Kate Raworth is found to be useful as a guide to public policy making that meets the core needs of all through economic activities that live within the means of the planet, or thrive in balance with the planet (Raworth, 2017). Amsterdam, for example, has adopted this global concept of the "doughnut" and turned it into a transformative action tool in the city (The Full Amsterdam Circular Strategy, 2020/2025).

5.2.2 ENERGY, ENVIRONMENT, AND ECONOMY

Energy, environment, and economy are highly interconnected (Tiba & Omri, 2017). For example, Gozgora et al. (2018) examined the relationship between energy consumption and economic growth for some Organisation for Economic Co-operation and Development (OECD) countries. And the efficient management of the energy–environment–growth nexus – (i) by reducing total energy consumption, and (ii) by developing economic and environmental policies that provide incentive mechanisms for renewable energy projects and which encourage the private sector to participate are identified as a must for sustainable growth (Dinç & Akdoğan, 2019).

Figure 5.1 shows the current energy system and the simplistic role of chemical conversions in the transformation of fossil, low-fossil, and possibly no-fossil based energy system.

It was the World Bank's 1999 energy-sector strategy *Fuel for Thought* that helped increase investment and lending for renewable energy (Ahmed, 1994; Anderson & Ahmed, 1995; World Bank, 1999). In 2010, the European Union launched a 10-year growth and job strategy called "Europe 2020" which set targets in key areas of employment, education, research and development, climate/energy, social inclusion and poverty reduction (European Commission, 2014). The Europe 2020 strategy is supported by seven "flagship initiatives" in innovation, the digital economy, employment, youth, industrial policy, poverty, and resource efficiency. And the Commission's 100 billion Euro research and innovation plan - the "Horizon Europe" program (Horizon Europe, 2020) identifies five selected mission areas including (i) the adaptation to climate change including societal transformation, (ii) cancer, (iii) climate-neutral or smart cities, (iv) healthy oceans, seas, coastal or inland waters, (v) soil, health and food.

Long ago, the Stockholm Environment Institute prepared global GHG and energy scenarios to identify the cost-competitive potential for phasing out the use of all fossil fuels to minimize emissions by adopting Greenpeace International's normative target of zero carbon emissions by the year 2100 (UNEP, 2019). However, recent research shows that by 2030, the world is planning on producing far more coal, oil,

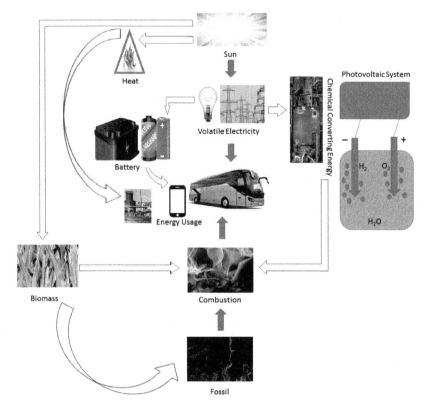

FIGURE 5.1 The current energy system and the simplistic role of chemical conversions in the transformation of fossil, low-fossil, and possibly no-fossil based energy system

and gas rather than abiding by the Paris Agreement (UNFCC, 2015). For example, Australian coal exports doubled and liquefied natural gas (LNG) exports tripled from 2000 to 2015 (Lazarus et al., 2019).

Asian giants like China, India, Japan, and South Korea are dependent on imported fuel. But they no longer wish to rely on unstable geopolitical equations and the pressure games or sanction threats from countries other than the two involved in exporting and importing. In recent decades, India and China have seen an enormous increase in polluted industrial zones and also an increase in the need for drafting suitable legislation on curbing pollution instead of lenient laws and traditional industrial practices. However, today, coal meets more than half of India's energy demand. And by the year 2040, its share of global energy demand is expected to double, which is the reason why India – once known as the coal king – is heavily investing in renewable sources (IEA, 2019; WEF, 2019). Renewables or renewable energy technologies have the potential to reduce hazards or eliminate pollution risks while providing a whole new range of entrepreneurial opportunities (Martinot, 2001) and a more planet and people-centered or inclusive growth model for a more resilient economy. India, home to 13 of the world's 20 most polluted cities, is the world's third-largest emitter of CO_2. Under the Paris Agreement (Climate action tracker, 2020), India has committed to generating 40% of its electricity from renewable sources and in addition to that has intended to reach a target of 450 GW of renewables by 2030 (World Bank brief, 2019). India has, in quick time, witnessed an unprecedented growth in its solar power industry and has also the lowest capital cost per MW globally for installing solar power plants (Muneer et al., 2005; Chandra et al., 2018; MNRE India, 2020). Loaded with over capacity of thermal power, coupled with only a tepid demand and a rising share of renewable energy, India is going to witness a marked shift toward an efficient supply and optimum generation mix.

The issue with regards to poor air quality in heavily industrialized developing nations needs to be addressed with appropriate policy making, while seeking the optimal balance between environmental quality and economic growth (Alberini et al., 1997); or it will become an issue (Chen et al., 2013; Ernst & Young, 2014; OECD, 2020) associated with increasing social unrest. On the bright side, the Global Commission set up by the International Renewable Energy Agency (IRENA) reported that China is set to become the world's renewable energy superpower owing to technology, with 29% of the renewable energy patents globally, well over 150,000 as of 2016 (Forbes, 2019).

5.2.3 Oil Diplomacy and the Petrodollar

Author Jane Kinninmont in her book *The GCC in 2020: The Gulf and its People* writes that the Gulf Cooperation Council (GCC)'s population explosion will continue while the labor market remains dependent on expat workers (Kinninmont, 2009). She adds that countries in the GCC are home to some of the world's youngest populations, and the future development of the region ultimately depends on the education and employment of these young people. And in view of the increase in industrial activities, energy consumption, and economic growth, there is a pressure on the existing refineries to seek new methods to optimize efficiency and throughput. Today, the demand for energy supply in the GCC region and also in all of the Middle East countries is increasing the expenditure of government utilities, and the energy supply capacity is dependent on capital investments in the oil and gas power sector. The increase in oil prices contributes to the decrease in the supply of some goods due to the increase in the costs of producing them and thereby aggravates the impact of oil shocks on the economy, at the regional as well as the global scale. And it is obvious that Middle East oil producers and locations or channels through which such business takes place attract immense attention. For example, almost 25% of the total global oil consumption and a third of the world's LNG make the Strait of Hormuz a very important strategic location and one of the world's busiest routes for international trade. Contemporary geopolitical strategies and situations like the Yom Kippur War (1973), the Iranian Revolution (1979), Operation Desert Shield/the Gulf War (1990–91) had triggered a jump in oil prices and negatively affected the increasingly urbanized and industrialized economies worldwide. Of late, the pandemic of the coronavirus disease in 2019 (Covid-19) drove down the oil price (IEA, 2020). Historical evidence shows that oil diplomacy or diplomatic deadlock with regards to fossil-fuel-based energy economics and the greed for possession of hydrocarbon reserves has led to economic sanctions as a control by fear strategy in those regions. And so, the mainstream news and media should not call the relationship between oil and international conflict an exaggeration (Belfer Center, 2013); there has always been this imminent threat of "resource wars" over possession of oil reserves while making changes to the leadership in the region, waging wars, depopulating the planet, defaming peaceful religion, creating paranoia across borders, and also inciting conflict between ethnicity and cultures in such strategically important locations.

In 1974, the United States of America and Saudi Arabia agreed to the use of the United States' dollar as the only currency for signing oil contracts and agreed in the year 1978 to petrodollar recycling. This cemented the oil–US dollar business relationship and since then any sign of a falling US dollar hurts the oil-exporting countries. Now the United States enforces its foreign policies by using the power of the petrodollar. In the 2000s, Iraq and Libya utterly failed in their separate attempts to challenge what they perceived as the United States' global hegemony through the petrodollar system. In fact, according to some accounts, the invasion of Iraq was due to the United States' rage over the United Nations' approval to Iraq's plan to begin using the euro as an alternative oil transaction currency. The euro being under attack from within (the eurozone crisis), there is still no good alternative world currency, except a tie-in to the gold standard, if all countries accept it! Countries like Venezuela and Iran have opted for signing oil deals in their own currencies. It would now be interesting to see if Saudi Arabia or other oil-producing nations accept China's call for a replacement of the US dollar as the only medium through which the world conducts oil transactions.

Of late, economists are finding a weakening link between the oil price and inflation. Apparently the US petrodollar is bound to lose its dominance (Robinson, 2012) as the world limits greenhouse gas emissions to fight global warming and shifts from oil to electric vehicles and solar or wind power generation (Bankruptcy of Our Nation, 2012; Hatfield, 2018; Amadeo, 2020; Robinson, 2020). The US is steadily losing its competitive edge in business, skills, and innovation with regard to renewable energy technologies to China and the European Union (Bosman & Scholten, 2013). But for some military complexes, strategists, and the business-monopoly political-mindset worldwide, making the current energy system sustainable or preferring renewable energy over fossil based energy is not only a technical but also an ideological challenge; they are more likely to choose strategies like keep going to war in the oil-producing nations or provoking geopolitical conflicts, depopulating the planet, imposing economic sanctions, or denying the "climate emergency" than one that delivers on the world's ambitious transition to cleaner energy which meets climate change goals and sustainable development. One can say this because some of the power structure lies in a persistent denial of the climate crisis and some of the industrial giants are trying to buy more time for bringing the emerging renewable energy sector under their crony capitalist exploitation and for taking over the sustainability campaign as it jeopardizes their economic monopoly. A new concept of philanthro-capitalism is also being floated around as if the increasingly unified activism and concerted international efforts tackling energy and climate crises are not doing enough. Large foundations or trusts are being established and sponsored by industrial tycoons, multinationals, and large companies in a way that the dividends only from their investments will keep fueling their current energy business that makes them richer; such a practice of philanthro-capitalism allows them to use not-for-profit/non-governmental organizations (NGOs) to take control of matters such as renewable energy, economy, entrepreneurship, innovation, climate change, or environmental protection, and even health. That's not philanthropy; it's a capitalist and democratic problem. *Planet of the Humans* – a documentary film, released free to the public on YouTube on the 50th Anniversary of Earth Day (April 22, 2020) – revealed the massive ecological impacts of a renewable-based energy system. In spite of the passivity around this issue, we know that it's just a transition phase; meeting the energy challenge is possible, and I see no fundamental reason to believe otherwise. Many coal-fired power plant owners have already retired their units in the USA (US EIA, 2019) and no matter how much the effort to shift away from fossil fuels is being criticized by the established politics on the industry-economics set-up, mixing efficient energy usage and conversions or inventing new affordable clean energy technologies would only contribute to the sustainability of the planet. In fact, researchers have recently hit back with an analysis of the scientific literature to debunk the myth (Heard et al., 2017) that 100% renewable energy systems are not possible. They claim that a shift to 100% renewables is technically feasible and economically viable (Brown et al., 2018).

5.3   PARADIGM SHIFT: UNDERSTANDING CHANGE IN THE ENERGY INFRASTRUCTURE

The book *Explosive Growth* was published on December 13, 2011 and in it the authors discussed "reshaping energy and infrastructure in our next 1000 days". Through a thorough analysis, Michael Rogol – a renowned consultant – warned the monopoly to adapt to a change in business practice, and he clearly showed how small businesses can take over an industrial sector dominated by giant corporations with assets of hundreds of billions of dollars as the economic feedback was bound to change the electricity system within 1000 days (Rogol with Rogol, 2011). As predicted, the electricity and renewable-based energy system has evolved to a level where oil isn't a precious commodity anymore, and ironically, in spite of oil crashing 321% into negative territory as demand evaporated (Forbes, 2020), many are still pushing for the old fossil-fuel-based growth model. Clayton Christensen in his book, *The Innovator's Dilemma: When New Technologies Cause Great Firms to Fall*, writes that the challenges mainly arise from: (i) new technology that is too expensive to threaten the old; (ii) new technology that does not meet established requirements; and (iii) new technology that does not fit the business model (Christensen, 2013). Belgium, Austria, and Sweden – in that order – became the first three European nations to eliminate coal completely from all sorts of heating and power generation (Europe Beyond Coal, 2020). And in Sweden, for example, researchers found that transitioning to a more sustainable energy system was considered very important and urgent in addressing global environmental and social challenges, while developing a theory-based evaluation framework that assesses and discusses both the robustness and transformative efforts of current policy and evaluation practices (Sandin et al., 2019). Jeremy Rifkin argues in his book *The Hydrogen Economy* that hydrogen could be the next generation's norm and could provide a plentiful and clean alternative to oil (Rifkin, 2003). The oil-producing nations (USEIA, 2020) are under significant economic pressure already, and even Middle Eastern economies are shifting towards revolutionizing the energy sector by choosing hydrogen as an alternative or at least as a crucial part of the global energy mix (The National, 2019). To some extent, in the current situation, hydrogen is extracted from fossil-based fuels. The conversion of carbon dioxide to valuable energy products is also an option offered by the chemistry research enterprise, though we are not doing enough there either. There is a need to achieve various (actually 17) goals of sustainable development, including Goal 4: Quality Education, and Goal 7: Affordable and Clean Energy. The concept of "scientific temper" (Nehru, 1989, p. 513) should also be added to such SDGs by which one may reject prejudice and embark upon a systematic search for the beauty of knowledge and truth. Scientific temper and a more open discussion may replace the partly strategic and socio-economic debate that creates hurdles for the generation and application of knowledge at a premature stage. Having said that, scientific temper not only lodges intellectual inquiry into analyzing, discussing, and then understanding or communicating issues of concern, but it also encompasses the adoption of a new understanding of previous conclusions in the face of new evidence. Activities that we advocate should mainly involve the co-evolution of innovation and entrepreneurship by identifying local problems and linking human beings to the global sustainability system.

Figure 5.2 shows the relationship between the energy system and sustainability.

As projected by International Energy Outlook, renewables are expected to be fast-growing energy sources worldwide, increasing by an average of 2.3% per year between 2015 and 2040 (USEIA, 2017). Renewable energy production is widely promoted across Europe and the tools aiming at reducing carbon emissions by 20%

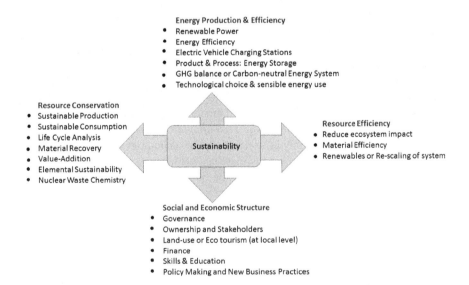

FIGURE 5.2   The energy system and sustainability

have been developed directly by the European Union (EU) and directed under the "Intelligent Energy Europe" program in spite of the disparity in priorities and support mechanisms in the member states (Haas et al., 2011; Intelligent Energy Europe, 2020). The government of the Republic of Ireland has also launched an ambitious climate action plan with the aim of reaching net zero-carbon emissions by 2050 (DCCAE, 2019). In 2013, the "New Climate Economy", a flagship project, was commissioned by the governments of Colombia, Ethiopia, Indonesia, Norway, South Korea, Sweden, and the United Kingdom to provide relevant help to governments, businesses, and society in making better-informed decisions in the quest to achieve economic prosperity while dealing with climate change. Its 2014 flagship report concluded that higher quality growth can be combined with strong climate action; and this new approach is expected to deliver higher productivity, more resilient economies, and greater social inclusion (New Climate Economy report, 2018).

5.3.1 Electricity

Electricity, the secondary source of energy, is the flow of charge or electrical power. It can be produced by converting coal, natural gas, solar energy, wind energy, or nuclear energy into electrical power. It is neither renewable nor non-renewable, but can be produced by both renewable and non-renewable primary sources of energy and can further be converted to other forms of energy such as heat or mechanical energy. The generation, storage, and regeneration of electricity is essential to solving energy problems. In the present-day scenario, electricity generation through thermo-mechanical systems (i.e., cooling or heating applications) still exists side-by-side with chemical energy conversion in its variants. Fuel cells or electrolysis or hydrogen generation would be much preferred as on-site or on-board energy applications. And there is much research endeavor being directed in this area. Solar energy is the best of our success stories; but the so-called elite or business monopoly in some countries are not helping the cause of shifting completely toward sun or wind energy. On the bright side, many have already started pairing wind or solar energy with traditional power to create a hybrid system. China tops the list in developing electricity from wind, followed in order by USA, Germany, and Spain. The Enel Group of companies and also Enel Green Power SpA and its subsidiaries, aiming to fully decarbonize their energy generation mix by 2050, built around 3029 MW of new renewable capacity all over the world in 2019 and thus bettered the previous year by 6.5%, adding around 190 MW (REVE magazine, 2020). Italian oil and gas giant Eni also announced a long-term energy shift away from fossil fuels by 2050. Morocco, a north African country that has relied for almost 99% of its energy needs on imported fossil fuels, successfully operates Africa's largest wind farm and has recently set a 52% renewable energy target by 2030 (Parkinson, 2016).

How energy can be stored is one of those life-changing questions to ask yourself today. In view of the Paris Agreement's long-term goal to keep the increase in the global average temperature to well below 2°C above pre-industrial levels, the International Energy Association (IEA) estimated that the world needs to increase the 176.5 GW of energy storage in 2017 to 266 GW by 2030. In a renewable energy world that sees energy storage technologies at center-stage, market players and policy makers are increasingly turning their attention to electrical energy storage technologies, and researchers and engineers cannot overlook the need for increased flexibility, a game where such technologies are attractive to power generators on account of the increased overall utilization of power system assets and which translate into higher average revenues and a low(er) risk of overcapacity. Climate change mitigation and energy storage solutions such as these address the issues of solar and wind power intermittency, capacity, and resilience of energy grids while making the grid more responsive, and at times responding quickly to large fluctuations, and most importantly, reducing the need to build backup power plants. Electrical energy storage technologies are broadly classified into:

1. Mechanical: (i) a pumped hydro sorage system (PHS); (ii) compressed air energy storage (CAES), and (iii) a flywheel (FES).
2. Electrochemical: (i) a secondary battery in the form of a lead-acid battery/ NaS/Li-ion battery, (ii) a flow battery with Redox flow/hybrid flow, and (iii) solid-state battery technology.
3. Electrical: (i) a (super) capacitor, and (ii) superconducting magnetic-SMES.
4. Thermochemical: solar fuels, solar hydrogen.
5. Chemical: hydrogen/fuel cell/electrolyzer.
6. Thermal: sensible/latent heat storage, including molten salt.

Today battery storage systems account for only around 4 GW of storage capacity, while the PHS system accounts for around 153 GW of storage worldwide. Battery technologies are quite expensive and PHS is largely constrained by the location of suitable sites. Universal access to energy sounds like a utopia and though batteries appear as a beautiful compromise, we know that they work well but: (i) on different scales (a few Wh to MWh, which are not always scalable), (ii) at a different efficiency (some are very inefficient, some not), (iii) at different costs (some with high $/kWh), and (iv) are difficult to combine energy and/or power.

5.3.2   A Shift Toward Digitalized Energy Technology and Sustainable Entrepreneurship

Digitalization by definition is the process of converting everything into digital format. Digitalization includes the adoption of digital technologies to modify a business model and integrate it into everyday life to create value from the process. Digitalization encompasses the use of novel technologies by exploiting advanced digital network dynamics and the giant digital flow of information (WPI, 2018; IGI Global, 2019). Digitalization, through sensors and predictive analysis, helps in optimizing performance in conventional as well as renewable energy plants. The development of smart power grids, for example, provide benefits such as better electricity transmission and integration of renewable energy systems. Digitalization influences the energy value chain and technologies like Big Data Analytics & Software, Digital Twins, and the open cloud-based Internet of Things that operating platforms such as Mind Sphere can use to boost efficiency in energy generation. Industrial entities use digitalization to improve manufacturing processes, enhance the quality of goods, and reduce the cost of production by reducing input and increasing output, whereas utility customers use digitally enabled electrification technologies (smart meters, power-save-mode gadgets) to monitor or lower consumption. On-site power or microgeneration is on the rise to supplement their respective needs rather than only relying on traditional centralized grid-connected power.

Many innovative ideas are progressing: (i) to generate and use energy more efficiently and bring about behavioral adaptations; (ii) to test the energy mix and reshape the power supply and demand by developing utility-scale energy-storage systems; (iii) to develop carbon capture and storage (CCS) technology for preventing CO_2, for example, from entering the atmosphere and for recycling carbon; (iv) to help produce renewable energy cost-effectively; (v) to store energy from solar cells, wind turbines, and manufacture rechargeable batteries; (vi) to bring in technologically viable set-ups; (vii) to combat climate change; and (viii) to wean ourselves off fossil fuel dependency or look for alternative energies other than coal, oil, and natural gas (Blanchard & Galí, 2007). Some useful examples of renewable energies include: (i) solar power (photovoltaics, active and passive solar heating); (ii) wind power (onshore and offshore); (iii) water (tidal energy, hydropower, wave power); (iv) geothermal energy; (v) biofuel (biomass, forest and agricultural waste stream, energy crops, sewage or landfill gas, etc.). An observation of mine in this respect is that, for sure, the fossil based energy system is highly complicated; the "few" are preferred over the many, and it is mainly centralized. The demand for decentralized energy infrastructure, the democratization of energy, and universal access to energy is higher than ever.

In 2019, Microsoft and Vattenfall in Sweden started a pilot program that offers customers the ability to match its power with renewables; this is done on an hour-by-hour basis by using smart meters to measure real-time energy consumption and guarantees of origin that verify the electricity is from solar, wind, or hydropower only (Vattenfall Sweden, 2019).

Google, the American multinational limited liability company (LLC) specialized in internet-related products and services, will now use renewable forecasting data from Danish startup "Tomorrow" to shift its own computing loads to better match its low-carbon electricity supply. This is a continuation of Google's white paper entitled, "The Internet is 24×7 – Carbon-free Energy Should Be Too".

Switzerland, under the Swiss Energy Strategy 2050, aims at replacing 30% of the electricity from nuclear plants with renewable energy. The "HyEnergy" project maps incident solar radiation through the country for every single hour of the year by using satellite measurements from machine learning technology and the statistical power of Big Data. Such a methodology is found to be useful in measuring renewable energy potential – for sun, wind, and geothermal – and provide a basis for future decision making in transitioning to decentralized local grids and energy distribution.

The practical value of institutional entrepreneurship is worth studying (Heiskanen et al., 2019), and within the energy field, it's interesting to see how various types of institutional entrepreneurships have promoted wind power "differently" in India and Finland (Jolly et al., 2016). Energy stakeholders are now almost forced to implement much more innovation-based business practices and entrepreneurial development with regard to clean energy and renewable energy based projects associated with better finance models describing a robust energy system to which many aspire. And it is obvious that, as a solution, a "low-carbon economy" or "clean electricity" are expected to contribute to the energy mix by reducing the dependency on "carbon-based" energy, for example, on internal combustion engine vehicles by way of employing well-planned and carefully operated variable and non-synchronous sources of power generation integrating solar photovoltaics and wind energy systems while the power grid is coupled with an energy storage facility that manages fluctuating renewable energy. Grids on a large scale and energy storage devices or batteries on a small scale on a single charge will help us to use renewable energy from the sun, for example, in transportation. Some leading civil societies and business entities are calling for a recovery from the environmental and economic crises (i) by adopting to a long-term growth model based on net-zero emissions, (ii) by supporting regulatory reforms and making wise decisions about the grid and storage investments – to be addressed together, stating that such a comprehensive approach would put the world on a stronger and cleaner footing and, some years from now, provide guidance to stakeholders built on a carefully planned sustainable business trajectory avoiding short-term, GHG emission-intensive, instant, money-making businesses or some sort of cover-up with a war-like situation, diplomatic-deadlock, and/or a pandemic. Of late, Repsol and British Petroleum have also committed to reduce GHG emissions to zero. Al Dhafra – the world's largest solar project – is planning on producing electricity and providing the lowest-ever energy tariffs (according to the Abu Dhabi Power Corporation, UAE). Recently, HydroWing and Tocardo, in an ambitious decarbonization plan, jointly announced tidal hydrogen generation, storage, and offtake. Innovation in green or sustainable chemistry and engineering will be key to transitioning to clean energy production and storage devices or technologies. I will now discuss some of the interesting chemistry research, case studies, and ventures that are helping us smoothly transition to renewable based energy systems.

5.4 CHEMISTRY TO COMBAT CLIMATE CHANGE AND THE ENERGY CHALLENGE

5.4.1 CHEMISTRY RESEARCH AND ENTERPRISE

The Energy Statistics Database maintained at the United Nations Statistics Division is a continued effort to provide a comprehensive picture of the energy sector worldwide and thereby contains integrated and updated comprehensive energy statistics on the production, trade, conversion, and final consumption of primary and secondary, conventional and non-conventional, and new and renewable sources of energy (UN The Energy Statistics Database, 2020). This dataset is sourced from more than 230 countries/territories and relates also to alcohol consumption by the chemical industry.

As per the law of the conservation of energy, science takes the view that energy can neither be created nor destroyed; rather it can only be transformed or transferred from one form to another. Though global warming is caused by the chemical conversion of carbonaceous materials into accumulated heat emissions or derived forms of

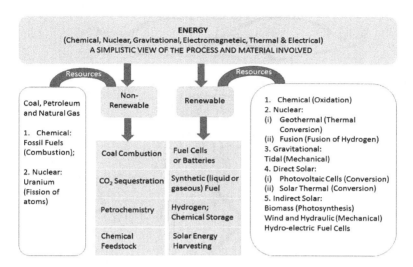

FIGURE 5.3 A simplistic view of the process and material involved in various different sources, resources, or forms of energy

energy, chemistry can play a critical role in controlling climate-forcing agents and energy loss, and the transition from a fossil-fuel-based energy system to a more sustainable one is impossible without the science of chemistry (see Figure 5.3).

Scientifically, chemistry is at the heart of petroleum products and/or their conversions. Synfuel is defined as a synthetic crude or gaseous and/or synthetic liquid product from the chemical conversion (AEO, 2006) of syngas (i.e., it is a mixture of carbon monoxide and hydrogen) from coal, natural gas, biomass, industrial, or other (municipal) solid waste feedstocks (Patel, 2007; Seo et al., 2018). The worldwide synfuel production capacity was around 240,000 barrels per day in accordance with the data obtained in July 2019. This synthetic fuel is refined by Fischer-Tropsch synthesis or coal liquefaction or methanol to gasoline conversion. Synfuel and syngas are both useful as an important chemical feedstock for further value-addition (Höök & Aleklett, 2010; Shaikh, 2014a).

Science, engineering, and chemistry in particular, intertwined with sustainability, will have to deliver a set of novel innovative energy technologies with high standards in stability, scalability, and sustainability. Chemistry can reduce losses from converting energy carriers. Chemistry can create energy materials for low carbon technologies (Islam et al., 2019). Though reduction of CO_2 and its valorization is still a material challenge on the energy efficiency side, chemical industries generate electricity and steam by using combined heat and power (CHP) plants. Such cogeneration plants are very efficient regarding fuel and in the supply of energy on an industrial scale. What we are looking at are some processes such as combustion and the syntheses of functional materials, for example. The chemical industry is therefore leading from the front in energy efficiency. On the utility level, the chemical industry provides smart solutions for energy efficient materials for building insulation, sealants and wraps in houses and factories, home appliances such as radiators or heating and cooling devices, solid-state lighting, electronics, health care products, textiles,

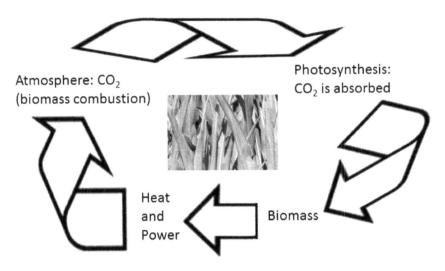

FIGURE 5.4   The carbon dioxide cycle through biomass

lightweight plastic auto parts that make car's fuel efficient, and aerospace products. Modern chemistry also seeks to establish a technology-neutral approach by carefully studying the "life-cycle assessment" (Herrchen & Klein, 2000) of products and processes, say for example, in the case of the biomass or valorization of an agricultural waste-stream (see Figure 5.4).

The Leiden Institute of Environmental Sciences, at the University of Leiden in the Netherlands, has the world's largest raw materials database and gathers essential knowledge about how these materials are used (Leiden University, 2020).

DuPont, an American multinational chemical company, through its business of innovation is committing to sustainability; its Solamet® – e.g., photovoltaic metallization pastes – is helping to make solar panels more energy efficient by increasing their power output by around 30% (Konno et al., 2010). Chemours Teflon EcoElite™ – biobased, durable, non-fluorinated, water-repellent, textile finish, protective clothes – reduces water and energy use for outdoor gear or sportswear (Brown et al., 2018). ExxonMobil Corp. technology is transforming algae into low-carbon clean energy. The Dow chemical company is producing silicones useful in solar panels and wind turbines which helps them resist solar radiation and/or extreme weather conditions.

While the development of lithium-ion batteries by notable chemists will provide energy storage in a fossil-fuel-free future (Castelvecchi & Stoye, 2019), the fluoride-ion battery, at the same time, is the potential next-generation, room temperature, rechargeable battery offering a very high energy-density device for application (Davis et al., 2018). Chemistry and the chemical industries can decide how to produce low-carbon or no-carbon energy, to store or distribute it, or to digitalize and consume it sensibly.

Under high-pressure industrial reactions, the conversion of CO_2 into clean fuels requires the right kind of solid catalysts; heterogeneous catalysis is a surface phenomenon, and such catalysts are useful in synthesizing fine chemicals and sometimes selective molecules or chemical building-blocks (Burri et al., 2007; Ross, 2011; Shaikh, 2014b).

Laboratory materials catalyze artificial photosynthesis, convert solar energy, and store it in the chemical bonds of organic molecules. Such catalytic systems outperform natural ones, as well as the demand for the development of suitable devices for synthetic as well as semi-synthetic strategies (Sokol et al., 2018). There is a widespread interest in electro-catalytic CO_2 reduction and using electrolysis to convert water and CO_2 into (carbon-containing) organic molecules, value added products, or fuels (Birdja et al., 2019; Qiao et al., 2014).

Technologies such as the CO_2 reduction over carbon-based electro-catalysts and the use of water and solar energy to recycle CO_2 into environmentally benign fuels over photo-catalysts address the issues of the global energy crisis and global warming simultaneously (Beydoun et al., 1999; Lubitz et al., 2008; Zhang & Reisner 2020). C.-T. Dinh et al. (2018) reported the use of thin copper-catalyst layers in a gas diffusion electrode leading a very selective and efficient electrochemical conversion of CO_2 to ethylene.

The recycling of CO_2 and water – the reverse of fuel combustion – yields liquid hydrocarbon fuel in non-biological processes using renewable or nuclear energy (Graves et al., 2011). This can enable a closed-loop carbon-neutral fuel cycle if atmospheric CO_2 is captured. In addition to that, CO_2 can be processed easily into methane, methanol, or dimethyl ether that burn cleanly in combustion engines (Catizzone et al., 2018). Methanol is a real "clean energy" option (European Commission, 2014; Araya et al., 2020; Methanol.org, 2020).

The steel industry is considered one of the highest CO_2 emitting industries. Swedish Steel India through its HYBRIT (Hydrogen Breakthrough Ironmaking Technology) initiative is working towards green chemistry replacing coking coke with hydrogen in steel manufacturing processes (Åhman et al., 2018).

Research that led to (i) understanding the photo-initiated charge-transfer processes on semiconductor interfaces and (ii) the practical application of dye-sensitized solar cell (DSCs) were really ground-breaking. This made a generation of scientists capitalize on the nanoscale for energy conversion since those days when nanotechnology was not a "buzzword". It's therefore worth noting here that the key scientific progress in solar cells and trendsetting energy applications could have won someone (O'Regan & Grätzel, 1991; Meyer, 2010) the Nobel Prize, but it hasn't (yet)! For the DSC to be turned into a successfully commercialized product, there is a need to address the current efficiency bottleneck (Graetzel, 1981; Peter, 2011). A lot needs to be done in developing methodologies to convert solar energy into chemical energy, making sustainable energy and the environment clean and energy production sustainable. Solar cells are set to be developed into more robust, reliable, and efficient energy technology for space applications as well, such as to power a Mars rover, robots, and also satellites, working under high energy radiation or harsh conditions in space (Kuendig et al., 2000; Cardinaletti et al., 2018; CESI, 2019; Espinet-Gonzalez et al., 2019).

Organic photovoltaics convert solar energy into electrical energy through light absorption (photon absorption and exciton formation), charge separation (exciton dissociation), charge transportation, and charge collection (by the electrodes). Researchers seek well-defined molecular design algorithms to concurrently optimize all these four processes (Yeh & Yeh, 2013). What we need now is a renewable energy storage system, so one can use energy when needed, irrespective of whether the wind is blowing or not, or whether the sun is shining or not.

Recent research on AquaPIM makes a case for the aqueous alkaline (flow-) battery technology for successfully producing long-lasting and low-cost grid batteries (Baran et al., 2019). AquaPIM is the polymer-based battery membrane based solely on readily available iron, zinc, and water. Lithium-ion batteries are useful in electric vehicles but the redox-flow batteries are found to possess even more of the much needed large energy capacity and to be able to discharge rapidly.

A scientific study published in 2015 made a radical suggestion on the combustion of metal particles to tap into the stored energy therein; in another approach aluminum was found to be reacting with water to produce hydrogen as a clean fuel (Bergthorson et al., 2015).

Reticular chemistry, the design principles and chemistry of the linking molecular building-blocks behind the syntheses of new classes of crystalline porous metal-organic frameworks (MOFs), zeolitic imidazolate frameworks (ZIFs), covalent organic frameworks (COFs), and related high surface area materials, have found potential applications in catalysis, electronics, gas separation, and onboard energy storage, while adsorbing hydrogen, methane, and CO_2 (Yaghi et al., 1998; Li et al., 1999; Yaghi & Li, 2011). In continuation with the idea of storing renewable energy, hydrogen is not only a combustion fuel (Nath & Das, 2007) but also an ideal renewable energy source. At standard pressure and temperature, hydrogen (H_2) is a colorless, odorless, and highly flammable gas. It can be stored as liquid. And water splitting into hydrogen could become an industrial version of photosynthesis in the near future (Walter et al., 2010). Producing electricity directly through an electrochemical process by feeding hydrogen gas into a fuel cell is a zero CO_2 emission technology. Thus, hydrogen is an energy carrier, and hydrogen fuel cells have the potential to power anything and build up a hydrogen economy by using electricity without toxic emissions (Bossel & Eliasson, 2003). Hydrogen could be the real game changer in the transportation sector; the hydrogen internal combustion engine vehicle (HICEV) uses a modified type of the gasoline-powered combustion engine, and the hydrogen fuel cell electric vehicle (FCEV) uses hydrogen electro-chemically, without combustion. The Hyndai Nexo, with its zero emission rating, is already available for sale in the UK while the Toyota Mirai and Honda Clarity are among other examples of hydrogen cars. Toyota partnered with BMW the "i Hydrogen NEXT" concept which is among the upcoming all-electric, no-combustion hydrogen car that one should be looking out for. Hydrogen – the most common element in the universe – is a very useful chemical input as well. Hydrogen gas reacts with CO_2 and nitrogen (N_2) or other gases to make methane or methanol (Chen et al., 2019) and ammonia (Schlögl, 2003) which can also be further transferred into fertilizers, plastics, or pharmaceuticals. Combining a clean electricity source with the electrolytic production of hydrogen gas is found to be economical and quantifies hydrogen generation as a solution to wind power intermittency (see Figure 5.5) (Glenk & Reichelstein, 2019).

About 95% of hydrogen produced globally is via the reforming of low-cost natural gas where methane reacts with steam under high (3–25 bar) pressure and high (700–1000°C) temperature over a nickel catalyst, leaving CO_2 as a waste stream. The process is strongly endothermic and is famously called steam methane reforming (SMR).

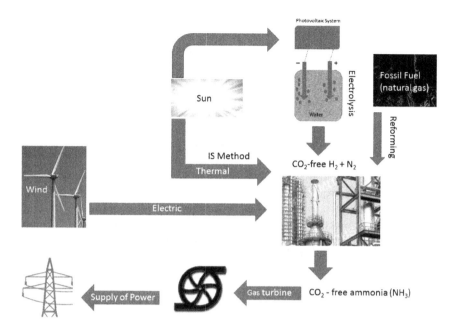

FIGURE 5.5 The energy chain utilizing CO_2 for the free ammonia synthesis for power generation

SMR:

$$CH_4 + H_2O \text{ (+heat)} \rightarrow CO + 3H_2$$

Auto-thermal reforming (ATR; using CO_2 and steam):

$$2\,CH_4 + O_2 + CO_2 \rightarrow 3\,H_2 + 3\,CO + H_2O$$

$$4\,CH_4 + O_2 + 2\,H_2O \rightarrow 10\,H_2 + 4\,CO$$

Water-gas shift reaction:

$$CO + H_2O \rightarrow CO_2 + H_2 \text{ (+small amount of heat)}$$

Partial oxidation of methane:

$$CH_4 + \tfrac{1}{2}O_2 \rightarrow CO + 2H_2 \text{ (+heat)}$$

Methanation (exothermic):

$$CO + 3H_2 \rightarrow CH_4 + H_2O, \ CO_2 + 4H_2 \rightarrow CH_4 + 2H_2O$$

SMR is more effective than ATR in producing hydrogen which is further used as a feedstock for fuel cell technology. A recent study reports electrified SMR that facilitates catalyst utilization and also the contact between heat source and reaction site to produce more hydrogen by reducing CO_2 formation (Wismann et al., 2019). Research in chemistry, reactor design, and engineering are therefore in high demand for greening fossil based energy and fine chemical technologies. Every chemical transformation can yield potential applications in energy management and thereby in developing molecular or material performance. In other words, research is needed to identify and investigate the chemical transformations that have high energy densities. We know that, for example, a chemical reaction, especially the reversible one, has the potential to convert, store, and manage or utilize thermal energy efficiently.

Water (H_2O) splitting by electrolysis produces H_2. On a small scale, researchers have succeeded in making hydrogen via electrolysis by converting electricity from the sun and/or the wind. This business can be grown. Since 2017, Nouryon, formerly AkzoNobel Specialty Chemicals, in Amsterdam, has been supplying hydrogen obtained as a byproduct of their chlorine production plant for operating hydrogen-powered buses at Frankfurt-Höchst industrial park in Germany. And Nouryon is also partnering with Tata Steel India, and the RISE institutes in Sweden to produce and use hydrogen renewably (Nouryon, 2018). Nouryon has teamed up with the Dutch energy company NUON to work toward incorporating hydrogen into the ammonia process. Nouryon in collaboration with the Dutch gas transporter Gasunie is investigating methanol synthesis using hydrogen and oxygen from electrolysis, plus CO, CO_2, and biomass (Nouryon, 2018). Aiming at Europe's largest green hydrogen project, the Shell oil company is partnering with Gasunie in the "North$_2$" initiative off the Netherlands and thereby plans to have in place 10 GW of wind turbines by 2040. In addition to that, Shell would like to advance the NortH$_2$ project with a large hydrogen electrolyzer at Eemshaven by linking up with Groningen Seaports.

Cooperatives have a different model of ownership than private business organizations Hansmann, 1996). Royal DSM, in the Netherlands, has a local initiative driven by citizen cooperatives and members of a green energy purchasing consortium to obtain energy from wind farms in south-west Holland. And Royal DSM deals with the Dutch energy supplier Eneco to aim at making 100% of electricity purchases in the Netherlands to come from renewable sources only (Ottewell, 2019).

The Haber-Bosch process is used for the industrial production of ammonia directly from nitrogen (N_2) and hydrogen (H_2), requiring H_2 that is obtained from coal or natural gas as a source under very high pressure and super-heated steam. But the factories emit a vast amount of carbon dioxide in the overall process. Hydrogen gas is fed to fuel cells. Fuel cells power vehicles. Fuel cells generate electricity from the energy stored in chemical bonds. Ammonia (NH_3) synthesized from sun, air, and water can offer the sustainable technology we need: ammonia can easily be cooled into liquid fuel and stored, and when needed it can be converted back into electricity or hydrogen gas (Zhou et al., 2017; Service, 2018). Malaysia, one of the largest supplier nations of palm oil, is developing various thermo-chemical processes for producing hydrogen rich gas from oil palm biomass (Mohammed et al., 2011).

The chemical recycling of CO_2 is near the top of the agenda for chemists (Centi & Perathoner, 2009). The Sabatier reaction, for example, is useful in catalytically transferring carbon dioxide into a carrier of chemical energy (Muller et al., 2013). Methane thus produced has immense chemical value and can be converted into methanol (Cui et al., 2018) or ammonia (Bai et al., 2018). Methane thereby holds a feedstock potential for the synthesis of organic chemicals and can also be fed into the existing natural gas network:

$$CO_2 + 4H_2 \rightarrow CH_4 + 2H_2O$$

Such technologies have once again brought catalysis to the center of chemistry in dealing with the climate challenge caused by fossil based energy. And topics such as this bring about a positive change in the energy–chemistry nexus through alternative energy technology and chemical material wealth creation for investing in sustainable entrepreneurship and also for considering environmental regulations that have started to influence present-day petroleum refining technologies. Chemical science will have to deliver much useful research with innovative clean energy technologies with plausible scalability for industrial applications (Perathoner et al., 2017). In an energy storage business, raw materials are very important. The EU's assessment of minerals shows that cobalt is critical and a conflict mineral. Though nickel replaces cobalt in many cases, it is only about three to four times more prevalent; finally, it's a well-known fact that lithium is not critical, but is highly strategic (Boissoneault, 2015; Drape, 2019; Malik, 2020). It would be interesting to see calcium (Nature news, 2019; Zhenyou et al., 2019) and other elements store solar and wind power and possibly replace lithium in batteries (Kavanagh et al., 2018).

5.4.2 PROGRESS IN FUTURISTIC ENERGY

Norway sees the future in renewables, especially in electricity from wind, sun, and water. It provides incentives to people who support the technological and business shift to renewables. Smart meters allow customers to harvest, store, and even sell solar energy back to power selling companies. And when it comes to Norway's smart-thinking business or wise investment decisions on renewable energy technologies developed by innovators, it's easy to believe that we can also stop driving other monopolized industrial economies on a one-way road to climate apocalypse. Norwegian company Ruden's iHEAT system is an entirely underground battery, developed using existing geological structures for storing heat from solar or wind power stations or waste incineration sites and then converting it into electricity. Oslo was honored with the European Green Capital award for 2019, on account of the many good works done by individuals and institutions in tackling climate change by conserving nature and restoring the waterway network. GHG emissions are cut by adopting electric-battery-based transportation, and the Oslo municipality aims for a circular waste system with well-functioning waste-to-energy, sorting, and biogas/fertilizer plants. High-lift heat pumps are used by Olyondo Technology for recycling heat from waste into high-temperature process steam for use in industries.

Fjords provide a structural advantage to Norway, and it is the largest hydro-powered nation in Europe and the seventh largest in the world. It is also the third largest exporter of natural gas and the eighth largest exporter of oil. Norway is trying to use the oil and gas sector for chemical commodities and feedstock conversion sustainably. It is now leading electric battery vehicles, the electric mobility revolution, and it has become a big market for electric cars and renewable-powered living. Equinor Norway and its subsidiaries such as Hywind Scotland and Hywind Tampen are among the first floating offshore wind farms in the world. The difficulty with regards to finding space is solved by building solar and wind farms on water. Ocean Sun is another company specializing in floating solar in Norway and will also be cooperating with Statkraft's Banja reservoir in Albania. Aquaculture industry inspired Ocean

Sun's patented technology of silicon solar modules installed on large floating structures. The Statkraft Company has Europe's largest virtual power plant which connects over 1300 solar and wind energy installations in Germany, and accumulates energy from flexible gas production, and solar, wind, and battery storage in the United Kingdom and operates successfully toward developing solar power capacity in India, Spain, and the Netherlands. Aiming at mitigating GHG emissions, Norway is planning on developing fully electrified domestic passenger planes by the year 2030. The Norwegian shipping industry has promised to develop carbon-free business as well. The location of northern Norway, in particular, due to its cold climate and accessibility to dark fiber offers a lot of scope for a robust data center such as Kolos – powered by 100% wind and hydropower.

On a philosophical note, the climate emergency doesn't describe the planet as a failure, but vividly shows a lack of our love for it. Industrialists need always to be honest and shouldn't mind sounding ignorant if someone points out to them the potential dangers for human health from manufacturing or disposal processes. Though global warming and pollution has caused a man-made climate crisis, it's not all downhill from here and I believe that our inherited knowledge and future scholarly sustainable enterprises will suffice to take on the energy and environmental challenges. Water vapor in the atmosphere can be used as a source of energy (Lax et al., 2020) and hydrogen can be produced by planting enzymes into sample algae (Kanygin et al., 2020). Today's actions to create sustainable energy solutions can present us with inclusive and positive socio-economic development and ensure our future (IRENA Report on a Roadmap to 2050, 2018). Science can help us assess where we stand and what needs to be done while progressing and achieving the SDGs (Global Sustainable Development Report of the UN, 2019). There is lot to learn from each other and to collaborate to make this transition to a renewable energy system possible, with a focus on no carbon footprint, our sustenance, and our sustain "ability". There are some good examples from where we may seek inspiration to treat resources with respect and restore them, as and when required, because the planet earth and its motherly resources are our only shared home.

5.5 CONCLUSION AND FUTURE DIRECTIONS

Chemistry – the study of matter, its properties, reactions, and interaction with energy or energy conversion – plays a strategic role in sustainable development (Shaikh. 2019). The science of chemistry gives a credible impression of achieving sustainability through various practical approaches, operational tools, and monitoring tools. Chemistry is not only confined to offering molecules and materials for energy storage, catalysts for easing reactions, and minimizing waste, but it also greatly influences life-cycle assessment, and resource and energy efficiency. I have highlighted the cross-sectional roles of chemistry in transforming the energy system; chemistry responds to "energy" and "climate" crises by offering non-nuclear alternative energy processes.

Chemistry can help us find ways to control and convert energy by bearing it "in chemical bonds", and also by seeking fine chemicals or value-added products, as it has done over the centuries by creating high-density fossil energy carriers through geochemical processes. As the eminent Professor Robert Schlögl once said, "Chemistry is at the centre of the energy challenge and the transformation of energy systems into a sustainable future will be impossible without chemical energy conversion". Chemistry can contribute to climate change goals by controlling the "energetic cost" of the conversion of energy carriers (Schlögl, 2016). Innovations in green or sustainable chemistry and engineering will be key to transitioning to eco-efficient and cost-effective energy. While development of clean energy generation technologies is on the horizon, recent research offers strategies for creating the energy mix and novel approaches of energy conversion. This includes early stage venture capitalists and clean energy technology incubators, focusing on end-use efficiency, demand control, and the decarbonization of the electricity sector irrespective of the source from where electricity is derived (Shaahid & El-Amin, 2009; Bumpus & Comello, 2020; OECD, 2020). Renewable electricity can also be stored as chemical energy in fuels. We have already succeeded in transforming primary energy from the sun into free energy, and there is a great opportunity ahead for industries to access and mix wind, water, and other overlooked renewable resources to develop a low carbon or carbon-neutral energy supply, energy carriers, smart materials, batteries, and relevant zero-emission energy technologies (Arico et al., 2005; Balaya, 2008; Manthiram et al., 2008).

The chemical storage of energy and hydrogen as the zero-emission fuel is one of the hot topics of research (Lu et al., 2011; Zhang et al., 2018, 2019) and there are potential technological pathways for producing hydrocarbons without using fossil fuels or biomass and also for mitigating the increasing concentration of CO_2 in the atmosphere and in the acidification of oceans (Barker & Ridgwell, 2012). Countries exporting waste to waste-to-energy plants abroad and treating the oceans as a CO_2 sink should rethink their approach to pollution prevention and waste management. Major anthropogenic activities, including excessive land-use change or cultivating practices, livestock, and deforestation, together with the direct combustion of fossil fuels, and industries contribute to the addition of heat-trapping greenhouse gases in the atmosphere (Mitchell 1989; Lal, 2004). The recycling of CO_2 and water – in the reverse of fuel combustion – yields liquid hydrocarbon fuel in non-biological processes using renewable or nuclear energy. This can enable a closed-loop carbon-neutral fuel cycle if atmospheric CO_2 could be captured (and stored).

Researchers continue to look for advancing electricity-based production or synthesis of useful chemical commodities like ammonia (Schlögl, 2003) electrochemically (Lazouski et al., 2020). A development in industrial chemistry demands, for example, looking into the electrochemical conversion of CO_2 into ethanol and ethylene as fuel and as a useful monomer respectively (De Luna et al., 2019, 2020). Progress is needed in catalysts and catalysis to be able to make chemical commodities by electrochemical processes and scale them up and make them quick and practical.

This chapter, though at the cross-roads of many disciplines, has not aimed at replacing specialized expertise on renewable energy, digitalization, and entrepreneurship in their separate domains, but it has attempted to give a genuine insight on how chemistry is at the heart of energy technologies and how it is important to better understand the key role chemistry has in the shift from the current fossil-based energy system to renewable-based energy projects, including the scope towards commercialization, decentralization, and the importance of sustainable entrepreneurship in addressing the energy challenge and climate crisis. In addition to that, readers may appreciate understanding the overlooked renewables and the strength that chemistry offers in energy conversion and materials wealth. This knowledge is intended for students, scholars, entrepreneurs, policy makers, as well as practitioners, including those with no formal training in chemical science and energy technologies.

Society is dependent on local policy makers, entrepreneurs, industry leaders, and scientists when it comes to discussing science, technological advancements, and their implications. Scientific literacy through education, awareness campaigns, regular dialogue, and communication provides a platform for the public (Shaikh, 2012) and policy makers for reforming the foundations of their knowledge of scientific activities and relevant policy developments and discussions about the risks and ethics of energy technologies. Education, especially within STEM (science, technology, engineering, mathematics), research, and entrepreneurial sustainable development are keys for dealing with these issues and without them it's impossible for us and future generations to identify the severity of the problems we face and troubleshoot them in time and wisely (Shaikh, 2015, 2016).

An observation of mine in this respect is that the fossil-fuel-based energy system is highly complicated; the "few" are preferred over the many, and it is mainly centralized. It's high time we started to look for a shared or public ownership of the energy sector, be it microgeneration or large-scale production, distribution, or storage. Innovation that was brought into our lives through fossil fuels has been found to be the cause of climate change; it needs to be greened now by cutting GHG emissions sustainably or replacing them with clean fuels. Our love for capitalism, industrialization, life sophistication, and growth should not lead us to lose our relationship with the planet. It was knowledge technology that brought energy to the doorsteps of mankind, and that's our common heritage! But later it was monopolized and developed into a tool for exploitation – exploitation of the people and also the planet to such an extent that one day we might lose the life-supporting system this planet offers. Scientific evidence shows us that it is high time we kept a check on the fossil-based energy system and our greed for industrialization in pursuit of material wealth by disregarding nature; but the mode of denial of the climate crisis is appalling.

To conclude, this chapter has provided a synopsis of the key findings, theory, and technological advances directed toward entrepreneurial sustainable development through chemistry, materials, renewable resources, or clean energy technologies, as well as concerted international efforts in policy making and practices in meeting today's energy challenge and climate change goals using digitalization wherever necessary and developing a robust economic system for entrepreneurs improving local energy delivery while undertaking carbon sequestration to secure CO_2 and prevent it from entering the atmosphere. The information and communication technology (ICT) sector contributes more than 2% of global carbon emissions. Efficient heat conduction through piped water or free air cooling along with the dark and cold climate of the northern countries offer much scope in managing the high performance computers of a data center and thereby to stop such data centers from gobbling up electricity (Jones, 2018).

As Albert Einstein, the genius scientist once said, "We cannot solve problems by using the same kind of thinking we used when we created them". And that's what the power structure would be doing if it keeps pushing the planet to the brink of collapse due to unsustainable industrial enterprise and considering depopulation as a strategy to manage resources and regions. Sorry if that's a lot of criticism; I hereby declare, this is not a part of any cynical, anti-capitalist, or regressive propaganda! The basic idea here, despite all the verbiage, is simple: There is an urgent need to institute a holistic approach, awaken humanism, and develop among people the scientific temperament – the spirit of inquiry and reform to carry us through this critical phase the planet is living through due to current resource consumption and the fossil-fuel-based energy system that is causing climate change and a race for resources and regions. It is also important to not let exploiters build a monopoly around a newly evolved renewable-based energy system.

The most highlighted fact is that the chemistry represents an interesting scope for futuristic energy. The literature survey and ideas noted above do not fully claim to conceptualize all the aspects of renewable-energy-based projects, digitalization, and sustainable entrepreneurship. In the review of the literature, the basic concept of fossil-based fuel, gas-to-liquid technology, pollution caused by the oil and gas industries, a shift to using direct or indirect solar energy, artificial photosynthesis, wind energy, hydrogen energy, and the various strategies for benign energy conversions have been discussed. This chapter has elucidated the pivotal role of chemical science intertwined with sustainability to deliver energy conversion methodologies with inventive steps and the possibility of commercialization at high standards of scalability, stability, and sustainability.

The chapter has illustrated the significance of a variety of energy conversions and their relationship to climate change from the chemistry point of view. The ideas noted and the many useful references cited do not fully claim to conceptualize all the aspects of chemistry for futuristic energy but might offer clues on new directions of research and also an evaluation of the sustainable entrepreneurial opportunities to address them in a state-of-the-art and in a what-if manner. Scientific activities and policy making can concurrently be pushed to prioritize and overcome the technical challenges in renewable energy generation, conversion, storage, distribution, or discharge. On the basis of a careful assessment of renewable-based energy projects, choices can be made in meeting the energy challenge, together with tackling the adverse effects caused by fossil-fuel-based energy technologies.

REFERENCES

Åhman, M., Olsson, O., Vogl, V., Nyqvist, B., Maltais, A., Nilsson, L. J., ... Nilsson, M. 2018. Hydrogen steelmaking for a low-carbon economy: A joint LU-SEI working paper for the HYBRIT project. EESS report 109, vol. EESS, 109 edn, Miljö- och energisystem, LTH, Lunds universitet, Lund, Sweden. e-ISBN: 978-91-86961-35-0, Category: Research, Published - 2018 Sep 10) and also Publisher: Envriomental and Energy Systems Studies, e-ISSN: 1102-3651). https://portal.research.lu.se/portal/en/publications/hydrogen-steel-making-for-a-lowcarbon-economy(e8289959-1fa9-48e0-87d6-e3f79cb93aac).html.

Ahmed, K. 1994. Renewable energy technologies: A review of the status and costs of selected technologies. World Bank Technical Paper No. 240, Washington DC, USA, World Bank.

Alberini, A., Cropper, M., Fu, T.-T, Krupnick, A., Liu, J.-T., Shaw, D., Harrington, W. 1997. Valuing health effects of air pollution in developing countries: The case of Taiwan. *Journal of Environmental Economics and Management* 34:107–126.

Amadeo, K. 2020. Trade Policy written by Kimberley Amadeo for the *Balance*. https://www.thebalance.com/what-is-a-petrodollar-3306358 (Accessed 14 April 2020).

Anderson, D., Ahmed, K. 1995. The case for solar energy investments. World Bank Technical Paper No. 279. Washington DC, USA, World Bank.

Annual Energy Outlook (AEO) 2006 with Projections to 2030. Washington, D.C. USA: Energy Information Administration. pp. 52–54. DOE/EIA-0383(2006). http://www.pseudology.org/gazprom/EnergyOutlook2006.pdf (Accessed 14 April 2020).

Araya, S. S., Liso, V., Cui, X., Li, N., Zhu, J., Sahlin, S. L., Kær, S. K. 2020. A review of the methanol economy: The fuel cell route. *Energies* 13(3):596. doi:10.3390/en13030596

Archer, D., Brovkin, V. 2008. The millennial atmospheric lifetime of anthropogenic CO_2. *Climatic Change* 90:283–297.

Arico, A. S., Bruce, P. Scrosati, B., Tarascon, J. M., Schalkwijk, W. V. 2005. Nanostructured materials for advanced energy conversion and storage devices. *Nature Materials* 4:366–377.

Bai, X., Tiwari, S., Robinson, B., Killmer, C. P., Hu, J. 2018. Microwave catalytic synthesis of ammonia from methane and nitrogen. *Catalysis Science & Technology* doi:10.1039/c8cy01355a.

Balaya, P. 2008. Size effects and nanostructured materials for energy applications. *Energy & Environmental Science* 1:645–654.

Bankruptcy of Our Nation (Revised and Expanded). 2012. New Leaf Press; Revised, Expanded edition (August 10, 2012), Language: English, ISBN-10: 0892217138.

Baran, M. J., Braten, M. N., Sahu, S., Baskin, A., Meckler, S. M., Li, L., ... Helms, B. A. 2019. Design rules for membranes from polymers of intrinsic microporosity for crossover-free aqueous electrochemical devices. *Joule* 3(12): P2968–P2985.

Barker, S., Ridgwell, A. 2012. Ocean acidification. *Nature Education Knowledge* 3(10):21.

Belfer Center. 2013. Policy Brief by *Jeff D. Colgan*. https://www.belfercenter.org/publication/oil-conflict-and-us-national-interests (Accessed 14 April 2020).

Bergthorson, J. M., Goroshin, S., Soo, M. J., Julien, P., Palecka, J., Frost, D. L., Jarvis, D. J. 2015. Direct combustion of recyclable metal fuels for zero-carbon heat and power. *Applied Energy* 160:368–382. doi:10.1016/j.apenergy.2015.09.037.

Berntsen, T., Fuglestvedt, J., Myhre, G., Stordal, F., Berglen, T. F. 2006. Abatement of greenhouse gases: Does location matter? *Climatic Change* 74:377–411.

Beydoun, D., Amal, R., Low, G., McEvoy, S. 1999. Role of nanoparticles in photocatalysis. *Journal of Nanoparticle Research* 1(4):439–458. doi:10.1023/A:1010044830871.

Birdja, Y. Y., Perez-Gallent, E., Figueiredo, M. C., Gottle, A. J., Calle-Vallejo, F., Koper, M. T. M. 2019. Advances and challenges in understanding the electrocatalytic conversion of carbon dioxide to fuels. *Nature Energy* 4(9):732–745.

Blanchard, O. J., Galí, J. 2007. The macroeconomic effects of oil price shocks: Why are the 2000s so different from the 1970s? Economics Working Papers 1045, Department of Economics and Business, Universitat Pompeu Fabra, revised Oct 2008. NBER Working Paper No. 13368. https://www.nber.org/papers/w13368.

Boissoneault, L. 2015. JSTOR daily. https://daily.jstor.org/salar-de-uyuni/ (Accessed 20 June 2020).

Bosman, R., Scholten, D. 2013. How renewables will transform commercial and (geo)political relations. *EnergyPost*. http://energypost.eu/renewables-will-transform-commercial-geopolitical-relations/ (Accessed 14 April 2020).

Bossel, U., Eliasson, B. 2003. Energy and the hydrogen economy. https://afdc.energy.gov/files/pdfs/hyd_economy_bossel_eliasson.pdf (Accessed 14 April 2020).

Brown, T. W., Bischof-Niemz, T., Blok, K., Breyer, C., Lund, H., Mathiesen, B. V. 2018. Response to "Burden of proof: A comprehensive review of the feasibility of 100% renewable-electricity systems". *Renewable and Sustainable Energy Reviews* 92:834–847.

Brundtland Commission. 1987. Our Common Future: Report of the World Commission on Environment and Development, United Nations' Commission (headed by Ms. Brundtland, the then Prime Minister of Norway) on Environment and Development, Published as Annex to General Assembly document A/42/427, Development and International Co-operation: Environment.

Bumpus, A., Comello, S. 2020. Emerging clean energy technology investment trends. *Nature Climate Change* 7:382–385. doi:10.1038/nclimate3306.

Burri, D. R., Shaikh, I. R., Choi, K.-M., Park, S.-E. 2007. Facile heterogenization of homogeneous ferrocene catalyst on SBA-15 and its hydroxylation activity. *Catalysis Communications* 8(4):731–735.

Butt, N., Beyer, H. L., Bennett, J. R., Biggs, D., Maggini, R., Mills, M., ... Possingham, H. P. 2013. Biodiversity risks from fossil fuel extraction. *Science* 342(6157):425–426. doi:10.1126/science.1237261 (Accessed April 14, 2020).

Cardinaletti, I., Vangerven, T., Nagels, S., Cornelissen, R., Schreurs, D., Hruby, J., ... Manca, J. V. 2018. Organic and perovskite solar cells for space applications. *Solar Energy Materials and Solar Cells* 182:121–127. doi:10.1016/j.solmat.2018.03.024.

Castelvecchi, D., Stoye E. 2019. Chemistry nobel honours world-changing batteries. *Nature* 574:308. doi:10.1038/d41586-019-02965-y.

Catizzone, E., Bonura, G., Migliori, M., Frusteri, F., Giordano, G. 2018. CO_2 recycling to dimethyl ether: State-of-the-art and perspectives. *Molecules* 23:31. doi:10.3390/molecules23010031.

Centi, G., Perathoner, S. 2009. Opportunities and prospectus in the chemical recycling of carbon dioxide to fuels. *Catalysis Today* 148(3–4):191–205.

CESI. 2019. https://www.cesi.it/space-solar-cells/ (Accessed 14 April 2020).

Chandra, Y. P., Singh, A., Kannojiya, V., Kesari, J. P. 2018. Solar energy a path to India's Prosperity. *Journal of the Institution of Engineers (India): Series C* 100(3):539–546. doi:10.1007/s40032-018-0454-6.

Chen, Y., Ebenstein, A., Greenstone, M., Li, H. 2013. Evidence on the impact of sustained exposure to air pollution on life expectancy from China's Huai river policy. *Proceedings of the National Academy of Sciences* 110(32):12936–12941. http://www.pnas.org/content/110/32/12936.full.pdf.

Chen, Y., Li, H., Zhao, W., Zhang, W., Li, J., Li, W., ... Si, R. 2019. Optimizing reaction paths for methanol synthesis from CO_2 hydrogenation via metal-ligand cooperativity. *Nature Communications* 10:1885. doi:10.1038/s41467-019-09918-z.

Christensen, C. M. 2013. *The Innovator's Dilemma: When New Technologies Cause Great Firms to Fail*. Harvard Business Review Press, Boston, Massachusetts, United States. ISBN: 142219602X.

Climate action tracker. 2020. https://climateactiontracker.org/countries/india/pledges-and-targets/ (Accessed 14 April 2020).

Cui, X., Li, H., Wang, Y., Hu, Y., Hua, L., Li, H., ... Bao, X. 2018. Room-temperature methane conversion by graphene-confined single iron atoms. *Chem* 4(8):1902–1910. doi:10.1016/j.chempr.2018.05.006.

Davis, V. K., Bates, C. M., Omichi, K., Savoie, B. M., Momčilović, N., Xu, Q., ... Jones, S. C. 2018. Room-temperature cycling of metal fluoride electrodes: Liquid electrolytes for high-energy fluoride ion cells. *Science* 362(6419):1144–1148. doi:10.1126/science.aat7070.

DCCAE. 2019. Government of the Republic of Ireland. https://www.dccae.gov.ie/en-ie/climate-action/publications/Pages/Climate-Action-Plan.aspx (Accessed April 14 2020).

De Luna, P., Hahn, C., Higgins, D., Jaffer, S. A., Jaramillo, T. F., Sargent, E. H. (2019). What would it take for renewably powered electrosynthesis to displace petrochemical processes? *Science* 364(6438):eaav3506. doi:10.1126/science.aav3506.

De Luna, P., Hahn, C., Higgins, D., Jaffer, S. A., Jaramillo, T. F., Sargent, E. H. 2020. Erratum for the review: What would it take for renewably powered electrosynthesis to displace petrochemical processes? *Science* 367(6482):eabb0992. doi:10.1126/science.abb0992.

Dekra Sustainability Magazine. 2017–18. https://www.dekra.com/media/dekra-sustainability-magazine-2018.pdf (Accessed 14 April 2020).

Dinç, D. T., Akdoğan, E. C. 2019. Renewable energy production, energy consumption and sustainable economic growth in Turkey: A VECM approach. *Sustainability* 11:1273. doi:10.3390/su11051273.

Dinh, C. T., Burdyny, T., Kibria, M. G., Seifitokaldani, A, Gabardo, C. M., García de Arquer F. P., ... Sargent, E. H. 2018. CO_2 electroreduction to ethylene via hydroxide mediated copper catalysis at an abrupt interface. *Science* 360(6390):783–787. doi:10.1126/science.aas9100.

Drape, R. 2019. National geographic magazine. https://www.nationalgeographic.com/magazine/2019/02/lithium-is-fueling-technology-today-at-what-cost/ (Accessed 20 June 2020).

Ernst & Young. 2014. Capitalizing on China's renewable energy opportunities. http://www.ey.com/Publication/vwLUAssets/EY-white-paperrenewable-energy-financing-in-china-en-26dec/$FILE/EY-white-paper-renewableenergy-financing-in-china-en-26dec.pdf (Accessed 14 April 2020).

Espinet-Gonzalez, P., Barrigón, E., Otnes, G., Vescovi, G. Mann, C., France, R. M., ... Atwater, H. A. 2019. Radiation tolerant nanowire array solar cells. *ACS Nano* 13(11):12860–12869. doi:10.1021/acsnano.9b05213.

Europe Beyond Coal. 2020. www.beyond-coal.eu (Accessed April 14 2020).

European Commission. 2014. Guidance for beneficiaries of European Structural and Investment Funds and related EU instruments. https://ec.europa.eu/regional_policy/sources/docgener/guides/synergy/synergies_beneficiaries.pdf (Accessed 14 April 2020).

Forbes. 2019. Freelance Journal article by *Dominic Dudley*, https://www.forbes.com/sites/dominicdudley/2019/01/11/china-renewable-energy-superpower/#68318b06745a (Accessed 14 April 2020).

Forbes. 2020. https://www.forbes.com/sites/brentanalexander/2020/04/21/40-oil-will-return-this-isnt-the-end-of-fossil-fuels/ (Accessed 14 April 2020).

Frederica, P. 2017. Pollution from fossil-fuel combustion is the leading environmental threat to global pediatric health and equity: Solutions exist. *International Journal of Environmental Research and Public Health* 15(1–16):17. doi:10.3390/ijerph15010016.

Glenk, G., Reichelstein, S. 2019. Economics of converting renewable power to hydrogen. *Nature Energy* 4:216–222. doi:10.1038/s41560-019-0326-1.

Global Sustainable Development Report. 2019. Independent Group of Scientists appointed by the Secretary-General, *Global Sustainable Development Report 2019: The Future is Now – Science for Achieving Sustainable Development*, United Nations, New York.

Gozgora, G., Laub, C. K. M., Luc, Z. 2018. Energy consumption and economic growth: New evidence from the OECD countries. *Energy* 153:27–34.

Graetzel, M. 1981. Artificial photosynthesis: Water cleavage into hydrogen and oxygen by visible light. *Accounts of Chemical Research* 14(12):376–384. doi:10.1021/ar00072a003.

Graves, C., Ebbesen, S. D., Mogensen, M., Lackner, K. 2011. Sustainable hydrocarbon fuels by recycling CO_2 and H_2O with renewable or nuclear energy. *Renewable and Sustainable Energy Reviews* 15(1):1–23. doi:10.1016/j.rser.2010.07.014.

Haas, R., Panzera, C., Rescha, G., Ragwitz, M., Reece, G., Held, A. 2011. A historical review of promotion strategies for electricity from renewable energy sources in EU countries. *Renewable and Sustainable Energy Reviews* 15:1003–1034.

Hansmann, H. B. 1996. *The Ownership of Enterprise*. Cambridge, MA: Belknap Press.

Hatfield, M. 2018. The Worrisome Deal: China and Saudi Arabia. *Harvard Political Review*. https://harvardpolitics.com/world/the-worrisome-deal-china-and-saudi-arabia/ (Accessed 14 April 2020).

Heard, B., Brook, B., Wigley, T., Bradshaw, C. 2017. Burden of proof: A comprehensive review of the feasibility of 100% renewable-electricity systems. *Renewable and Sustainable Energy Reviews* 76:1122–1133. doi:10.1016/j.rser.2017.03.114.

Heiskanen, E., Kivimaa, P., Lovio, R. 2019. Promoting sustainable energy: Does institutional entrepreneurship help? *Energy Research & Social Science* 50:179–190. doi:10.1016/j.erss.2018.11.006.

Herrchen, M., Klein, W. 2000. Use of the life-cycle assessment (LCA) toolbox for an environmental evaluation of production processes. *Pure and Applied Chemistry* 72(7):1247–1252.

Höök, M., Aleklett, K. 2010. A review on coal-to-liquid fuels and its coal consumption. *International Journal of Energy Research* 34:848–864. doi:10.1002/er.1596.

Horizon Europe. 2020. https://ec.europa.eu/info/horizon-europe-next-research-and-innovation-framework-programme_en (Accessed 14 April 2020).

IEA. 2019. https://www.iea.org/reports/world-energy-investment-2019 (Accessed April 14 2020).

IEA. 2020. https://www.iea.org/articles/the-global-oil-industry-is-experiencing-shock-like-no-other-in-its-history (Accessed 14 April 2020).

IEA Flagship Report. 2018. https://www.iea.org/weo2018 (Accessed April 14 2020).

IGI Global. 2019. https://www.igi-global.com/chapter/the-shift-towards-a-digital-business-model/220028 (Accessed 14 April 2020).

Intelligent Energy Europe. 2020. https://ec.europa.eu/energy/intelligent/projects/ (Accessed 14 April 2020).

IPCC. 2014. https://ar5-syr.ipcc.ch/topic_summary.php (Accessed April 14, 2020).

IRENA. 2018. Global energy transformation: A roadmap to 2050. https://www.irena.org/-/media/Files/IRENA/Agency/Publication/2018/Apr/IRENA_Report_GET_2018.pdf.

Islam, M. S, Bruce, P. G, Catlow, C. R. A., Nelson, J. 2019. Energy materials for a low carbon future. *Philosophical Transaction of the Royal Society: A Mathematical, Physics and Engineering Sciences* 377(2152):20190219. doi:10.1098/rsta.2019.0219.

Johannesburg Declaration. 2002. Johannesburg Declaration on Sustainable Development at the World Summit on Sustainable Development, in Johannesburg, South Africa, 26 August to 4 September 2002. sustainabledevelopment.un.org and http://www.un-documents.net/jburgdec.htm (Accessed 14 April 2020).

Johnson, M. P., Schaltegger, S. 2019. Entrepreneurship for sustainable development: A review and multilevel causal mechanism framework. *Entrepreneurship Theory and Practice*. doi:10.1177/1042258719885368 (Accessed 14 April 2020).

Jolly, S., Spodniak, P., Raven, R. P. J. M. 2016. Institutional entrepreneurship in transforming energy systems towards sustainability: Wind energy in Finland and India. *Energy Research & Social Science* 17:102–118. doi:10.1016/j.erss.2016.04.002.

Jones, N. 2018. How to stop data centres from gobbling up the world's electricity. *Nature* 561:163–166. doi:10.1038/d41586-018-06610-y.

Kanygin, A., Milrad, Y., Thummala, C., Reifschneidera, K., Bakera, P., Marcob, P., ... Redding, K. E. 2020. Rewiring photosynthesis: A photosystem l-hydrogenase chimera that makes H_2 in vivo. *Energy & Environmental Science*. doi:10.1039/C9EE03859K.

Kavanagh, L., Keohane, J., Garcia Cabellos, G., Lloyd, A., Cleary, J. 2018. Global lithium sources—Industrial use and future in the electric vehicle industry: A review. *Resources* 7(3):57. doi:10.3390/resources7030057.

Kinninmont, J. 2009. The GCC in 2020: The gulf and its people. A report from the economist intelligence unit sponsored by the Qatar financial centre authority. https://www.academia.

edu/534995/The_GCC_In_2020_The_Gulf_And_Its_People (Accessed 14 April 2020) and https://www.arabianbusiness.com/gcc-population-seen-growing-53m-by-2020-10831. html (Accessed 14 April 2020).

Konno, T., Kitagaki, T., Kojo, H. 2010. U.S. Patent No. 7,767,254 B2.

Kuendig, J., Goetz, M., Niquille, X., Shah, A., Vaccaro, S., Mosig, J., ... Fernandez, E. 2000. Thin-film silicon solar cells for space applications: Radiation hardness and applications for an integrated Solant (solar cell-antenna) module. *Conference Record of the Twenty-Eighth IEEE Photovoltaic Specialists Conference - 2000 (Cat. No.00CH37036)*, Anchorage, AK, USA, pp. 1079–1082. doi:10.1109/PVSC.2000.916074.

Lal, R. 2004. Soil carbon sequestration to mitigate climate change. *Geoderma* 123:1–22.

Lax, J. Y., Price, C., Saaroni, H. 2020. On the spontaneous build-up of voltage between dissimilar metals under high relative humidity conditions. *Scientific Reports* 10(1):7642–7650. doi:10.1038/s41598-020-64409-2.

Lazarus, M., Verkuijl, C., Yehle, E. 2019. Closing the fossil fuel production gap. Stockholm Environment Institute, Stockholm. https://www.sei.org/wp-content/uploads/2019/09/closing-the-fossil-fuel-production-gap-brief.pdf (Accessed 14 April 2020).

Lazouski, N., Chung, M., Williams, K., Gala, M. L., Manthiram, K. 2020. Non-aqueous gas diffusion electrodes for rapid ammonia synthesis from nitrogen and water-splitting-derived hydrogen. *Nature Catalysis*. doi:10.1038/s41929-020-0455-8.

Lee, S., Speight, J. G., Loyalka, S. 2007. *Handbook of Alternative Fuel Technologies*. CRC Press. Leiden University, 2020. https://www.universiteitleiden.nl/en/research-dossiers/keeping-the-planet-liveable/a-smarter-approach-to-energy-and-raw-materials (Accessed April 14 2020).

Li, H., Eddaoudi, M., O'Keeffe, M., Yaghi, O. M. 1999. Design and synthesis of an exceptionally stable and highly porous metal-organic framework. *Nature* 402:276–279.

Lu, L., Ren, N., Zhao, X., Wang, H., Wu, D., Xing, D. 2011. Hydrogen production, methanogen inhibition and microbial community structures in psychrophilic single-chamber microbial electrolysis cells. *Energy & Environmental Science* 4:1329. doi:10.1039/c0ee00588f.

Lubitz, W., Reijersea, E. J., Messinger, J. 2008. Solar water-splitting into H_2 and O_2: Design principles of photosystem II and hydrogenases. *Energy & Environmental Science* 1:15–31.

Malik, J. 2020. *Tharunka* magazine, University of New South Wales, Australia. https://tharunka.arc.unsw.edu.au/elon-musk-the-bolivian-coup-innocent-businessman-or-opportunistic-neo-coloniser/ (Accessed 20 June 2020).

Manthiram, A., Murugan, A. V., Sarkar, A., Muraliganth, T. 2008. Nanostructured electrode materials for electrochemical energy storage and conversion. *Energy & Environmental Science* 1:621–638.

Martinot, E. 2001. Renewable energy investment by the World Bank. *Energy Policy* 29(9):689–699. doi:10.1016/S0301-4215(00)00151-8.

Methanol.org. 2020. https://www.methanol.org/energy/ (Accessed 14 April 2020).

Meyer, G. J. 2010. The 2010 millennium technology grand prize: Dye-sensitized solar cells. *ACS Nano* 4:4337–4343.

Mitchell, J. F. B. 1989. The greenhouse-effect and climate change. *Reviews of Geophysics* 27:115–139.

MNRE India. 2020. Ministry of new & renewable energy. 2020. Physical Progress (Achievements). https://mnre.gov.in/the-ministry/physical-progress (Accessed 14 April 2020).

Mohammed, M. A. A., Salmiaton, A., Wan Azlina, W. A. K. G., Mohammad Amran, M. S., Fakhru'l-Razi, A., Taufiq-Yap, Y. H. 2011. Hydrogen rich gas from oil palm biomass as a potential source of renewable energy in Malaysia. *Renewable and Sustainable Energy Reviews* 15(2):1258–1270. doi:10.1016/j.rser.2010.10.003.

Muller, K., Städter, M., Rachow, F., Hoffmannbeck, D., Schmeiber, D. 2013. CO_2-methanation by catalytic conversion. *Environmental Earth Sciences* 70:3771–3778.

Muneer, T., Asif, M., Munawwar, S. 2005. Sustainable production of solar electricity with particular reference to the Indian economy. *Renewable and Sustainable Energy Reviews* 9(5):444–473. doi:10.1016/j.rser.2004.03.004.

Nath, K., Das, D. 2007. Production and storage of hydrogen: Present scenario and future perspective. *Journal of Scientific & Industrial Research* 66:701–709.

Nature news. 2019. https://www.nature.com/articles/d41586-019-02634-0 (Accessed 20 June 2020).

Nehru, J. 1989. *The Discovery of India* (Centenary ed.). Oxford: University Press.

New Climate Economy Report. 2018. (a). https://newclimateeconomy.report/2018/ (b). http://newclimateeconomy.report/ (Accessed 14 April 2020).

Nouryon. 2018. https://www.nouryon.com/news-and-events/features-overview/hydrogen-enabling-the-transition-to-green-chemistry/ (Accessed 14 April 2020).

O'Regan, B., Grätzel, M. 1991. A low-cost, high-efficiency solar cell based on dye-sensitized colloidal TiO2 films. *Nature* 353(6346):737–740.

OECD. 2020. Renewable energy (indicator). doi:10.1787/aac7c3f1-en. https://data.oecd.org/energy/renewable-energy.htm (Accessed 14 April 2020).

Ottewell, S. 2019. https://www.chemicalprocessing.com/articles/2019/chemical-makers-turn-to-renewable-electricity/ (Accessed 14 April 2020).

Parkinson, G. 2016. Renew economy. https://reneweconomy.com.au/new-low-for-wind-energy-costs-morocco-tender-averages-us30mwh-81108/ (Accessed 14 April 2020).

Parmesan, C. 2006. Ecological and evolutionary responses to recent climate change. *Annual Review of Ecology Evolution and Systematics* 37:637–669.

Perathoner, S., Gross, S., Hensen, E. J. M., Wessel, H., Chraye, H., Centi, G. 2017. Looking at the future of chemical production through the European roadmap on science and technology of catalysis the EU effort for a long-term vision. *ChemCatChem* 9(6):904–909. doi:10.1002/cctc.201601641.

Peter, L. M. 2011. The grätzel cell: Where next? *The Journal of Physical Chemistry Letters* 2(15):1861–1867. doi:10.1021/jz200668q.

Qiao, J., Liu, Y., Hong, F., Zhang, J. 2014. A review of catalysts for the electroreduction of carbon dioxide to produce low-carbon fuels. *Chemical Society Reviews* 43:631–675.

Raworth, K. 2017. *Doughnut Economics: Seven Ways to Think Like a 21st-Century Economist.* Chelsea Green Publishing.

REVE. 2020. https://www.evwind.es/2020/01/09/enel-green-power-starts-the-year-with-a-bang-beating-its-previous-record-with-over-3-gw-of-renewable-capacity-built-in-2019/72997 (Accessed 14 April 2020).

Rifkin, J. 2003. *The Hydrogen Economy: The Creation of the Worldwide Energy Web and the Redistribution of Power on Earth,* TarcherPerigee; 1 edition (25 August 2003). ISBN-10: 1585422541.

Robinson, J. 2012. *Bankruptcy of Our Nation.* New Leaf Press; revised expanded edition.

Robinson, J. 2020. Follow the Money. https://followthemoney.com/preparing-for-the-collapse-of-the-petrodollar-system-part-1/ (Accessed 14 April 2020).

Rogol, M. with Rogol, S. H. 2011. *Explosive Growth.* Live Oak Book Company. (December 13, 2011), USA.

Ross, J. R. H. 2011. Heterogeneous catalysis: Fundamentals and applications. Elsevier B. V. doi:10.1016/C2009-0-19388-1.

Sandin, S., Neij, L., Mickwitz, P. 2019. Transition governance for energy efficiency - insights from a systematic review of Swedish policy evaluation practices. *Energy Sustainability and Society* 9:17.

Schlögl, R. 2003. Catalytic synthesis of ammonia—a "never-ending story"? *Angewandte Chemie International Edition* 42:2004–2008.

Schlögl, R. 2016. Sustainable energy systems: The strategic role of chemical energy conversion. *Topics in Catalysis* 59:772–786. doi:10.1007/s11244-016-0551-9.

Seo, Y.-C., Alam, M. T., Yang, W.-S. 2018. Gasification of municipal solid waste, gasification for low-grade feedstock, chapter in IntechOpen. doi:10.5772/intechopen.73685. https://www.intechopen.com/books/gasification-for-low-grade-feedstock/gasification-of-municipal-solid-waste (Accessed 14 April 2020).

Service, R. F. 2018. https://www.sciencemag.org/news/2018/07/ammonia-renewable-fuel-made-sun-air-and-water-could-power-globe-without-carbon (Accessed 14 April 2020).

Shaahid, S. M., El-Amin, I. 2009. Techno-economic evaluation of off-grid hybrid photovoltaic-diesel battery power systems for rural electrification in Saudi Arabia-A way forward for sustainable development. *Renewable & Sustainable Energy Reviews* 13:625–633.

Shaikh, I. R. 2012. Scientific literacy in public: Another look at science education and communication and its relationship to society. *The Online Educational Research Journal*www.oerj.org.

Shaikh, I. R. 2014a. Development of some eco-efficient and selective chemical processes and products using sustainable and green chemistry approaches, Ph.D. thesis submitted (www.JJTU.ac.in) in July–August 2014; viva-voce is not conducted yet due to parochial attitude, intolerant, religious fanatic agenda of the RSS and terrorism of the said University. A patent application was also filed at the Indian Patent Office. Methanol Carbonylation to Acetic Acid, Ordinary Indian Patent Application Number: 11049/MUM/2015.

Shaikh, I. R. 2014b. Organocatalysis: Key trends in green synthetic chemistry, challenges, scope towards heterogenization and importance from research and industrial point of view. *Journal of Catalysts* (Review; Article ID 402860). http://www.hindawi.com/journals/jcat/aip/402860/.

Shaikh, I. R. 2015. Delivering impact through STEMS: The imperative. PIC-STEMS. *Proceedings of the Indian Conference on Science, Technology, Engineering, Mathematics & Sustainability*. No. 2, 1(2015):12–26.

Shaikh, I. R. 2016. Geoengineering: The planet salvage operation? *PIC-STEMS (Proceedings of the Indian Conference on Science, Technology, Engineering, Mathematics & Sustainability)* 1(2016):11–17.

Shaikh, I. R. 2019. Development of some eco-efficient and selective chemical processes and products using sustainable and green chemistry approaches. 2(4). doi:10.32474/DDIPIJ.2018.02.000145.

Sokol, K. P., Robinson, W. E., Warnan, J., Kornienko, N., Nowaczyk, M. M., Ruff, A., ... Reisner, E. 2018. Bias-free photoelectrochemical water splitting with photosystem II on a dye-sensitized photoanode wired to hydrogenase. *Nature Energy* 3:944–951. doi:10.1038/s41560-018-0232-y.

Stephan Schmidheiny with World Business Council for Sustainable Development. 1992. *Changing Course: A Global Business Perspective on Development and the Environment.* Cambridge, USA: MIT Press.

The Full Amsterdam Circular Strategy. 2020/2025. https://www.circle-economy.com/insights/the-amsterdam-city-doughnut-a-tool-for-transformative-action (Accessed 14 April 2020).

The National UAE news article. 2019. https://www.thenational.ae/business/energy/uae-in-prime-position-as-hydrogen-power-revolution-accelerates-1.831617 (Accessed 14 April 2020).

The Ocean Portal Team, Smithsonian Institution. 2018. https://ocean.si.edu/conservation/pollution/gulf-oil-spill (Accessed 14 April, 2020).

Tiba, S., Omri, A. 2017. Literature survey on the relationships between energy, environment and economic growth. *Renewable and Sustainable Energy Reviews* 69:1129–1146.

UN General Assembly. 2015. https://www.un.org/sustainabledevelopment/sustainable-development-goals/ (Accessed April 14 2020).

UN The Energy Statistics Database. 2020. http://data.un.org/Data.aspx?d=EDATA&f= cmID%3ADL (Accessed 14 April 2020).

UNCED report. 1992. Report of the United Nations Conference on Environment and Development (UNCED). Rio Declaration and Agenda 21, Rio de Janeiro, Brazil. Available online as "Agenda 21" at https://sustainabledevelopment.un.org/content/documents/Agenda21.pdf (Accessed 14 April 2020).

UNEP. 2011. Towards a green economy: Pathways to sustainable development and poverty eradication. http://web.unep.org/greeneconomy/resources/green-economy-report (Accessed 14 April 2020).

UNEP. 2019. https://www.unenvironment.org/resources/emissions-gap-report-2019 (Accessed 21 June 2020).

UNESCO - United Nations' Educational, Scientific and Cultural Organization: Declaration on the responsibilities of the present generations towards future generations. 12 November 1997. http://portal.unesco.org/en/ev.php-URL_ID=13178&URL_DO=DO_TOPIC&URL_SECTION=201.html (Accessed 14 April 2020).

UNFCC. 2015. https://unfccc.int/process-and-meetings/the-paris-agreement/the-paris-agreement (Accessed 14 April 2020).

United Nations. 2017. Promoting entrepreneurship for sustainable development: A selection of business cases from the Empretec network. United Nations. https://unctad.org/en/PublicationsLibrary/diaeed2017d6_en.pdf (Accessed 14 April 2020).

USEIA. 2017. https://www.eia.gov/outlooks/ieo/pdf/0484(2017).pdf (Accessed 14 April 2020).

USEIA. 2019. https://www.eia.gov/todayinenergy/detail.php?id=40212 (Accessed 14 April 2020).

USEIA. 2020. https://www.eia.gov/tools/faqs/faq.php?id=709&t=6 (Accessed 14 April 2020).

Van Hise, J. R. 2008. CO_2 removal rate in earth's atmosphere. *Research Journal of Chemistry and Environment* 12:14–16.

Vattenfall Sweden. 2019. https://group.vattenfall.com/ (Accessed 14 April 2020).

Walter, M. G., Warren, E. L., McKone, J. R., Boettcher, S. W., Mi, Q., Santori, E. A., Lewis, N. S. 2010. Solar water splitting cells. *Chemical Reviews* 110(11):6446–6473. doi:10.1021/cr1002326.

WEF. 2019. https://www.weforum.org/agenda/2019/05/india-is-investing-more-money-in-solar-power-than-coal-for-first-time/ (Accessed 14 April 2020).

WHO. 2016. WHO's ambient air quality database. http://www.who.int/phe/health_topics/outdoorair/databases/cities/en/ (Accessed April 14, 2020).

Wismann, S. T., Engbæk, J. S., Vendelbo, S. B., Bendixen, F. B., Eriksen, W. L., Aasberg-Petersen, K., … Mortensen, P. M. 2019. Electrified methane reforming: A compact approach to greener industrial hydrogen production. *Science* 364(6442):756–759. doi:10.1126/science.aaw8775.

World Bank. 1999. Fuel for thought: Environmental strategy for the energy sector. Sector strategy paper prepared jointly by the Environment Department; the Energy, Mining, and Telecommunications Department; and the International Finance Corporation, Washington DC, USA. www-esd.worldbank.org (Accessed 14 April 2020).

World Bank brief. 2019. https://www.worldbank.org/en/about/partners/brief/india-developing-solar-energy-meet-rising-demand-electricity (Accessed 14 April 2020).

WPI - Worcester Polytechnic Institute. 2018. https://digitalcommons.wpi.edu/iqp-all/5185/ (Accessed 14 April 2020).

Yaghi, O. M., Li, H., Davis, C., Richardson, D., Groy, T. L. 1998. Synthetic strategies, structure patterns, and emerging properties in the chemistry of modular porous solids. *Accounts of Chemical Research* 31:474–484.

Yaghi, O. M., Li, Q. 2011. Reticular chemistry and metal-organic frameworks for clean energy. *MRS Bulletin* 34(9):682–690. doi:10.1557/mrs2009.180.

Yeh, N., Yeh, P. 2013. Organic solar cells: Their developments and potentials. *Renewable and Sustainable Energy Reviews* 21:421-431. doi:10.1016/j.rser.2012.12.046.

Zachos, J., Pagani, M., Sloan, L., Thomas, E., Billups, K. 2001. Trends, rhythms, and aberrations in global climate 65 Ma to present. *Science* 292:686–693.

Zhang, J. Z., Reisner, E. 2020. Advancing photosystem II photoelectrochemistry for semi-artificial photosynthesis. *Nature Reviews Chemistry* 4:6–21. doi:10.1038/s41570-019-0149-4.

Zhang, P., Guo, Y., Chen, J., Zhao, Y. R., Chang, J., Junge, H., ... Li, Y. 2018. Streamlined hydrogen production from biomass. *Nature Catalysis* 1:332–338. doi:10.1038/s41929-018-0062-0.

Zhang, P., Guo, Y., Chen, J., Zhao, Y. R., Chang, J., Junge, H., ... Li, Y. 2019. Author correction: Streamlined hydrogen production from biomass. *Nature Catalysis* 2:828. doi:10.1038/s41929-019-0337-0.

Zhenyou, L., Fuhr, O., Fichtner, M., Zhao-Karger, Z. 2019. Towards stable and efficient electrolytes for room-temperature rechargeable calcium batteries. *Energy & Environmental Science* 12:3496–3501. doi:10.1039/C9EE01699F.

Zhou, F., Azofra, L. M., Ali, M., Kar, M., Simonov, A. N., McDonnell-Worth, C., ... MacFarlane, D. R. 2017. Electro-synthesis of ammonia from nitrogen at ambient temperature and pressure in ionic liquids. *Energy & Environmental Science*. 10:2516–2520. doi:10.1039/C7EE02716H.

6 Sustainable Development of Marine Fisheries in Pakistan

Abdul Baset

CONTENTS

6.1 INTRODUCTION

The growth in the world population has resulted in increased consumption of animal protein, while seafood from fisheries resources are important to overcome this demand of the human population (FAO, 2017). Fisheries resources not only provide cheap animal protein, but also play a significant role in the national economy and influence human development and welfare, such as in terms of providing employment, production, and trade. Employment in the fisheries sector has grown rapidly, especially in Asia, where over 85% of the world's fisherfolk live. The fisheries resources are mainly concerned with self-renovating. Therefore, if these resources are well studied in terms of stock assessment and biological parameters, and well managed, then their effects could be huge (Pinzón-Espinosa, 2018) in overcoming the demand for seafood and fisheries.

The growing demand for seafood requires an increase in food production in a sustainable manner. The sustainability of fish populations can be affected by human activities due to the different management schemes, including direct effects and indirect effects. A direct effect on production can be the harvesting impacts of management systems. Indirect changes in key environmental variables like temperature, salinity, wind, and ocean currents can also significantly change the abundance, distribution, and availability of fish populations, either directly or by affecting prey or predator populations (Breitburg et al., 2018). Overfishing is common in the exploitation of most of the wild population, in which sustainable exploitation of the important fisheries resources is difficult but necessary.

Traditionally, fish stocks have been managed as solitary units, but nowadays the interest in associating stocks with other constituents of the ecosystem, such as climate and fisheries catch levels, has been considered as an ecosystem-based approach to fisheries management (Maxwell, 2019). This approach to fisheries management acknowledges that fisheries are part of the environment and cannot be managed in isolation (Laugen et al., 2014) and requires recognition of the ecosystem's dynamics and structure. Indicators of ecosystem change need to be understood in order to measure the impacts of fishing, climate change, and other factors across ecosystems and to provide management guidance at an ecosystem level (Maxwell, 2019). Fisheries management aims to regulate exploitation in commercially important stocks of fish to ensure their long-term sustainability. To manage these stocks successfully, an understanding of stock assessment and the dynamics of the population is necessary.

Fish population dynamics play a vital role in the management of fisheries resources. The objectives of these dynamics are helpful for the proper management and conservation of fish stocks. It helps to regulate the current level of catch and fishing effort, and thus the level of abundance of major fish species (Costello et al., 2016). Fish population dynamics identify all the factors that affect the fish and its ecology, such as growth, food, reproduction, mortality, and breeding grounds, breeding season, and nursery grounds of major fish stocks. The maximum sustainable yield (MSY) concept of the major fish stocks aims to increase productivity and obtain the highest potential economic benefits in the long run (Zhou et al., 2010). Population dynamics of fish species are important to provide basic knowledge to maintain the fish stock at a sustainable yield. It helps us to develop the skills of separating fish stock, the estimation of total fish production, growth measurement, recruit measurement, measurement of variation in recruitment, mortality measurement (Z, M, and F, etc.), controlling over-fishing, and also avoiding under fishing (Hilborn and Walters, 2013). Fish population dynamics is a significant tool for fisheries sustainability in Pakistan.

Pakistan is located in the northern Arabian Sea with a coastline that extends from Iran to India, to about 1150 km from its south-east border with the Sindh coast (68° 10′E), and north-west to the Baluchistan coast (61° 30′E). The Exclusive Economic Zone (EEZ) of Pakistan spreads over an area of 240,000 sq. km (Minton et al., 2015). The Baluchistan coast has generally a rocky bottom and a rough shelf which is about 772 km long. The Jiwani, Gawader, Pasni, Kalmat, Ormara, and Sonmiani Bays are the major fish landing areas along the Baluchistan coast. This has a rough bottom with a narrow and rough shelf, therefore trawling is not possible. The coast along the Sindh is about 348 km long and the bottom is generally sandy, muddy, or a mix. The flow of fresh water from the Indus River makes this area more productive. This region has creeks with a forest of mangroves where it creates an ecosystem that supports the biodiversity that is beneficial in the nursery grounds and provides a good habitat for fisheries resources (Whitfield, 2017). The mangrove forests of the Sindh coastline are the sixth largest in the world.

6.2 FISHING GEAR

Several types of fishing gear were used before the independence of Pakistan, includ-ing gillnets, cast nets, hand lines, long lines, boat seines, beach seines, and trawl nets. However, the gillnet is the most common way to catch fish either by small or large boats. Initially, they were made of cotton twins, but since the 1960s were replaced by nylon nets. More than 23,000 registered vessels of different sizes operate the fishing, from small to medium sized boats, to large launches and trawlers. Of these vessels about 15,000 operate in Sindh while the remaining 8000 are in Baluchistan province. There are 30 fishing vessels (20 stern trawlers and 10 tuna longliners) that operate the EEZ of Pakistan (FAO, 2017).

Typically, small boats as well as large vessels are used to catch demersal fish fauna by using gillnets. Locally it is known as a ruch and is considered to be the main type of fishing gear. The small sized fishing boats operate in shallow water for differ-ent demersal fish fauna, including catfish, sharks, trevallies, croakers, pomfrets, and mullet. The gillnet is also used for small pelagic and large pelagic fish (FAO, 2017).

6.2.1 Thukri

Thukri is a bottom set of gillnets and is used to catch shrimps. It is used by small boats of size 6–8 m and also used for hand casting. However, shrimps are mainly caught by using trawling in Pakistan. The length of shrimp trawlers is usually 15–20 m. For pelagic fish fauna, including mackerel and sailfish, large gillnets are used (Psomadakis et al., 2015).

6.2.2 Katra

Katra is a type of encircling net with a length ranging from 150 to 200 m and a depth of 24 to 30.6 m. Usually, it is used along the coast of Sindh province for small pelagic fish resources such as sardinellas (especially Indian oil sardines and anchovies). Line fishing gear is also used to capture demersal fish fauna including eels, sea bream, and croakers. More thukri is a type of gillnet used for mullet and patri thukri is an entangled net used for crabs (Khan and Khan, 2011). Various types of nets with different mesh sizes are used, for example, the mesh size of 13–17 cm is used for tuna fishery (Psomadakis et al., 2015).

6.2.3 POPLATE

A poplate plastic net is a polyamide monofilament net with a mesh size of 155 mm over a range of 2–4 km, commonly used for the catch of demersal fish fauna, including pomfrets, used by doonda fishing boats within a 16–20 m depth. A stretched mesh size at 78 mm is set at the bottom and is about 2–4 km long for targeting mackerel (Psomadakis et al., 2015).

6.2.4 PHAT

A phat is also a type of gill net and is used for shrimp catching. A bin is a type of casting net to obtain fish along the coastline. Fishing in the creeks uses a dori. A Rach lara is a gill net used for medium sized boats (25–35 ft) and launches, which fish in the buffer zone (between 12 and 35 nautical miles from shore). This gear in Pakistani waters is made in Pakistan and locally assembled (Psomadakis et al., 2015).

6.3 FISHING CRAFT

Marine fishing after independence in 1947 was entirely small scale and non-mechanized. In 1958, the first landing center was constructed in Karachi, though the progress of the fishing industry was comparatively slow. Fishers along the Sindh coast started mechanization of their traditional fishing vessels in 1958. Nowadays, there are more than 23,000 vessels of different sizes engaged in fishing, including small and medium-sized boats, large launches, and trawlers. The size of small boats ranges from 18 to 25 ft in length and they operate in coastal waters. The medium-sized boats, ranging in length from 25 to 35 ft, operate within a 12-mile territorial limit. The larger launches and trawlers operate offshore in deeper waters. There are two types of fishing vessels working at the same time (Khan and Khan, 2011).

6.3.1 MECHANIZED DOCKED BOATS

Since 1956 mechanized boats were gradually introduced. In 1983 the number of powerboats, operating in Pakistani marine waters, was just 1173, though this increased significantly with the passage of time. By 2020 the number of vessels in the marine waters of Pakistan was 23,000. The Sindh coast of Pakistan is wider than Baluchistan and has a smooth sandy or sandy muddy bottom in which these trawlers are operated. The number of trawlers has reached 3026 in Pakistan. These trawlers are wooden made having a local traditional and modern design. The number of sailboats and gillnetters are 6550 and 3408, respectively. The keel of these boats varies between 8.5 and 18 m. The normal keel length in boats is 11.6 m (Durch, 2016).

Fishermen have structurally modified their boat's "Hora" (a Pakistani boat) to 10–23 m for trawling purpose, either changing their boat to have a pointed or straight stern. The average length of the hora is 15 m. The boats are 2–7 m wide, whereas their depth varies from 0.8 to 3 m. The average width and depth is 4 and 1.9 m correspondingly. The boats are diesel driven with a power ranging from 22 to 170 hp. The average engine power is 125 hp.

The wooden gillnetters used for fishing have pointed ends at both sides. Their length varies between 13 and 35 m. These power gillnetters use diesel, and their engine power varies between 120 and 250 hp. Commercial fishing is allowed in the EEZ of Pakistan, 20–35 nm, with government permission. Foreign vessels also operate in this area. On the other hand, small scale fisheries are confined to 12 nm from the coastal belt.

6.3.2 MECHANIZED SAILBOATS

Powered sailboats including hora boats are used for fishing in Pakistani marine waters. These wooden boats operate with a diesel engine, usually 2–3 hp. Their size is generally smaller than gillnetters and recently they have decreased in number because of their small size. Mechanized sailboats made of wood number 9322 in Pakistan.

Since late 1980, specially crafted boats made from scrapped ships are also used for fishing. These boats are locally known as Doondha and operate at a 20 m depth. They use engines of 22–23 hp and are 7.5–10.5 m long. Mostly, they are operated out of Karachi; however, some are also based in Gadani, Pasni, and Gawader.

6.4 POPULATION OF FISHERMEN

The fisheries sector in Pakistan has an importance for employment. Fishermen work on operational crafts in marine as well as inland waters. Employment in fisheries sector-related industries include fish harbors, curing yards, and fish farming processing plants. The marine fishing sector supports 400,000 people directly, and about 1 million indirectly. In fisheries and related sectors, the employment of fishermen is categorized in terms of full-time employees, part-time employees, and casual (FAO, 2017).

6.5 THE PREFERABLE SEASONS FOR FISHING

Some fish species can be caught throughout the year, but summer and autumn are the preferable seasons for fishing many species, especially from June to October in Pakistani marine waters, since more fish with commercial importance are caught during these months (Islam and Wahab, 2005). These commercially important fish include grunts, grunter's threadfins, croakers, and catfish. Shrimp fishery shows a catch pattern throughout the year. However, during the months of June and July the

Sindh government has put a ban on the commercial catching of shrimps due to their breeding season (Abbas et al., 2013).

6.6 CONSUMPTION OF SEAFOOD

Seafood is a preferred food worldwide, and it is proven that fish are a source of essential amino acids. Pakistani people eat fish, especially the people who live in coastal areas, though they are not very fond of other seafood. Most people do not show any tendency to eat seafood such as shrimps, mussels, oysters, shells, crabs, and lobsters. This is perhaps because of a religious point of view; they prefer to eat lamb, poultry, and vegetables (Roy, 2012).

Fish and fishery products in Pakistan were 2.5 kg/capita in 1961 according to Qureshi (1961). The consumption of fish in the coastal city of Karachi was estimated by Ahmad (1970) as 5 kg in 1970, and on the Makran coast as 10 kg, while it is only 1 kg in all other parts of the country. The Makran coast is undeveloped and obtaining other food is expensive, which may be the reason for the high consumption of seafood, which is slowly increasing. In major cities of the country, 30% of the consumption of seafood is from marine catches. However, only 45–50% of the total catch is used for consumption while about 22% of it is exported. In addition to this, 28% of the catch is used to prepare various fishery products (Nasim, 2010).

6.7 MAJOR FISHING GROUNDS AND FISH LANDING SITES

The location of Pakistan in the northern Arabian Sea has a coastline extending from Iran to India, to about 1150 km from its south-east border with the Sindh coast (68° 10′E) and to north-west of the Baluchistan coast (61° 30′E). The EEZ of Pakistan spreads over an area of 240,000 sq. km and is very important and significant (Minton et al., 2015). The marine areas facilitate the largest fisheries in the country. Large pelagic fishery resources include shark, mackerel, and tuna, which are found in offshore waters, which are considered a major fishing ground, while small pelagic fishes are hunted mainly in basic fishing grounds whereas offshore waters are particularly good for the capture of large pelagic fish species. Small pelagic fishery resources include mullet and silver wittings. In Sindh, due to the Indus delta, a number of fishing settlements in the creeks are considered important fishing grounds as well as a breeding ground for many marine and estuarine fish species. However, there are some other small rivers that open out into coasts, including enclosed and semi-enclosed bays, which are also important for small scale fishing. These grounds act as a fishery resource reservoir from which many commercially important fish fauna are harvested (Abbas et al., 2013).

Pakistan's marine waters are divided into three different zones from a strategic and management point of view: The first is used for small-scale fisheries and is measured from the coastline up to 12 nm out to sea; the next area ranges from 12 to 20 nm and is regarded as a buffer zone. The second zone is from 20 to 35 nm, where fishing is carried out using medium-sized vessels. In the third zone, ranging from 35 to 200 nm, only industrial fishing can operate, using large-sized fishing vessels and trawlers (Potgieter, 2012).

There are a total of 22 fish landing sites in Pakistan of which seven are major fish landing sites; of these, three are in Sindh province and four are in Baluchistan. Karachi, the Korangi fishing harbors, and Ibrahim Haidery are major fish landing sites in Sindh province. Pasni and Gawader are fish harbors, while Gaddani and Damb are major fish landing sites in Baluchistan province. The most important major fish landings are in Karachi in Sindh, while Gawader and Pasni are in Baluchistan (MFD, 2014).

The topographical features of the coast vary in Pakistan. Thus, considering its characteristic features, the coastline is divided into two regions. The first belongs to the Sindh coast which is situated at the south-eastern side, whereas the second region belongs to the Baluchistan or Makran coast which is located on the north-western side. The Makran coast is long at 772 km as compared to the Sindh coast at 348 km (Baset et al., 2017a) and stretches from the Iranian border on one side and the Hub river on the other. This coast contains many bays such as Ormara, Kalmat, Sonmiani, Gawader, Pasni, and Jiwani. These bays serve as main fishing areas in Baluchistan. This coast has a rough bottom and is mostly rocky; the continental slope is from 10 to 30 miles off the coast. Therefore, it is impossible to practice bottom trawling here.

From the Hub River to the Indian border, an area of 348 km, is the Sindh coast. As compared to the Baluchistan coast, this one has more fish breeding grounds thus is rich in aquatic fauna and very fertile because of the Indus delta. It has a bottom of sand and mud. There are many creeks along its length. Mangroves associated with the creeks act as the best nursery grounds. Therefore, various types of aquatic fauna, particularly finfish, shellfish, and shrimps, dwell there (Panhwar et al., 2012). The major areas of the shelf expand to about 80 miles into the Indus delta. Like the Baluchistan coast, the Sindh coast also has many major fishing grounds such as Khobar, Jhari, Phitti, Gharo, Korangi, Hydri, Keti Bandar, Hajamro, and Khuddi. There are four major fish harbors (two in Sindh and two in Baluchistan province) in Pakistan. Korangi fish harbor is operated by the Federal Ministry of Port and Shipping. Karachi fish harbor is the biggest and oldest one in Pakistan. This large fishing harbor houses nearly 4000 fishing vessels. Karachi fish harbor is operated by the provincial government of Sindh. About 95% of fish and fishery products are exported through Karachi fish harbor, where about 90% of the total fishery is landed. Because of the congestion at the harbor, it was strongly felt that another harbor nearby should be developed, so the Asian Development Bank provided a loan in 1984 for Korangi fish harbor, the construction of which was completed in 1992. Gawader fish harbor, Pasni fish harbor, Korangi fish harbor, and Karachi fish harbor are major fish landing sites in Pakistan. Various departments of government operate these harbors, such as Gawader fish harbor by the Federal Ministry of Communication and Pasni fish harbor by the Government of Baluchistan. Recently, due to the China–Pakistan Economic Corridor (CPEC), Gawader fish harbor will be developed further (FAO, 2017).

However, there are also small landing sites in Baluchistan as well as Sindh. In Baluchistan, the coastal development authority operates a small landing site at Gadani fish landing jetty, while in Sindh, the Karachi local government operates Ibrahim Hydri.

6.8 MAJOR MARINE RESOURCES

Pakistani marine waters are subtropical and much richer in biodiversity as compared to temperate waters. Commercially important fish (finfish and shellfish) species are found in these marine waters, including about 250 species of demersal fish, 50 species of small pelagic fish, 20 species of large pelagic fish, and 15 species of medium-sized pelagic fish. Furthermore, there are 15 species of commercial shrimp, 12 of squid/cuttlefish/octopus, and five of lobster. The majority of them are edible and commercially important species (FAO, 2017).

A large number of demersal and pelagic fish are commercially important in terms of market value or in terms of quantity and are landed at different sites in the country. The main finfish are mullet, croakers, groupers, snappers, pomfrets, seabream, eels, thread-fin breams, sea catfish, ribbonfish, travellies, sharks, and rays. In shellfish, shrimps include *Penaeus indicus*, *P. merguiensis*, *P. monodon*, *P. japonicus*, *P. semisulcatus*, *P. penicillatus*, *Metapenaeus affinis*, *M. stebingii*, *M. brevicornis*, *M. monoceros*, *P. stylifera*, *P. hardwickii*, *P. sculptilis*, and *Metapenaeopsis stridulans*.

The Sindh coastline has a large freshwater flow from the River Indus, which has many creeks, estuaries, and mangrove areas which serve as the primary nursery ground for many marine finfish and shellfish species. The reason is that the Sindh coast is very fertile and is rich in fish biodiversity. Major finfish and shellfish resources are from the Sindh coast. Most of the fishing grounds are also along the Sindh coastline (Psomadakis, 2015).

6.9 MAJOR THREATS TO MARINE BIODIVERSITY

The most critical problem for Pakistan's marine biodiversity are global climate change, which results in the flow of water from the Indus River being decreased due to the construction of barrages over the river for agricultural use, which has affected the mangrove home of aquatic fauna. Illegal, unreported fishing, and overexploitation are also major threats to marine biodiversity. Destructive fishing gear, including the gujjo for bottom trawling, the katra seine net, and the bhoola tidal trap net, is also damaging habitation. Likewise, the seawater is becoming more turbid as a result of pollution. This tainted water contains a low level of oxygen which causes a serious threat to marine biota (Inam et al., 2007).

Global climate change affects directly and indirectly marine biodiversity in Pakistan. In the future, this effect is expected to continue (Roessig et al., 2004). Climate change may result in salinity increase, oceanic temperature variation, and increase in the acidity and pH with the passage of time. Under the influence of climatic conditions, the sea surface temperature (SST) will increase with time (from 20 to 30°C). In addition to the temperature, sea salinity also varies around the year (from 35.4 to 36.7 PPT). Salinity, near the coastal belt, depends upon the water flow from rivers which come from glacier melting or rainfall (from 36.5 to 37.2 ppt). From July to September the salinity drops because more freshwater flows into the sea from the river. Conversely, during March to the peak of June a rise in the salinity of seawater is observed because less water flows into the sea during these months (Psomadakis et al., 2015). The salinity ranges from 36.2 to 36.5 ppt in the northern part of the

Arabian Sea. Some factors, like increasing pollution from urban development and the low freshwater flow of the Indus River into the sea, cause salinities to increase, which affects the marine ecosystem (FAO, 2007).

The continuing reduction of the Indus water flow causes the shrinking of the Indus estuary system. The mangroves serve as the nursery grounds of various fish and shrimp species. The Sindh coast is blessed with a mangrove forest system, due to its sandy and muddy nature, whereas the Baluchistan coast has mangrove forest at Miani Hor and in other places, but it is not developed because of its uneven and rocky nature. The population of mangroves is decreasing around the coastline of Pakistan due to the low rainfall, the reduction of the Indus water flow, and the use of those mangroves for other purposes (FAO, 2017). A number of mangrove species have become extinct; there are now only three species, that is, *Avicennia marina*, *Ceriops tagal*, and *Rhizophora mucronata* (Psomadakis, 2015).

6.10 COMMERCIALLY IMPORTANT FISH FAMILIES

Demersal fisheries are the key resource of Pakistani waters, followed by pelagic fisheries. Important fish families are: *Clupeidae, Sillaginidae, Haemulidae, Sciaenidae, Scombridae, Serranidae, Lutjanidae, Carangidae, Ariidae, Chirocentridae, Engraulidae, Lethrinidae, Nemipteridae, Serranidae, Sparidae, Terapontidae, Istiophoridae, Monacanthidae, Polynemidae, Sillaginidae, Sphyraenidae, Coryphaenidae, Cynoglossidae, Drepaneidae, Muraenesocidae, Stromateidae, Penaeidae, Palinuridae, Portunidae, Sepiidae, Loliginidae, Alopiidae, Carcharhinidae, Triakidae, Rhinidae, Rhinobatidae, Dasyatidae, Gymnuridae, Myliobatidae, and Mobulidae.* The Pakistani marine fishery depends on these fish families (Psomadakis et al., 2014; Gausmann, 2018).

6.11 FISH STOCK ASSESSMENT SURVEYS

The mismanagement and unrecognized status of exploited fisheries also result in the overfishing and depletion of stocks. Stock assessment of resources provides meaningful information on the exploitation of these living aquatic resources so as to exploit them for a maximum sustainable yield. Obtaining such a yield after evaluating the stock's condition, along with optimum fishing modes, is an important step in the execution of management strategies for rejuvenation of valuable stock. The mismanagement of a stock may result in commercially valuable species changing undesirably in term of biological characteristics and productivity (Friedlander et al., 2015).

Several fishery resource surveys have been conducted for the purpose of fish stock assessment in Pakistani waters by the Marine Fisheries Department of Pakistan in cooperation with foreign agencies. The first provisional survey was conducted, using a wood and a shrimp trawler named *Ala* and *Machhera* respectively, in 1960. Later on, the fish stock from deeper waters was assessed by Pruter (1964) and Hida and Pereyra (1966). After this, during the period 1966–1969, another survey was conducted through a commercial shrimp trawler named *Machhranga*. In 1970, two Russian scientists (Gololobov & Grobe, 1970) worked on the stock assessment of

Pakistani waters and used the Marine Fisheries Department's vessel. In 1973, the stock of demersal fish species was estimated by Zupanovic & Mohiuddin (1973) in Pakistani waters. In 1975–1976 and 1977 the Norwegian research vessel *Dr. Fridtjof Nansen* and a Japanese research vessel (*Shoyo Maru* in 1975–1976) conducted surveys that found a huge stock of giant squid (*Symplectoteuthis ovaliniensis*). Appleyard et al. (1981) analyzed the data obtained from another survey which was done in the EEZ of Pakistan using the research vessel *Nauka* in 1969. Abidgaard et al. (1984) used two research vessels *Machhera* and *Tehkik* and studied the biomass of demersal fish resources from 0 to 200 m in depth.

The latest surveys were conducted using the Iranian research vessel *Ferdows-I* and local fishing vessel *Mehboob-e-Madina*. The Norwegian research vessel *Dr. Fridtjof Nansen* in 2009–2010 surveyed demersal and pelagic species in Pakistani waters. The FAO project (Fisheries Resources Appraisal in Pakistan) funded these surveys under the Marine Fisheries Department, Karachi, Government of Pakistan.

6.12 FISHERIES MANAGEMENT IN PAKISTAN

Fishery in Pakistan contributes 1% of the total GDP of the country, which needs to improve its fisheries sector by adopting modern techniques of management for the protection and conservation of its fisheries resources (Nazir et al., 2015). Fisheries management in Pakistan is a great challenge for the managers of present and future generations. They have to address the effects of the natural environment, pollution, climate factors, and fishing. However, it seems that management has failed to manage the fisheries sector of Pakistan because of the common evidence of overexploitation (Baset et al., 2017b).

A big challenge for marine fishery in Pakistan is the illegal fishing by unregistered vessels, particularly small boats, which affect the accuracy of the catch effort data. The second challenge is to establish marine protected areas (MPAs) to safeguard juveniles especially in the nursery grounds of finfish and shellfish so that they can breed at least once in their lifetime (FAO, 2007).

The other challenges are the ban of destructive fishing methods, the closing season, control of fishing licenses, regulations on gear type, and mesh size. However, there is no satisfactory implementation of those management measures. The control, surveillance, and monitoring system are able to sustain the fisheries sector for the upcoming period, which requires that stockholders and fisheries research institutions work collaboratively. When fishery managers are unaware of the status and potential of the resources under their responsibility, they are unlikely to act at the right time or to make the right decisions.

The statistics department of the marine fisheries of Pakistan rarely publish the figures for the fishery stock and catch, hence data are not accessible for assessment by scientists and experts. The fishery statistics data should be published yearly so that experts support the fishery managers to observe the effects of fishing and other factors to describe their past and current status, and to plan for the better management of a fish stock for the future, and to make predictions about how a fish stock will respond to current and future management measures (Soncini-Sessa et al., 2007).

The most important thing for a successful fisheries management is to maintain over the long term where clear management policies are implemented by a proactive management process. The changing resource conditions will be a big problem, but a necessary step in the long term for fishery managers if achieving agreement on the management actions to be taken. Pre-agreed harvest control rules may be good for fishery as they should reduce the impact of short-term political factors in making difficult decisions about future catches. Many fisheries stakeholders used to plan annually to maintain their benefits from fishery, but long-term strategies may also be difficult to implement in a political world (Soncini-Sessa et al., 2007).

6.13 FISHERIES PRODUCTION IN PAKISTAN

6.13.1 TOTAL FISHERIES PRODUCTION

Pakistan has rich fisheries resources with a total production of 24,337,449 MT from 1950 to 2017 where the average production rate was 357,903 MT per year. During this period, captured fisheries production was 21,981,192 MT and from aquaculture was 2,356,257 MT. In 1999, it obtained the highest fish production of 677,606 MT, while the lowest was reported during 1950–1963, with an average production rate of 64,237 MT per year. For the first time, an increase was seen in production during 1964–1975, with an average of 1,705,935 MT yearly, where the growth rate was recorded at 187 MT per year. Over 20 years (1984–2003) fish production showed an average of 530,916 MT yearly. Remarkably, the growth rate has increased over the last ten years (2008–2017) with an average rate of 617,275 MT per year. In the period of 1950–2017, the growth rate per year was 5.95% where the lowest was calculated at 19.6% in 1974 and the highest was 109% in 1953 (Figure 6.1) (Shah et al., 2018).

6.13.2 CAPTURE FISHERY PRODUCTION

Capture fishery has a significant role in the fish production of Pakistan from 1950 to 2017 with an average of 320,504 MT per year. Between 1950 and 1959, per year production was estimated as 56,599 MT; however, the lowest production rate was recorded in 1950. After the first decade, an increase was recorded, where the highest catch was recorded as 654,530 MT in 1999 (Figure 6.2). Capture fishery contributed 98% yearly from 1950 to 1999 in total fish production, while during 2000–2017, it was calculated as 80% per year. The highest percentage of capture fishery in total fish production was recorded as 98% in 1950, while the lowest was calculated as 76.5% in 2017. The reduction of capture fishery participation means aquaculture production increases year-wise. However, Pakistan needs to pay more attention to protect fishery resources by decreasing vessel licenses (Shah et al., 2018).

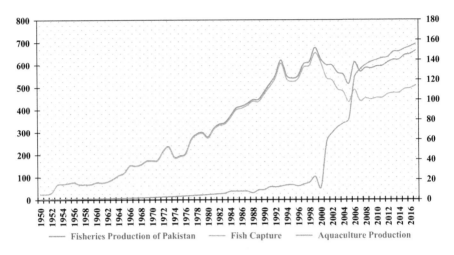

FIGURE 6.1   Time scale series data of total fisheries production from 1950 to 2017 in Pakistan

Source: Shah et al., 2018.

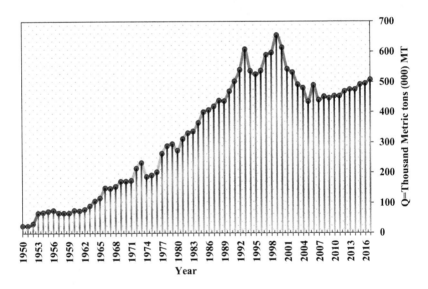

FIGURE 6.2   Capture fishery production in Pakistan from 1950 to 2017

Source: Shah et al., 2018.

6.13.3   Aquaculture Production

Aquaculture has contributed to fish production in Pakistan since 1950. Aquaculture contributed only 144,208 MT from 1950 to 1990, where the lowest contribution was recorded in 1950. After then the growth rate was estimated at 5460 MT per year for the next three decades (1991–2017), where the highest was estimated at 8430 MT per

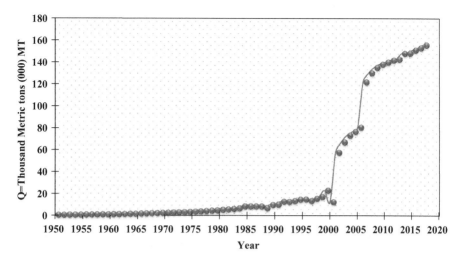

FIGURE 6.3   Aquaculture production in Pakistan from 1950 to 2017

Source: Shah et al., 2018.

TABLE 6.1
Country-wise export of sea food (USD thousands)

Country	2015–2016	2014–2015	2013–2014
Vietnam	79,532	88,102	91,802
Thailand	48,821	6,700	40,463
China	44,020	63,816	45,113
U.A.E	38,141	41,816	52,489
Korea	25,199	18,386	18,321
Kuwait	13,509	8,831	13,659
Malaysia	11,731	11,846	19,339
Japan	9,404	5,419	6,602
Egypt	8,926	11,713	16,096
Indonesia	5,633	7,291	8,892
Sir Lanka	4,994	4,424	4,437
Hong Kong	3,704	4,320	6,097
Qatar	3,435	2,096	4,437
United Kingdom	2,983	2,594	1,499
All others	24,837	71,881	40,212
Total	**324,869**	**349,235**	**369,458**

Source: Trade Development Authority of Pakistan.

year (Figure 6.3). However, Pakistan needs also to give more positive attention to the improvement of aquaculture to compete with the demand for seafood (SBP, 2016).

6.13.4   EXPORT FISHERIES PRODUCTION

Pakistan exports fisheries products to more than 30 countries, of which 14, including Vietnam, Thailand, and China, are the chief importers (Table 6.1). According to the

TABLE 6.2
Export of sea food from Pakistan

Year	Value (USD thousands)
2009–2010	226,936
2010–2011	296,182
2011–2012	320,182
2012–2013	317,652
2013–2014	369,458
2014–2015	349,235
2015–2016	324,869

Source: Trade Development Authority of Pakistan.

Trade Development Authority of Pakistan, the highest export occurred in 2015–2016 which earned USD324,869,000 (Table 6.2) (SBP, 2016).

6.13.5 IMPORT FISHERIES

The import of fisheries products is much less than the exports of Pakistan; however, from 1980 to 2002 it was 24.2 MT (USD54,600) and from 2003 to 2013 it was estimated at 1468 MT (USD2,622,000) per year (SBP, 2016).

6.14 CONCLUSION

Pakistani marine waters are very fertile for every type of commercially important fishery (finfish and shellfish). Fisheries management in Pakistan is a great challenge for the managers for present and future generations. They have to address the effects of the natural environment, pollution, climate factors, and the effect of fishing. However, it seems that they have failed because of the common evidence of overexploitation, which results in capture fisheries production in Pakistan decreasing with the passage of time. Fisheries overexploitation creates not only ecosystem damage but also decreased biomass production.

There are some limitations in the research and management, such as insufficient monitoring of commercial fisheries catches, the fact that sustainable fisheries need adequate monitoring, there are out-of-date assessments of fishery stocks, and that no work was done after the FAO project (Fisheries Resources Appraisal in Pakistan) in 2009–2010. The continued and timely appraisal of marine fisheries is very necessary for the country's sustainable fisheries. There is limited availability of qualified human resources engaged with commercial fisheries management. There is limited availability of training institutions, and fisheries management do not get much benefit from them. The limited practice and ignorance of mariculture is directly proportional to the overexploitation of marine capture fisheries. There is limited investigation about offshore deep-water fishing grounds which could provide alternative resources for excess fishing efforts along the coastline. The continuing reduction of the Indus water flow causes the shrinking of the estuary system, an increase in salinity, and a decrease in the population of mangroves which serve as the nursery grounds of various fish and shrimp species. Coastal pollution is one of the biggest challenges for marine fisheries management, which needs to establish institutions where people can be trained in modern, low-cost, pollution-control technology.

The unique contribution of this study is very significant in that it may provide some guidelines for marine fisheries and management so that they may take some serious steps for its betterment. We suggest the following recommendations for marine fisheries management.

1. The Pakistan fisheries managers should take steps to reduce the catch because in this way we can control fishing effort, the number of fishing vessels, and trawler mesh size. So we can save and protect the growth of small fish resources. We should provide them with the time to breed once in their lifespan.
2. The government must ensure the improvement of the institutional structure. The Pakistan Marine Fisheries Department (MFD) should establish institutions under the federal government as well as provincial administration that chiefly focuses on aquaculture. The Department of Mariculture and Marine Aquaculture should be engaged for the benefit of marine fisheries and plants.
3. Fishery management should promote full realization of optimum yields as originally envisioned in the Pakistan National Fisheries Policy (2007) by ensuring that harvests do not reduce stock abundance below levels that can sustain maximum yields over a long period. For currently overfished stocks, harvest levels must allow rebuilding of stock over specified periods to a level that can support sustainable maximum yields.
4. Fishery management should control entry into and the wasteful deployment of capital, labor, and equipment in marine fisheries.
5. The federal government must ensure that research institutions (the Marine Fisheries Department, the Sindh Fisheries Department, etc.) improve data collection, analysis, and distribution to make evaluations and policy decisions and to develop a long-range goal of an ecosystem management framework that ensures sustainable levels of natural resources and that minimizes the effects of human actions both on the ecosystem as a whole and on the individual components of it – for example, species' habitats and the food-web structure.
6. The government must ensure the reduction of the bycatch or discard problems within fisheries to reduce illegal and banned fishing in the spawning seasons, and checkups for registration and net size – so that fish at least spawn once in their lifespan.
7. A major national program should be developed to determine what habitats are critical for fish reproduction and growth and how they can be protected. Such a program would bring the problem of degradation of fish habitats to national attention, and would provide a means of coordinating measures to achieve adequate protection.
8. The CPEC is a great prospect for Pakistan and China to establish joint institutions and research centers for marine fisheries management and cooperation in marine affairs and their development. The institutions should develop projects for the control and protection of fisheries resources, to establish marine aquaculture farms in Sindh and Baluchistan provinces.

REFERENCES

Abbas, G., Khan, M.W., and Ahmed, M., 2013. *Coastal and Marine Fisheries Management in SAARC Countries*. SAARC Agriculture Centre, Dhaka, Bangladesh.

Baset, A., Liu, Q., Hanif, M.T., Liao, B., Memon, A.M., and Mohsin, M., 2017a. Estimation of maximum sustainable yield using production modeling: A stock appraisal of Indian Oil Sardine (*Sardinella longiceps*) from Pakistani Waters. *Pakistan Journal of Zoology*, 49(2), pp. 521–528.

Baset, A., Qun, L., Pavase, T.R., Hameed, A., and Niaz, Z., 2017b. Estimation of maximum sustainable yield of *Scomberomorus* species fish stocks in Pakistan using surplus production models. *Indian Journal of Geo Marine Sciences*, 46(11), pp. 2372–2378.

Breitburg, D., Levin, L.A., Oschlies, A., Grégoire, M., Chavez, F.P., Conley, D.J., Garçon, V., Gilbert, D., Gutiérrez, D., Isensee, K., and Jacinto, G.S., 2018. Declining oxygen in the global ocean and coastal waters. *Science*, 359(6371).

Costello, C., Ovando, D., Clavelle, T., Strauss, C.K., Hilborn, R., Melnychuk, M.C., Branch, T.A., Gaines, S.D., Szuwalski, C.S., Cabral, R.B., and Rader, D.N., 2016. Global fishery prospects under contrasting management regimes. *Proceedings of the National Academy of Sciences*, 113(18), pp. 5125–5129.

Durch, W., 2016. *Constructing Regional Security: The Role of Arms Transfers, Arms Control, and Reassurance*. Springer, New York.

FAO, 2007. *The State of World Fisheries and Aquaculture 2006*. Food and Agriculture Organization, Rome.

FAO, 2017. *The Future of Food and Agriculture: Trends and Challenges*. Food and Agriculture Organization of the United Nations, Rome.

Friedlander, A.M., Nowlis, J., and Koike, H., 2015. Improving fisheries assessments using historical data. In *Marine Historical Ecology in Conservation: Applying the Past to Manage for the Future*, p. 91.

Hilborn, R. and Walters, C.J. eds., 2013. *Quantitative Fisheries Stock Assessment: Choice, Dynamics and Uncertainty*. Springer Science & Business Media, Dordrecht.

Inam, A., Clift, P.D., Giosan, L., Tabrez, A.R., Tahir, M., Rabbani, M.M., and Danish, M., 2007. The geographic, geological and oceanographic setting of the Indus River. In *Large Rivers: Geomorphology and Management*, pp. 333–345.

Islam, M.S. and Wahab, M.A., 2005. A review on the present status and management of mangrove wetland habitat resources in Bangladesh with emphasis on mangrove fisheries and aquaculture. In *Aquatic Biodiversity II*, pp. 165–190. Springer, Dordrecht.

Khan, S.R. and Khan, S.R., 2011. Fishery degradation in Pakistan: A poverty–environment nexus? *Canadian Journal of Development Studies/Revue canadienne d'études du développement*, 32(1), pp. 32–47.

Laugen, A.T., Engelhard, G.H., Whitlock, R., Arlinghaus, R., Dankel, D.J., Dunlop, E.S., Eikeset, A.M., Enberg, K., Jørgensen, C., Matsumura, S., and Nusslé, S., 2014. Evolutionary impact assessment: Accounting for evolutionary consequences of fishing in an ecosystem approach to fisheries management. *Fish and Fisheries*, 15(1), pp. 65–96.

Maxwell, K., 2019. Informing Ecosystem-Based Fisheries Management from an Indigenous Perspective: The Mōtū Kahawai Fishery. Thesis, Victoria University of Wellington.

MFD, 2014. *Handbook of Fisheries Statistics of Pakistan*. Compiled by Marine Fisheries Department (MFD), Karachi, Pakistan.

Minton, G., Reeves, R., Collins, T., and Willson, A., 2015. Report on the Arabian Sea Humpback Whale Workshop: Developing a collaborative research and conservation strategy.

Nasim, A., 2010. Enterprises based fisheries sector study and strategic plan for interventions at enterprise's level to enhance quality production. UNIDO final report.

Nazir, K., Yongtong, M., Kalhoro, M.A., Memon, K.H., Mohsin, M., Kartika, S., 2015. A preliminary study on fisheries economy of Pakistan: Plan of actions for Fisheries Management in Pakistan. *Canadian Journal of Basic and Applied Sciences*, 03, pp. 7–17.

Panhwar, S.K., Liu, Q., Khan, F., and Siddiqui, P.J., 2012. Maximum sustainable yield estimates of Ladypees, Sillago sihama (Forsskål), fishery in Pakistan using the ASPIC and CEDA packages. *Journal of Ocean University of China*, 11(1), pp. 93–98.

Potgieter, T.D., 2012. Maritime security in the Indian Ocean: Strategic setting and features. *Institute for Security Studies Papers*, 2012(236), pp. 24–24.

Psomadakis, P.N., Osmany H.B., and Khan, W., 2014. Important coastal fishery species of Pakistan. A pocket guide. FAO FishFinder Programme. FAO, Rome. 63 cards + CD-ROM.

Psomadakis, P.N., Osmany, H.B., and Moazzam, M., 2015. *Field Identification Guide to the Living Marine Resources of Pakistan*. Food and Agriculture Organization of the United Nations, Rome.

Roessig, J.M., Woodley, C.M., Cech, J.J., and Hansen, L.J., 2004. Effects of global climate change on marine and estuarine fishes and fisheries. *Reviews in Fish Biology and Fisheries*, 14(2), pp. 251–275.

Roy, P., 2012. Pakistan A Crisis of Legitimacy and Liberalization 1980–2010.

SBP, 2016. State Bank of Pakistan—Annual Report—2015–16.

Shah, S.B.H., Mu, Y., Abbas, G., Pavase, T.R., Mohsin, M., Malik, A., Ali, M., Noman, M., and Soomro, M.A., 2018. An economic analysis of the fisheries sector of Pakistan (1950–2017): Challenges, opportunities and development strategies. *International Journal of Fisheries and Aquatic Studies*, 6(2), pp. 515–524.

Soncini-Sessa, R., Weber, E., and Castelletti, A., 2007. *Integrated and Participatory Water Resources Management—Theory*. Elsevier.

Whitfield, A.K., 2017. The role of seagrass meadows, mangrove forests, salt marshes and reed beds as nursery areas and food sources for fishes in estuaries. *Reviews in Fish Biology and Fisheries*, 27(1), pp. 75–110.

Zhou, S., Smith, A.D., Punt, A.E., Richardson, A.J., Gibbs, M., Fulton, E.A., Pascoe, S., Bulman, C., Bayliss, P., and Sainsbury, K., 2010. Ecosystem-based fisheries management requires a change to the selective fishing philosophy. *Proceedings of the National Academy of Sciences*, 107(21), pp. 9485–9489.

Gausmann, P., 2018. Synopsis of global freshwater occurrences of the bull shark (*Carcharhinus leucas* Valenciennes 1839, Carcharhinidae) with comments on the geographical range. Unpublished report.

7 Factors Affecting Consumer Purchase Behavior in the Digital Marketing Arena in Kolkata, India

Biswajit Roy and Sudin Bag

CONTENTS

7.1 INTRODUCTION

Traditional marketing uses conventional marketing strategies like personal selling and advertising through traditional media like TV, newspapers, and bill boards. It mostly uses hardcopy material and opts for interpersonal communication, which makes it static. On the other hand, digital marketing uses sales and promotion of goods and services through electronic media. It is viewed as more interactive and dynamic in nature, though it suffers from drawbacks like lack of security and reliability. Indeed, digital marketing relies on the use of electronic devices such as computers, tablets, smart phones/cell phones, digital bill boards, and game consoles, which are deployed to engage with consumers and other business partners (Das and Lall, 2016).

Managers as well as entrepreneurs are now in a dilemma with regard to choosing either traditional or digital marketing – and they are attempting to find the perfect balance between them (Shukla, 2010). This study provides some guidelines for them so that they can make a balance between these two ways of marketing.

It is important to underline that digital marketing practices are increasing more and more rapidly with the introduction of numerous technologies in this era of digitalization (Issa, 2014). Busy daily life activities are also pushing citizens to opt for online marketing processes, though traditional marketing remains a must for several areas or sectors.

To define digital marketing (electronic marketing), it is commonly argued that it arises from the use of information and communication technology (ICT) and numerous technologies applied to traditional marketing (Keserwani, 2014). This implies that people using digital marketing processes should benefit from a technological awareness and easy access to technological equipment. Governments are also encouraging people to use digital marketing techniques. However, due to a lack of availability of technology and public awareness, the traditional ways of marketing still have a lot of preference among public organizations. In fact, in some areas, these are the only marketing processes that people can avail themselves of. More especially, it is found through observation that, due to the involvement of technology and comfort associated with the use of computing devices and applications, youth (college goers) have preference towards digital marketing more than aged people. Rapid technological advancements over the last two decades have dramatically changed consumers' expectations (Khan and Chavan, 2015; Shaikh et al., 2020). Fulfillment of customer expectation is the first priority for the business to gain competitive advantages in the market (Bag and Sen, 2012). Digital services help to deliver the best services to mitigate customers' expectations (Hoffman and Novak, 2017). The expansion of the digital platform makes customers well aware of the techniques employed by firms while doing business (Slintak, 2019; Zhidebekkyzy et al., 2019). A website can be used to mitigate customers' expectations by detailing the correct information about products, prices, offers, and delivery (Egger, 1999; Khare, 2016). Keeping this in mind, the current study considers college goers in conducting an opinion based survey. An analysis was then done on the collected data. It was ensured that the respondents selected for conducting this research had made at least one transaction digitally so that they had experienced both traditional and digital marketing processes. After that, the respondents were asked to select either the digital or traditional marketing process for their next transactions – 20% of them notably expressed an opinion toward the benefits of traditional marketing over digital marketing. This attracts our involvement toward the understanding of the pros and cons of digital and traditional marketing through a literature review.

7.2 LITERATURE REVIEW

Researchers from different parts of the world have attempted to identify the factors that play a significant role on the adaptation of either traditional marketing or digital marketing practices. Consumers show their preferences toward either traditional or digital marketing or both processes based on the factors they prefer the most. Joseph (2008) has identified that factors that attract people toward traditional marketing processes are after-sales service and support for customers. Also use of digital marketing requires some technical knowledge and devices that are not available to many. But according to the researcher, the factors for which traditional marketing is not favored by many is that it is found to be often expensive and time consuming.

Digital marketing is more of a recent concept in comparison to traditional marketing. Researchers from different parts of the world have attempted to address factors that play an important role in consumers' minds at the time of their adaptation. Kumar and Dange (2012) in their paper "A Study of Factors Affecting Online Buying Behavior: A Conceptual Model" attempted to understand buying motives by identifying the internal and external factors of online marketing. They discussed about online security, privacy, trust, and trustworthiness in the context of digital marketing.

In India a significant study was made by Ganapathi (2015). In his paper entitled "A Study on Factors Affecting Online Shopping Behavior of Consumers in Chennai", this researcher applied exploratory factor analysis and thereafter a regression analysis and concluded that convenience, security, website features, and time saving were positively and significantly influencing the purchasing decisions of consumers at the 1% level. The study suggested that online retailers need to ensure that the online shopping process on their websites is designed to be as simple and convenient as possible for online consumers to shop online. In addition to this, online retailers might ensure an efficient delivery service to their consumers.

In another significant study, Franco and Bulomine Regi (2016) identified some of the factors to which customers are attracted regarding digital marketing, such as convenience, time saving, options, ease of comparison, ease of finding reviews, and coupons and deals. Consumer awareness about the internet and digital media makes them develop interests, which influence their online purchase of products and services (Burman and Aggrawal, 2015; Ray et al., 2020). However, customers are scared of using digital marketing due to some of the factors related to for example privacy and security, quality, hidden costs, delay in receiving goods, need to access the internet, and lack of personal interaction, for which traditional marketing is preferred.

Little research has explored the relative importance of the factors that influence both digital and traditional marketing. Kulathunga et al. (2018), in their paper entitled "Factors Affecting Online Purchase Intention: A Study of Sri Lankan Online Customers", investigated this issue from a technological perspective. According to them, the factors are perceived usefulness, perceived ease of use, website content, and trust. The findings reveal that trust has a mediating effect on perceived usefulness, perceived ease of use, and website content on the online purchase intention related to retail shopping in Sri Lanka.

To perform the comparative study on the factors, which have an impact on the customers' choice toward using either the traditional or the digital purchasing process, it is important to choose respondents who have a considerable exposure to both processes. In this regard, it is customary to choose the youth group of customers who have experienced both marketing processes. Jukariya and Singhvi (2018) in their paper "A Study of Factors Affecting Online Buying Behavior of Students", considered the inclination and interest of the youth population toward online shopping. They identified transaction security and multiple payment options, personal privacy and security, product price and quality, the speed of access, and after-sales service as some major factors that affect students' online shopping behavior. According to the study the factor which has the greatest impact on online buying is personal privacy and security followed by transaction security and multiple payment options and convenience and saving time.

This study hence selected the youth population and considered some of those common factors that may have an impact on choosing either the traditional or the digital marketing process.

7.3 RESEARCH METHODOLOGY

The process of collecting data aims at identifying the main influential factors on consumer purchase behavior in the digital marketing arena. This is followed by a literature review of prior research that addresses similar research problems.

7.3.1 RESEARCH DESIGN

Exploratory research design is followed in this study to find the influencing factors by using statistical methods. The study also includes some deeper psychological considerations that even the respondents may not be aware of regarding their liking or disliking of the marketing process they are exposed to.

7.3.2 SAMPLING TECHNIQUE

Being the essential part of the business research process, the sample selected meaningful respondents of a representative population by following the Simple Random Sampling procedure. The underlying psychology behind selecting respondents was to choose those who have experienced both the digital and traditional ways of marketing, since it was observed during the sample selection that young persons are those who frequently purchase goods and services and hesitate to select either the digital or the traditional marketing technique. Students were thus interrogated in this study. Data were collected from January to March 2019. The sample frame of the gathered information is discussed below.

7.3.3 SAMPLE FRAME

The demographics of respondents (included in the sample) are depicted in Table 7.1. A total of 160 interviewees were questioned for the purposes of the present study.

7.3.4 TYPES OF DATA COLLECTED IN THE STUDY

The primary data had been directly collected from the consumers by applying a data collection method through an interview. For this purpose, a self-designed, structured, undisguised questionnaire was used in the present study, via the face-to-face mode of interview undertaken in Kolkata. Then, the secondary data were gathered from different journals or websites for the background study.

TABLE 7.1
Statistical description of the sample

Features

Age Group	18–21 yrs	22–25 yrs	25–28 yrs
Sample	77	59	24
Profession	**Students**	**Service/Business**	**Others**
Sample	125	27	8
Gender	**Male**	**Female**	**Other**
Sample	96	64	—
Location	**North Kolkata**	**Central Kolkata**	**South Kolkata**
Sample	38	27	95

Note: Total number of respondents: 160.

TABLE 7.2
List of variables and their abbreviations used in the research

Sl_No.	Variables	Statements of variables
1	Fact1	Digital marketing is a faster process which is helpful
2	Fact2	Digital marketing needs a lot of technical knowledge
3	Fact3	Digital marketing is not very user friendly
4	Fact4	There is no scope for error correction
5	Fact5	Less time is required for shopping reliably
6	Fact6	Digital marketing is based on the performance of the computer I am allotted to
7	Fact7	Computerized payment makes digital marketing difficult to rely on
8	Fact8	Proper training is required before using digital marketing
9	Fact9	I simply don't like the digital marketing process over the traditional process
10	Fact10	Digital marketing is time saving
11	Fact11	Digital marketing is more difficult since I am not comfortable with computers
12	Fact12	I get tensed the whole day of the expected delivery date of products
13	Fact13	The digital marketing process is a bit difficult since I don't know how to revise the order
14	Fact14	Digital marketing is a smart process since it needs less effort
15	Fact15	The digital marketing process is not preferred since it needs a lot of payment option setting effort from the customer side
16	Fact16	The digital marketing process is not preferred since delivery time is at least one day for purchased products
17	Fact17	Digital marketing is good since delivery address details are automatically submitted with little effort
18	Fact18	Digital marketing is good because it is error free
19	Fact19	The digital marketing process is less reliable

7.3.5 Research Variables

A total of 19 variables were selected in the research based on which the opinions of the respondents were judged. The list of the variables considered in the research is given in Table 7.2.

7.4   DATA ANALYSIS AND MAIN RESULTS

The base of the present study depends on knowing how many of the respondents are inclined toward digital or traditional marketing techniques after applying both of them.

Table 7.3 reports that, among 160 respondents, about 80% of them are in favor of digital marketing.

Figure 7.1 depicts the same findings. Indeed, the distribution of liking a specific type of marketing by interviewees is important. For more precision, all the respondents were asked about their opinion toward the marketing process to reveal that they are either in favor of or against the variables they were exposed to. Based on their responses, a factor analysis was applied to identify the factors that were important to choosing a specific type of marketing practice.

Table 7.3 shows the primary checking used for identifying whether the data captured during the survey were conducive to the factor analysis or not.

Table 7.4 shows that the Kaiser–Meyer–Olkin (KMO) value of 0.610 is good enough to suggest that the sample size is adequate to conduct the factor analysis. At the same time Bartlett's test result is found to be less than 0.05 and thus is significant for conducting factor analysis.

Table 7.5 reports that the 19 variables were selected for judging the perception of respondents toward choosing a specific marketing practice. Those variables were

TABLE 7.3
The marketing process

Marketing technique	Frequency	Percentage
Digital	128	80.0
Traditional	32	20.0
Total	160	100.0

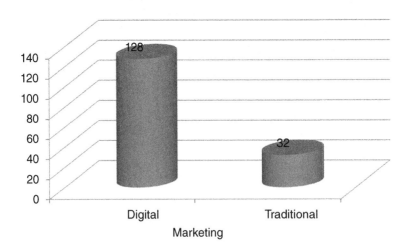

FIGURE 7.1   Distribution of opinion toward preferring a specific marketing practice

TABLE 7.4
KMO and Bartlett's test

KMO measure of sampling adequacy		0.610
Bartlett's test of sphericity	Approx. Chi-square	1651.326
	Df	171
	Sig.	.000

TABLE 7.5
Total variance explained

Component	Initial eigen values			Extraction sums of squared loadings		
	Total	% of variance	Cumulative %	Total	% of variance	Cumulative %
1	4.652	24.484	24.484	4.652	24.484	24.484
2	2.377	12.509	36.992	2.377	12.509	36.992
3	1.948	10.253	47.245	1.948	10.253	47.245
4	1.674	8.812	56.058	1.674	8.812	56.058
5	1.545	8.133	64.191	1.545	8.133	64.191
6	1.178	6.198	70.389	1.178	6.198	70.389
7	1.030	5.422	75.811	1.030	5.422	75.811
8	.925	4.870	80.681			
9	.751	3.954	84.635			
10	.693	3.649	88.283			
11	.578	3.040	91.324			
12	.379	1.993	93.316			
13	.326	1.718	95.034			
14	.276	1.451	96.485			
15	.176	.924	97.409			
16	.174	.918	98.327			
17	.141	.743	99.070			
18	.120	.630	99.700			
19	.057	.300	100.000			

Note: Extraction method: principal component analysis.

reduced to seven factors that can explain the 75.811% of the total variance. Such a percentage is acceptable for the current study.

The next step was to identify the member variables of these factors. Table 7.6 shows the factor loading and that factor 1 clearly includes five member variables, factors 2 and 3 encompass three member variables each, and the remaining four factors consist of two member variables each.

The factors, based on the common characteristics of the member variables, are reported in Table 7.7.

Table 7.7 completes the analysis of the first research objective.

The next step consists in finding out the comparative importance of each identified factor. For this purpose, the artificial neural network technique for multilayer perception was applied.

TABLE 7.6
Rotated component matrix[a]

	Components						
Factors	**1**	**2**	**3**	**4**	**5**	**6**	**7**
Fact2	.821						
Fact8	.809						
Fact15	.737						
Fact19	.641						
Fact6	.598						
Fact4		.758					
Fact9		.683					
Fact1		−.576					
Fact16			.790				
Fact10			.656				
Fact13			.651				
Fact18				.814			
Fact3				−.723			
Fact12					.857		
Fact11					.552		
Fact5						.833	
Fact7						.708	
Fact14							.891
Fact19							.463

Notes: Extraction method: principal component analysis; rotation method: varimax with Kaiser normalization.
[a] Rotation converged in 12 iterations.

TABLE 7.7
Factor loadings

Factor variables*	Factor name
Fact2	Technical orientation
Fact8	
Fact15	
Fact19	
Fact6	
Fact4	Fast process
Fact9	
Fact1	
Fact16	Time
Fact10	
Fact13	
Fact18	User friendliness
Fact3	
Fact12	Comfort ability
Fact11	
Fact5	Reliability
Fact7	
Fact14	Effort
Fact17	

* Details of the factor list can be found in Table 7.2.

TABLE 7.8
Case-processing summary (multilayer perception)

		N	Percent
Sample	Training	106	66.3
	Testing	54	33.8
Valid		160	100.0
Excluded		0	
Total		160	

TABLE 7.9
Network information

Input layer	Covariates	1	Technical orientation	
		2	Fast process	
		3	Time	
		4	User friendliness	
		5	Comfort ability	
		6	Reliability	
		7	Effort	
	Number of units[a]			7
	Rescaling method for covariates		Standardized	
Hidden layer(s)	Number of hidden layers			1
	Number of units in hidden layer 1[a]			4
	Activation function		Hyperbolic tangent	
Output layer	Dependent variables	1	Marketing	
	Number of units			2
	Activation function		Softmax	
	Error function		Cross-entropy	

[a] Excluding the bias unit.

Table 7.8 provides the case processing summary, demonstrating that about 66% of the sample is used for training of the neural network, whereas 34% of it is employed for the testing purpose.

Table 7.9 reports that the identified seven factors are as pursued covariates in the neural network, except for one hidden layer considered in the analysis via four units. The activation function used in the analysis is "Hyperbolic tangent", knowing that the dependent is the "Marketing" process based on two different levels: the digital and the traditional ones.

The neural network (as shown in Figure 7.2) is used for depicting the relationship between input variables with the hidden layer and the impacts of the hidden layer on the output variable. The neural network can also be employed for predicting the effects of input variables on the output one. As shown in Figure 7.3, it can be observed that other than seven input variables there is a bias element that has a strong relationship with the hidden layer H (1:2). Besides, the hidden layer has a strong effect on the perception of respondents over "Digital marketing" output. It can be also deduced that the hidden layer H (1:3) mainly influences traditional

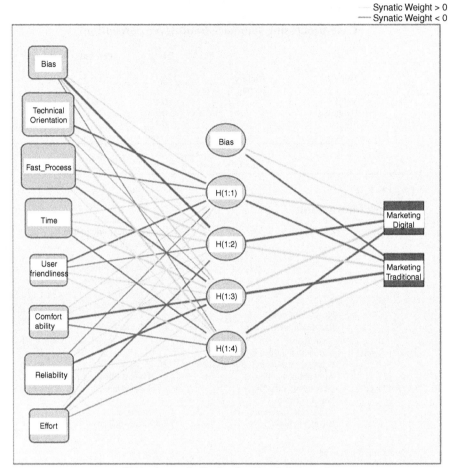

Hidden layer activation function: Hyperbolic tangent

Output layer activation function: Softmax

FIGURE 7.2   The neural network

marketing, as the thick line describes a strong synaptic weight. Moreover, the hidden layer H (1:3) has a strong association with the variables "Comfortability" and "Reliability".

Therefore, according to consumers, traditional marketing is more comfortable and reliable in comparison with digital marketing. Similarly, the neural network reports that requirement of "Technical orientation" attracts traditional marketing more than digital marketing. Meanwhile, "User friendliness" is another factor that attracts people toward traditional marketing as most of the respondents think that the digital marketing process is not so much "User friendly" oriented. Interestingly, interviewed customers think that traditional marketing is a "Faster process" when compared with digital marketing. On the other hand, digital marketing is mainly linked to the two

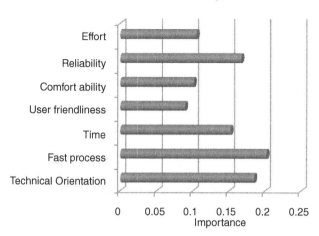

FIGURE 7.3   Relative importance of factors

TABLE 7.10
Model summary

Training	Cross entropy error	.043
	Percent incorrect predictions	0.0%
	Stopping rule used	Training error ratio criterion (.001) achieved
Testing	Training time	0:00:00.03
	Cross entropy error	.032
	Percent incorrect predictions	0.0%

Note: Dependent variable: marketing.

---

factors, because it saves "Time" and requires less "Effort" (as per the synaptic weights seen in the thick line: the thicker is the line, the stronger is the relation). Similarly, from Table 7.10, the same findings were obtained, knowing that the parameter estimates were calculated in numeric terms.

From the model summary (Table 7.11) depicts that the training and testing error of the model is almost negligible. So, the model thus can be used reliably for prediction purposes.

Table 7.12 outlines the details of classification for both the training and testing phases of the model. During the training phase, 86% of customers' responses are given for digital marketing, whereas 20% of responses are attributed in favor of the traditional marketing approach.

For testing purposes, 42% of customers' responses were in favor of digital marketing, whereas 12% of them were for the traditional marketing process.

TABLE 7.11
Parameter estimates

Predictor		Predicted						
		Hidden layer 1				Output layer	
		H(1:1)	H(1:2)	H(1:3)	H(1:4)	DM	TM
Input layer	(Bias)	.634	−3.132	1.625	−.445		
	Technical orientation	−1.575	1.050	−.028	2.033		
	Fast process	−.764	2.136	−1.370	.051		
	Time	.875	1.393	1.922	−.960		
	User friendliness	−1.242	−.551	.780	1.153		
	Comfort ability	.603	.293	−1.932	−.753		
	Reliability	−.489	.185	−1.898	.493		
	Effort	.144	−1.126	.387	−.464		
Hidden layer 1	(Bias)					1.103	−1.616
	H(1:1)					2.263	−1.712
	H(1:2)					−2.583	3.456
	H(1:3)					3.691	−3.452
	H(1:4)					−1.962	2.037

Note: DM = Digital marketing; TM = Traditional marketing.

TABLE 7.12
Classification

Model phase	Marketing approach	Digital	Traditional	Percent correct
Training	Digital	86	0	100.0
	Traditional	0	20	100.0
	Overall percent	81.1%	18.9%	100.0
Testing	Digital	42	0	100.0
	Traditional	0	12	100.0
	Overall percent	77.8%	22.2%	100.0

Note: Dependent variable: Marketing.

Table 7.13 reports the relative importance of the seven independent variables in numeric terms and Figure 7.3 depicts the same in the form of a bar chart. The normalized importance of Table 7.13 is plotted in Figure 7.3 to visualize the importance of each factor affecting the marketing process.

As per Figure 7.3, the most important factor which has an impact on the marketing process is "Fast process" which was discussed previously in Figure 7.2, that this

TABLE 7.13
Independent variable importance

Variables	Importance	Normalized importance (%)
Technical orientation	.185	91.7
Fast process	.202	100.0
Time	.152	75.2
User friendliness	.089	44.3
Comfort ability	.101	50.3
Reliability	.167	82.6
Effort	.105	51.9

variable belonging to the neural network is strongly associated with the traditional marketing process. "Technical orientation", the next most important factor, is also found to be in favor of traditional marketing. The other factors that have effects on choosing a marketing process are "Reliability", "Time", "Effort", "Comfortability", and "User friendliness".

So, it can be concluded that customers prefer the marketing process to be fast, and that it is least associated with technical difficulty. Then, the questions arising in the mind of the customer turn around reliability, the time required to purchase, the required effort, the comfort of the entire process, and user friendliness.

7.5 DISCUSSION OF RESULTS

The key findings of this study are: (1) about 80% of the respondents prefer digital marketing; (2) the preference for the digital or traditional marketing process is based on seven factors, that is, Technical orientation, Fast process, Time, User friendliness, Comfort ability, Reliability, and Effort. (3) The results show that among all the identified seven factors, consumers give the most importance to the factor "Fast process" followed by the "Technical orientation" of the customer that influences them toward either adoption of refusal of either the digital or traditional way of purchase. (4) It is also found in the study that customers prefer the traditional marketing process mainly due to two out of seven identified factors, that is, "Comfortability" and "Reliability". On the other hand, digital marketing is found to be preferred as it requires less "Time" and "Effort". Among the recent studies, Kumar and Dange (2012) found entirely different aspects of the factors in the context of digital marketing. Those ones are related to online security, privacy, trust, and trustworthiness. But, since the present study considers both the traditional and digital marketing aspects, all these factors are not exactly addressed here, though a factor named "Technical orientation" is considered. The analysis of this study is closely related to the research made by Ganapathi (2015). In his paper entitled "A Study on Factors Affecting the Online Shopping Behavior of Consumers in Chennai", the researcher applied the exploratory factor analysis and thereafter the regression analysis, and identified that the factors – convenience, security, website features, and time saving – have significant influence on customers' minds. This study also applied the exploratory factor analysis to extract the possible factors. However, unlike the work of

Ganapathi (2015), this chapter applied the "artificial neural network (multilayer perceptron)" to identify the factors that have a great influence on customers' minds. The factor "Time saving" found in this chapter is also similarly found. Other significant research carried out by Franco and Bulomine Regi (2016) identified some of the factors for which customers are attracted toward digital marketing, that is, Convenience, Time saving, Options, Ease to comparison, Ease to find reviews, and Coupons and deals. This study also found the 'Time saving feature' to be an influential factor in adopting digital marketing. In addition to this, Franco and Bulomine Regi (2016) and Javadi et al. (2012) mentioned that "Privacy and security" is one of the main factors for users of the traditional marketing process. On the other hand, the current study addresses the same factor called "Reliability". This research also found results similar to that of "A Study of Factors Affecting Online Buying Behavior of Students" (Jukariya and Singhvi, 2018). The latter identified the factor that has the greatest impact on online buying that includes personal privacy and security, and convenience and saving time, whereas the present study mentioned the factors "Reliability", "User friendliness", "Comfortability", and "Time" as the saving factors for the adaptation of digital marketing practices.

7.6 CONCLUSION

In this digital world, people purchase their products using two different processes: the traditional and the digital. The factors that they consider for purchasing products include "Fast process", the requirement for "Technical orientation", "Reliability" of the marketing process, "Time" required for completing the entire marketing process, "Effort" required by the customers for purchasing goods and services, whether the customers are "Comfortable" with the process or not, and "User friendliness" of the entire marketing process. It is found from the present study that customers pay a lot of attention to the process that they consider as "Faster", followed by the requirement of the "Technical orientation" of the process. In fact, customers are scared of the process that requires a "Technical orientation". It is also found from the results of this study that the factor that favors the digital marketing approach is "Time" (required for the process or any time purchase motivation). Moreover, the "Effort" required to purchase is favorable to customers. On the other hand, the reasons behind why people prefer the traditional way of marketing are linked to the fact that they think that it is a "Fast process". Indeed, they believe that they can purchase products easily from local stores and do not need to wait for at least one day. Customers were also found to prefer traditional marketing as, according to them, it does not need any "Technical orientation". The other reasons behind preferring the traditional marketing process are "Reliability", "Comfortability", and "User friendliness".

7.7 RESEARCH CONTRIBUTIONS AND FUTURE IMPLICATIONS

In recent years, people have been purchasing their products using two different processes, either the traditional or off line way and the digital or online way. The factors that they consider for purchasing products have been identified by this study. Furthermore, the results of this study have enabled us to discover the factors that favor the online or digital marketing approach as the best one. On the other hand, the reasons behind why people prefer the off line or the traditional way of marketing can also be discovered from the findings of this research.

Organizations are presently facing the dilemma of choosing either the off line (traditional) or the online (digital) marketing procedure. They are in search of finding the perfect balance between them (Shukla, 2010). This present study provides a guideline to making a fit between both ways of marketing. It can also help to identify the reasons why customers prefer digital or traditional processes for purchasing goods. It enables marketers and managers as well to identify the most influential factors for both the online (digital) and offline (traditional) ways of purchasing. The present study can also provide significant research outlines for researchers and business practitioners to formulate business strategies and plans. The neural network method used in this research can also be employed for predictive analyses and for a better understanding of consumer purchase behavior. The study can further be continued for different sample sizes to upgrade comprehension of consumer behavior.

REFERENCES

Bag, S. and Sen, S.S. (2012), Kolkata metro railway and customer satisfaction: An empirical study, *International Journal of Multidisciplinary Research*, Vol. 2, Issue 3, pp. 165–176.

Burman, D. and Aggrawal, A. (2015), Factors affecting buying behavior in E-commerce in India: A review of literature, *International Journal of Business Quantitative Economics and Applied Management Research*, Vol. 2, Issue 2, pp. 123–135.

Das, S.K. and Lall, G.S. (2016), Traditional marketing VS digital marketing: An analysis, *International Journal of Commerce and Management Research*, Vol. 2, Issue 8, pp. 5–11.

Egger, F.N. (1999), Human factors in electronic commerce: Making systems appealing, usable & trustworthy, in *Graduate Students Consortium & Educational Symposium, 12th Bled International E-Commerce Conference*, Bled, Slovenia, June 1999.

Franco, C.E. and Bulomine Regi, S. (2016), Advantages and challenges of e-commerce customers and businesses: An Indian perspective, *International Journal of Granthaalayah: A Knowledge Repository*, Vol. 4, Issue 3, pp. 102–113.

Ganapathi, R. (2015), A study on factors affecting online shopping behavior of consumers in Chennai, *Journal of Management Research and Analysis*, Vol. 2, Issue 2, pp. 123–126.

Hofacker, C.F. (1999), *Internet Marketing*. Dripping Springs: Digital Springs, Inc.

Hoffman, D.L. and Novak, T.P. (2017), Consumer and object experience in the Internet of Things: An assemblage theory approach, *Journal of Consumer Research*, Vol. 44, Issue 6, pp. 1178–1204.

Issa, T. (2014), Online shopping and human factors, in Lacka, E., Chan, H., and Yip, N. (eds.), *E-Commerce Platform Acceptance*, pp. 131–150. Cham, Switzerland: Springer.

Javadi, M.H.M., Dolatabadi, H.R., Nourbakhsh, M., Poursaeedi, A., and Asadollahi, A.R. (2012), An analysis of factors affecting on online shopping behavior of consumers, *International Journal of Marketing Studies*, Vol. 4, Issue 5, pp. 28–39.

Joseph, P.T. (2008), *E-Commerce: An Indian Perspective*, 3rd ed. New Delhi: PHI.

Jukariya, T. and Singhvi, R. (2018), A study of factors affecting online buying behavior of students, *International Journal of Current Microbiology and Applied Sciences*, Vol. 7, Issue 1, pp. 56–69.

Keserwani, H. (2014), A study of E-tailing challenges and success factors in India—An exploratory study, *Integral Review—A Journal of Management*, Vol. 7, Issue 2, pp. 92–103.

Khan, A. and Chavan, C.R. (2015), Factors affecting on-line shoppers' behavior for electronic goods purchasing in Mumbai: An empirical study, *International Journal in Management and Social Science*, Vol. 03, Issue 03, pp. 37–49.

Khare, A. (2016), Consumer shopping styles and online shopping: An empirical study of Indian consumers, *Journal of Global Marketing*, Vol. 29, Issue 1, pp. 40–53,

Kim, J.-E. and Kim, J. (2012), Human factors in retail environments: A review, *International Journal of Retail & Distribution Management*, Vol. 40, Issue 11, pp. 818–841.

Kulathunga, D., Athapaththu, C., and Jayani. (2018). Factors affecting online purchase intention: A study of Sri Lankan online customers, *International Journal of Scientific & Technology Research*, Vol. 7, Issue 1, pp. 120–128.

Kumar, V. and Dange, U. (2012), A study of factors affecting online buying behavior: A conceptual model, *SSRN Electronic Journal*.

Ray, N., Mukherjee, T., and Bag, S. (2020), A study on online shopping behavior in Kolkata, West Bengal, *Our Heritage*, Vol. 68, Issue 1, pp. 7738–7751.

Shaikh, A.A., Alharthi, M.D., and Alamoudi, H.O. (2020), Examining key drivers of consumer experience with (non-financial) digital services—An exploratory study, *Journal of Retailing and Consumer Services*, Vol. 55, p. 102073.

Shukla, T. (2010), Factors affecting 'Internet marketing' campaigns with reference to viral and permission marketing, *The IUP Journal of Management Research*, Vol. 9, Issue 1, pp. 26–37.

Slintak, K. (2019), A new concept of management, *Montenegrin Journal of Economics*, Vol. 15, Issue 1, pp. 199–213.

Zhidebekkyzy, A., Kupeshova, S., and Yesmurzayeva, A. (2019), Project management in Nanotechnology: A systematic literature review, *Montenegrin Journal of Economics*, Vol. 15, Issue 3, pp. 227–244.

8 Motivational Factors Behind the Participation Efforts in Crowdsourcing with the Moderating Effect of Trust in Digital Crowdsourcing Platforms

Muhammad Wasim Akram, Shaza Mahar, Aslan Amat Senin, and Kamariah Ismail

CONTENTS

8.1   INTRODUCTION

Xu et al. (2015) coined the term "crowdsourcing" to mean a way of capitalizing the talent of the individual. Crowdsourcing is about the involvement of the workforce through online communities (Xu et al., 2015). It is a form of open calling to solve a problem and complete a task for an organization. Crowdsourcing is an opportunity for organizations to collect workforce intelligence, skill sets, and their ability without hiring them – for mutual benefit (Bakici, 2020). Therefore, it is important for organizations to evaluate key factors that motivate a workforce to participate in crowdsourcing that is overlooked in the literature. Past studies revealed that workforce involvement in crowdsourcing relies on intrinsic and extrinsic motivation (Yang, 2019). Porter (2010) revealed that extrinsic motivation significantly influences workforce participation in crowdsourcing. Extrinsically motivated individuals perform their task for rewards and benefits (e.g., wealth, fame, grades, degree, and bonuses); Zhao and Zhu (2014) identified four antecedents of this motivation: (1) external, (2) identified, (3) interjected, and (4) integrated.

Identified motivation is about the individual's perspective on the task's importance and his or her performance; in return he or she will be rewarded, which is also important for his or her future development and growth (Wang et al., 2019). Integrated motivation is an autonomous motivation compatible with the ethics and opinions of the individual (Feller et al., 2012). Interjected motivation is about individual performance to avoid guilt or to achieve respect from colleagues or to sustain a position in an organization (Heo and Toomey, 2015). Similarly, Porter (2010) revealed that intrinsic motivation significantly influences an individual to participate in crowdsourcing. Such individuals have high self-confidence and are willing to accept challenges without management support (Župič, 2013). Trust is also a significant factor that influences individual participation in crowdsourcing (Feller et al., 2012). Trust gives confidence to participants that his or her idea will not be utilized in any malpractice, that no fraudulent activity will be carried out, and that the participants will be rewarded fairly.

Crowdsourcing is highly motivated in developed economies globally. Developed countries engage and motivate organizations to participate in crowdsourcing to revolutionize businesses. It is a means of interaction between organizations and society (Xu et al., 2015). Crowdsourcing is a modern approach that has gained support due to global technological advancement (Heo and Toomey, 2015). However, organizations in developing economies like Pakistan are still not ready to adopt crowdsourcing, though several research studies have been conducted in developed countries in this particular context (Zhao and Zhu, 2014; Lee et al., 2015; Choy and Schlagwein, 2016). Comparatively, crowdsourced organizations perform better than other organizations (Xu et al., 2015). But still, it remains unclear what the main factors are among all those motivational factors that influence participants to engage in crowdsourcing and motivate them to solve organizational problems. The literature also overlooks this particular problem which is only discussed by a few studies in the context of developed economies (e.g., Zheng et al., 2011; Boons et al., 2015; Deng et al., 2016).

Few risks are associated with being a participant of crowdsourcing because they feel that the idea might not be accepted by the crowdsourced organization or their idea might use in the malpractice or might use in any fraudulent activity or might not be paid fairly (Afuah and Tucci, 2012; Feller et al., 2012). Therefore trust is another important factor that can moderate the relationship between motivation and individual participation in crowdsourcing (Porter and Donthu, 2008; Jarvenpaa et al., 2010; Kim, 2014). The current study examines the role of motivation in influencing crowdsourcing for open innovation because the latter is important for any organization to compete in the market and maintain competitiveness. Different types of motivation are used by the current study, which are discussed in the literature. However, this study develops a conceptual model using self-determination theory as an underpinning so as to empirically test the literary association between motivation and individual participation in crowdsourcing through the moderating role of trust in the context of health care.

8.2 LITERATURE REVIEW

8.2.1 DIGITAL CROWDSOURCING

Luz et al. (2015) investigate crowdsourcing's task orientation. Crowdsourcing is referred to as the task accomplished by selected agents by outsourcing the organization in the form of an open call by an indeterminate large group of workers in return for small financial rewards (Doan et al., 2011). The social community has a significant impact on the ideas, life, and behavior of an individual (Konstas et al., 2009). According to Porter (2010), online communities decrease geographical gaps and promote the combination of ideas and behavior based on the global sector. The combination of ideas and behavior is called "collective intelligence" (Luo et al., 2009). Crowdsourcing is used to solve the problem of the system which is provided by the owners of the organization or system (Doan et al., 2011; Javadi and Aghaie, 2014). Zhao and Zhu (2014) evaluate the research on crowdsourcing with respect to the current scenario and future research. Crowdsourcing is the web-based phenomenon which attracts the attention of users as well as participants from previous decades.

8.2.2 SELF-DETERMINATION THEORY

According to Brabham (2008a) there is some kind of motivation behind the time and effort of the individual which is not always money, but may be enjoyment, that is, intrinsic reward. The result is that crowdsourcing workers need some motivational factor to remain part of an organization, such as a crowdsourcing agent (Zhao and Zhu, 2014). This kind of motivational factor is identified and external, namely monetary, enjoyment, and reputation. The study also shows that there is a wide area for future research and development in the concept of crowdsourcing. Liu (2014) studies the crisis within the framework of crowdsourcing and shows that it is the old phenomenon which gains more attention during the previous year. Crowdsourcing that happens during the crisis are volunteer based, yet other financial incentives and rewards are used to engage crowd workers and their continuous participation.

The study also shows that motivation has a significant impact on the quality of the task, which means the extrinsic or intrinsic motivation and the type of crowdsource workers affect the speediness and the superiority of the crowdsourcing tasks. Extrinsic motivation and intrinsic motivation are used as independent variables, whereas crowdsourcing is used as the dependent variable. The participant of crowdsourcing may have a diverse range of motivational factors, which leads to different rewards. According to the view of the participant, the crowdsourcing system provides the opportunity for the individual to work with small, medium, and large organizations where they enhance their skill, creativity, and also acquire work experience. According to Brito (2008), the organization which is non-profit uses crowdsourcing to solve their business problems.

Recently crowdsourcing has attracted more attention, and organizations now realize the importance of their business and launch many campaigns to grow it (Whitla, 2009; Rouse, 2010). But there are some challenges that arise while implementing crowdsourcing, like effective rewards, handling submissions, loss of mechanisms, and generating the trust (Jain, 2010). Kazman and Chen (2009) illustrate that crowdsourcing provides the norms and rules of the problems that exist in the real world. It is an application of an open source system (Howe, 2008). The result shows that the organization uses outsourcing to solve problems by a large number of unknown participants. It also shows that the organization may use its personal crowdsourcing or may use another entity for staging it.

8.2.3 INTRINSIC MOTIVATION

This study asserts that motivation is an important factor in the participation of crowdsourcing workers. It uses extrinsic and intrinsic motivation as an IV; crowdsourcing is used as the dependent variable (DV). Chesbrough (2003) introduces the new phenomenon of open source. According to Hossain (2012a), intrinsic motivation plays an important role in the involvement of the crowd. It is often seen that when the tasks are complex, extrinsic motivation is more dominant than the intrinsic rewards. Intrinsic motivation refers to the playful task which increases the participation contribution for fun, interest, or excitement, whereas extrinsic motivation can only be monetary (in the form of money or gifts) or non-monetary reward. The results show that intrinsic motivation and extrinsic motivation are important motivational drivers in gamified crowdsourcing.

8.2.4 Extrinsic Motivation

According to Jeppesen and Lakhani (2010), the motivation behind crowds is mostly extrinsic. The research shows that even people with a low education can be involved in crowdsourcing and earn money by accomplishing micro-tasks for the organization (Eagle, 2009). These motivational factors vary due to the difference in the task or the difficulties in the task. This study also shows that crowdsourcing is only successful when the organization that outsources their system understands the motivational factor of individual, for which they are involved in crowdsourcing participation. This chapter also shows the applications of crowdsourcing, which are idea generation, micro-tasking, open source software, and so on.

Kavaliova et al. (2016) illustrate the innovation of crowdsourcing and the development of the product. The purpose of the paper is to explore the way of employing gamification used by the company, that is, the element of the game and the designing techniques to motivate the participant to be involved in the crowdsourcing task. They used intrinsic motivation and extrinsic motivation as an independent variable, with crowdsourcing used as a dependent variable. According to Neyer et al. (2009), and Whitla (2009), crowdsourcing has a collaborative and a competitive nature that provides the opportunity for the organization to access and identify the skill, knowledge, experiences, and resources that are beyond its borders. By using crowdsourcing the company starts to increase the saving of cost because in many cases it has been seen that extrinsic rewards are not common (Howe, 2006a).

8.2.5 Trust

Trust is a multidisciplinary idea which has been clarified in different ways, as indicated by alternate points of view. While financial analysts bolster the perspective on trust as the consequence of a balanced estimation of expenses and advantages, clinicians will in general think about it as a characteristic of people, while sociologists support its properties of system relations (Appelbaum et al., 2000), despite the fact that researchers have utilized various depictions of trust (Kramer and Tyler, 1999). We concentrated our examination on trust in the board as moderator. Trust, truth be told, assumes a significant role in the connection between motivational variables and crowdsourcing, and it has been the subject of an impressive discourse in human resource management (HRM)writing.

8.2.6 Development of Hypotheses

8.2.6.1 Impact of External Motivation on the Crowdsourcing Participation Effort

Lee et al. (2015) introduce the feasibility of adopting the crowdsourcing system to solve problems of innovation of the business through the attributes and design of tasks. The study consists of 152 responses which were collected through a mailed out questionnaire. Correlation and regression analysis is used as a statistical tool. The study shows that task design is one of the major factors that has an impact on the participation and engagement of workers in crowdsourcing. According to Brabham (2008b) and Vukovic and Bartolini (2010), crowdsourcing can be used to decrease costs, solve the problems of the business, and also shorten the development cycle of the product. To obtain the innovative solution and to solve the problems of the business, the organization outsources their process by involving a third party (Verona et al., 2006).

Fuller et al. (2012) evaluate that extrinsic motivation and intrinsic motivation have a certain impact on the quality and the quantity of the solution of the innovative problem. Thuan et al. (2014) implement the Nexus model to make crowdsourcing successful for business. Crowdsourcing has been attracting the attention of business organization to collect crowds of people to harvest skill, labor, expertise, and knowledge. Zhao and Zhu (2014) illustrate that conclusions can be driven either collaboratively or independently. Hetmank (2013) identifies the different components of crowdsourcing applications as: user, task, contribution, and workflow organization. Therefore it was hypothesized that:

H1: There is a significant impact of external motivation on the participation efforts in crowdsourcing.

8.2.6.2 Impact of Introjected Motivation on Crowdsourcing Participation Efforts

Heo and Toomey (2015) explain the different visual factors that impact crowdsourcing so as to share information by motivating them. According to Hau and Kim (2011), intrinsic motivation is the individuals' desire to engage in the community due to their personal satisfaction and fun, whereas extrinsic motivation engages the individual by giving them monetary rewards. External rewards are controlled by external power rather than by offering a lower degree of autonomy (Ryan and Deci, 2000a). Introjected motivation is a type of internalized motivation though it is classified as a type of extrinsic motivation (Malhotra and Galleta, 2003). According to Ryan and Deci (2004), the identified motivation makes a personal connection with external rewards. The individual who experiences autonomy and capability can show more self-determined incentive in a crowdsourcing atmosphere (Zheng et al., 2011). The study shows that the individual experiences introjected motivation when he or she performs the task, not because he or she wants to do it, but because he or she is afraid of shame or wants to avoid the guilt of having not performed it. Hence it was hypothesized that:

H2: There is a significant impact of introjected motivation on the participation efforts in crowdsourcing.

8.2.6.3 Impact of Identified Motivation on Crowdsourcing Participation Efforts

Zhao and Zhu (2014) studied the impact of extrinsic motivation and intrinsic motivation on participating in the contest of crowdsourcing by considering 455 responses to a survey. The study chose correlation analysis to check the significance of the hypothesis. The initial and extensive growth of crowdsourcing depends on the mass involvement of the participants (Gregg, 2010). The independent variable used is extrinsic and intrinsic motivation, whereas crowdsourcing is used as the dependent variable.

Crowdsourcing is the latest technology that has been rapidly implemented by firms to solve the problem of their organization (Archak and Sundararajan, 2009). The participation of the individual is the hot topic nowadays, which can be also be studied by reviewing relative concepts like open source and collective intelligence (Ke and Zhang, 2010). According to Chan (2011), a certain type of motivation is present behind the participation of the individual, which is intrinsic and extrinsic. According to Krishnamurthy (2006), both types of motivation are important for participation. The authors of this study further divide extrinsic motivation into four categories, namely, external, introjected, identified, and integrated. The participant involved in crowdsourcing may work hard to achieve monetary rewards or to obtain good career opportunities in the future. This study shows that the contest of crowdsourcing is the type of crowdsourcing that is used by many companies for solving problems and for making decisions. Therefore it was hypothesized that:

H3: There is a significant impact of identified motivation on the participation efforts in crowdsourcing.

8.2.6.4 Impact of Integrated Motivation on Crowdsourcing Participation Efforts

Olson and Rosacker (2013) examine the participation in open source software (OSS) and crowdsourcing, by considering a sample of 118 respondents. Crowdsourcing refers to the participation of workers to accomplish organizational goals and objectives. The variable used in this study was the independent one of integrated motivation; crowdsourcing was the dependent variable. It has been proven from past history that creativity is necessary for the accomplishment of economic, industrial, and agricultural growth, and also for information (Florida, 2003). According to Cook (2008), the participation of individuals has been increasing more and more. These motivational factors are aligned with the external motivation, such as money, flexitime, learning, stability, and security. This study also shows that sometimes participants are found to be individualistic, assigning great importance to reputation. Crowdsourcing is the application of OSS, linking the organization with software that is beyond its boundaries (Howe, 2006b). According to Raymond (2001), money does not play a role when the participant is knowledgeable and experienced. OSS is good under some conditions, like openness, trust, transparency, tact, and professionalism (Agerfalk and Fitzgerald, 2008). Cook (2008) shows that the contribution of the user generates various different systems in the crowdsourcing environment. The study shows that the appreciation of peer assessment, self-marketing, and career opportunities in the future are some of the main factors that motivate participation in crowdsourcing. Therefore it was hypothesized that:

H4: There is a significant impact of integrated motivation on the participation efforts in crowdsourcing.

8.2.6.5 Impact of Intrinsic Motivation on Crowdsourcing Participation Efforts

Wang et al. (2013) illustrated the perception of crowdsourcing annotations for the processing of natural languages. Annotations are a kind of task that are designed to entertain workers. According to Ahn and Dabbish (2008), fun is the main motivation that contributes to crowdsourcing. It is often seen that some people are involved in the online system to crowdsource so as to generate profits (Snow et al., 2008; Iperiotis, 2010). The study shows that fun, profit, and altruism are the internal motivational factors. Fun is the most important motivator.

Royo and Yetano (2015) study crowdsourcing as an instrument of e-participation, by considering 223 valid responses of participants. According to Schweitzer, Buchinger, Gassmann and Obrist (2012), members in crowdsourcing interact with each other to develop more creative ideas. According to Nam (2012), crowdsourcing should be evaluated by using the three pillars, namely: participation, transparency, and collaboration. Župič (2013) examines the social media that enables crowdsourcing, based on 23 in-depth two-way communication. According to Wasko and Faraj (2000), the interaction of the individual with an online community increases the knowledge of community members. There are four types of motivation which are involved in crowdsourcing, namely: community love, innovative skills, earning money, or signaling for freelance workers (Brabham, 2010; Ridge, 2014). Crowdsourcing is similar to OSS development, having only a minor difference; both are enabled by the Web. Therefore it was hypothesized that:

H5: There is a significant impact of intrinsic motivation on the participation efforts in crowdsourcing.

8.2.6.6 Moderating Role of Trust in Crowdsourcing Platforms on Crowdsourcing Participation Efforts

There exist two standards of writing literature while examining trust in innovation, that is, interpersonal trust and system-like trust (Lankton et al., 2015). Since crowdsourcing demonstrates a high level of humanness, for example, volition in assessment and reward (Feller et al., 2012), as per Lankton et al. (2015), human-like trust in innovation seems relevant to our study. The past literature explains that trust matters in the face of risks and prevailing vulnerabilities (Jarvenpaa et al., forthcoming) and encourages future practices (e.g., Kim, 2014). Trust is viewed as a pertinent factor in vulnerabilities innate in online activities (Jarvenpaa et al., forthcoming). In crowdsourcing, a firm's opportunistic behaviors prevail (Afuah and Tucci, 2012). In this matter, trust is a significant factor that helps to solve participation in our research. The literature suggests that trust is an important and critical factor for the success of crowdsourcing and that it may influence the participation of solution makers (Feller et al., 2012). However, there is an absence of research that has empirically tested its effects. As trust reduces the need to act in a self-defensive manner and encourages risk-taking behavior (Jarvenpaa et al., 1998), we predict that trust in crowdsourcing will urge members to participate in it, even when partaking in risks and opportunistic practices.

Few have investigated the precursors of trust in online exercises (e.g., Ridings et al., 2002; Porter and Donthu, 2008). Doorman and Donthu (2008) identify that apparent efforts to provide quality and substance and to cultivate embeddedness (e.g., looking for the opinions of individuals) help manufacturers to trust in firm-supported virtual networks. With regard to crowdsourcing, past writing proposes that guaranteeing problem solvers of being compensated appropriately (Feller et al.,

2012) and moderating the expenses of cooperation (Yang et al., 2008) should impact trust. Consequently, we assert that money related reward, loss of learning influence, and intellectual exertion will influence trust.

Therefore it was hypothesized that:

H6: Trust in crowdsourcing platforms positively moderates the relationship between external motivation and participation effort in crowdsourcing.

H7: Trust in crowdsourcing platforms positively moderates the relationship between introjected motivation and participation effort in crowdsourcing.

H8: Trust in crowdsourcing platforms positively moderates the relationship between identified motivation and participation effort in crowdsourcing.

H9: Trust in crowdsourcing platforms positively moderates the relationship between integrated motivation and participation effort in crowdsourcing.

H10: Trust in crowdsourcing platforms positively moderates the relationship between intrinsic motivation and participation effort in crowdsourcing.

8.2.7 CONCEPTUAL FRAMEWORK

The participation effort in crowdsourcing platforms is the dependent variable, whereas extrinsic motivation and intrinsic motivation are our independent variables. Extrinsic motivation is further divided into four types of motivation, namely: external, integrated, identified, and introjected (Zhao and Zhu, 2014), while trust in crowdsourcing platforms is the moderator between independent and dependent variables. To identify the impact of different facets of motivation on the participation of crowdsourcing, the following model has been developed for this study (Figure 8.1).

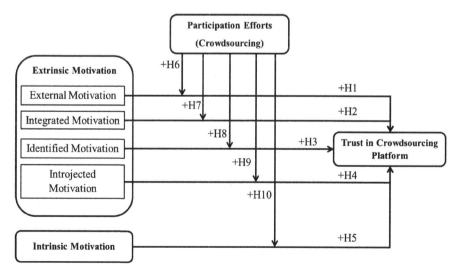

FIGURE 8.1 Theoretical model

Deci and Ryan (2008) develop a theoretical model, the self-determination theory, in which they discuss the motivational factor of the people or workers involved in a certain job. The self-determination theory evaluates the activities that are performed by the individual to attain the long term goal. They divide human motivation into two types, namely autonomous motivation and controlled motivation. Autonomous motivation is the combination of extrinsic motivation (i.e., identified motivation and integrated motivation) and intrinsic motivation. Controlled motivation includes the two types of extrinsic motivation (i.e., external motivation and introjected motivation). According to the theory, autonomous motivation and controlled motivation energize and motivate people to perform the task.

The importance of the task is represented by the amount of money paid to the crowdsource worker in return for their involvement in the online system (Zhao and Zhu, 2014). Introjected motivation involves the employee completing the task for the organization and resolving the problem of the business. It shows that people complete their task, not for external rewards like money, awards, and bonuses, but that they are involved in the task to gain self-esteem and to avoid punishment or guilt for not completing the task (Foss et al., 2009). Identified motivation helps people to identify and evaluate the importance of their particular task and performance. It is the type of extrinsic motivation which motivates people to perform for future benefit (Vukovic et al., 2010).

This interest may be related to other aspects of their life or job activities (Gagne and Deci, 2005). Intrinsic rewards motivate people to do the task for the organization for their enjoyment, entertainment, fun, or to pass time. People who are motivated intrinsically have been shown not to be involved due to their desire for money or external motivation; rather they work for the organization if they find it challenging or have a creative task. The people involved in intrinsic motivation are satisfied by their job task, which motivates and energizes them to fulfil their role by giving their best performance. Intrinsically motivated people try to involve themselves in an enjoyable task (Isen and Reeve, 2005).

8.3 RESEARCH DESIGN

Data was collected through paper and pencil survey. A convenience sampling technique was used in order to identify the respondents. A total of 250 individuals were approached to obtain a response to the adapted questionnaire, whereas 208 useful responses were further used for the analysis. The response rate was 83.2%. The respondents were asked to answer the questions on a five-point Likert scale which ranged between "strongly disagree = 1" and "strongly agree = 5"; the five items of external motivation and the seven items of introjected motivation were from the studies of Amabile (1994) and Zhao and Zhu (2014). Similarly, seven items of identified motivation, four items of integrated motivation, and thirteen items of intrinsic motivation were used from the studies of Amabile (1994), Allen and Meyer (1996), Bergami and Bagozzi (2000), and Zhao and Zhu (2014). Three items of participation effort were adopted from the studies of Yeo and Neal (2004), Ke and Zhang (2010), and Zhao and Zhu (2014), and four items of trust were adapted from the study of Kankanhalli et al. (2005). For face and content validity the relevant practitioners and academics were consulted for the review of the questionnaire; modification was done accordingly. Based on their review, a pilot study based on 40 respondents was conducted. The scale was found to be internally consistent. For the full measurement, empirical testing, structural equation modelling was used through smart partial least square (PLS).

8.4 RESULTS

8.4.1 DEMOGRAPHICS

Table 8.1 presents the demographics of the sample surveyed in terms of frequencies and percentages. Gender was sub-categorized into the Male and Female population. Of all the surveyed respondents, we didn't find any females participating in crowdsourcing at careem. The second demographic dimension was Age, which was divided into four subcategories: 10 respondents were in the age group 21–25 years, which constituted 4.8% of the total respondents; 80 respondents were in the age group 26–30 years, constituting 38.5%; 84 respondents were in the age group 31–35 years, which made up 40.4%; 4 respondents were in the age group 36–40, constituting 1.9%; and 30 respondents were of age 40+, constituting 14.4%. Hence the majority belonged to the age group 26–35 years. The third survey dimension was Qualification. As it turned out, 93 respondents had a high school degree, which constituted 44.7% respondents; 115 had a bachelor's degree, which made up 53.3%; and none of them had a master's degree. The fourth survey dimension in terms of demographics was Experience which included respondents with: less than 1 year's experience, 1–3 years, 3–5 years, and more than 5. All of the 208 respondents had experience of less than a year since this sector is new in Pakistan.

8.4.2 VALIDATION OF MEASUREMENT MODEL

In model assessment, Smart PLS 2.0 was used to confirm the reliability and validity of the study. The loading and cross-loading of questionnaire items was examined

TABLE 8.1
Demographics

Demographics	Category	f	%
Age	20–25	10	4.8
	26–30	80	38.5
	31–35	84	40.4
	36–40	4	1.9
	41+	30	14.4
	Total	208	100
Education	High school	93	44.7
	Bachelors	115	55.3
	Masters	0	0
	Total	208	100
Gender	Male	208	100
	Female	0	0
	Total	208	100

by the researcher to capture any problems, which is a prerequisite of the measurement model.

As argued by Hair et al. (2010, p. 14) convergent validity is checked when all the items have factor loadings higher than 0.5. Regarding this study, out of 43 items, 42 had such factor loadings, as shown in Table 8.2.

Table 8.2 exhibits average variance extracted (AVE) and Cronbach's alpha values of all constructs. The research shows that at least a 0.70 value should be acceptable for composite reliability and 0.50 of AVE (Fornell and Larcker, 1981; Hair et al., 2014). As shown in the table, the Cronbach alpha was calculated to measure internal

TABLE 8.2
Confirmatory factor analysis

Items	Loadings	α	CR	AVE
External motivation				
EXTER1	0.90	0.93	0.94	0.76
EXTER2	0.94			
EXTER3	0.76			
EXTER4	0.87			
EXTER5	0.90			
Identified motivation				
IDENT1		0.97	0.97	0.86
IDENT2	0.94			
IDENT3	0.99			
IDENT4	0.84			
IDENT5	0.99			
IDENT7	0.91			
Integrated motivation				
INTEG1	0.86	0.93	0.95	0.86
INTEG3	0.96			
INTEG4	0.96			
Intrinsic motivation				
INTRI1	0.89	0.96	0.96	0.76
INTRI2	0.88			
INTRI3	0.82			
INTRI5	0.90			
INTRI7	0.91			
INTRI10	0.81			
INTRI11	0.84			
INTRI13	0.93			
Introjected motivation				
INTRO1	0.58	0.87	0.91	0.72
INTRO5	0.96			
INTRO6	0.89			
INTRO7	0.93			
Participation in crowdsourcing				
PARTI1	0.65	0.77	0.86	0.67
PARTI2	0.90			
PARTI3	0.89			
Trust in crowdsourcing platforms				
TRUST1	0.95	0.92	0.94	0.84
TRUST2	0.98			
TRUST4	0.82			

consistency (George and Mallery, 2003), which provides the rule that $\alpha > 0.9$ is excellent, $\alpha < 0.8$ is good, and $\alpha < 0.7$ is acceptable. As far as this research is concerned, Table 8.2 indicates all items have Cronbach alpha more than 0.6, so this indicates good consistency. For all items, AVE is greater than 0.5, which shows the reliability of the measurement model.

A correlation matrix was introduced to ensure external consistency of the model, based on the correlation between the latent variables. The constructs were compared with the square root of the AVEs. As shown in Table 8.3 all the correlations between the constructs are lower than the square root averages (AVEs), which are bolded in diagonal.

8.4.3 STRUCTURAL MODEL

In PLS, the structure model provides internal analysis of the immediate relationships among the variables of the research study and their t-values with respect to path coefficients. Following the standard guidelines of Hair et al. (2014), the bootstrapping technique was performed (with 500 testing emphases for 209 cases/perceptions) to obtain beta estimations of the coefficient and t-values, of which more than 1.64 is viewed as significant, which is additionally utilized for deciding on the proposed hypothesis. The fundamental reason for this investigation is to concentrate on model assessment with an examination of direct connections and also to test the hypothesized connections among the variables in the basic model. In this research six hypothesis which have direct connections were tried. Out of six, three were demonstrated to be supported and three were rejected. Table 8.4 outlines that all hypothesis that were upheld and acknowledged have t-values more than 1.64 and that the speculations which are rejected had t-values under 1.64. It was completely clarified in Table 8.3 which demonstrates the impact of all IV on DV.

To compute the directing impact, the analysts ran a PLS calculation to obtain the beta coefficient estimations which are given in Table 8.5. Regarding the theory testing, the scientists ran a bootstrapping strategy to check whether trust in crowdsourcing platforms moderates the connection between on the one hand external, identified, integrated, introjected, and intrinsic motivation, and on the other hand participation effort in crowdsourcing As shown in Table 8.5, out of five hypotheses, four are significant and one is insignificant.

TABLE 8.3
Correlation matrix

Variables	1	2	3	4	5	6	7
External motivation	1						
Identified motivation	−0.88	1.00					
Integrated motivation	−0.87	0.90	1.00				
Intrinsic motivation	0.84	−0.91	−0.92	1.00			
Introjected motivation	0.46	−0.53	−0.48	0.42	1.00		
Participation effort	0.20	−0.27	−0.19	0.20	0.83	1.00	
Trust	−0.87	0.99	0.92	−0.92	−0.52	−0.25	1.00

TABLE 8.4
Structural model results

Path	O	M	SD	S.E	t-Stat	Result
External Motivation → Participation Effort	0.01	0.00	0.19	0.19	0.04	Rejected
Identified Motivation → Participation Effort	0.04	0.11	0.49	0.49	0.07	Rejected
Integrated Motivation → Participation Effort	0.46	0.41	0.23	0.23	2.01	Supported
Intrinsic Motivation → Participation Effort	0.37	0.31	0.19	0.19	1.89	Supported
Introjected Motivation → Participation Effort	1.00	0.99	0.07	0.07	14.15	Supported

TABLE 8.5
Moderation effect analysis

Path	O	M	SD	S.E	t-Stat	Result
External Motivation * trust → Participation Effort	0.47	0.01	0.41	0.41	1.14	Not supported
Identified Motivation * trust → Participation Effort	1.39	1.02	1.11	1.11	1.65	Supported
Integrated Motivation * trust → Participation Effort	0.28	0.18	0.84	0.84	2.10	Supported
Intrinsic Motivation * trust → Participation Effort	0.10	0.00	0.20	0.20	2.88	Supported
Introjected Motivation * trust → Participation Effort	0.39	0.34	0.28	0.28	1.92	Supported

In addition, the R^2 value is increased to 0.798 by introducing the moderating variable which is trust between IVs and DV.

8.5 DISCUSSION OF RESULTS

This overall study shows the impact and relationship of external, identified, integrated, introjected, and intrinsic motivation on participation efforts in crowdsourcing. There have been several research works, and each researcher's input assists organizations to recognize the significance of the study. The study was conducted in the service sector of Islamabad to analyze the independent and dependent variables. The study shows that external, identified, integrated, introjected, and intrinsic motivation increase the participation efforts in crowdsourcing. By analyzing the impact of the independent variables on the dependent variables, the current study reveals that: H1 is not supported by the results, which shows that external motivation has no significant impact on participation efforts in crowdsourcing because trust and awareness proved to be a noteworthy variable of interest in an organizational scenario and has been defined as a distinct period. The concept of crowdsourcing seems new in the Pakistan context and people are not fully aware of it. The identified motivation evaluates the personal connection with the external rewards. The individual who experiences autonomy and capability can show more self-determined incentive in the crowdsourcing atmosphere (Zheng et al., 2011). The study shows that when the individual is motivated by introjected motivation so he or she performs the task, not because he or she wants to do it, but because he or she is afraid of shame or seeks to avoid the guilt of not having performed it. Hence, H2 is confirmed.

In the current study, identified motivation shows no significant impact on participation in crowdsourcing efforts. Identified motivation is the extent to which a person identifies with the benefits of the performance for their own future development. Pakistani society focuses on short-term goals rather than long-term plans. Hence H3 is rejected. Empirical evidence is provided for a positive relationship between integrated motivation and participation in crowdsourcing efforts, providing empirical support for the theoretical claim that integrated motivation is a predictor of such participation (Agarwal and Fitzgerald, 2008). Thus, H4 is supported by the results.

H5 is also supported by the results, showing that in crowdsourcing the monetary rewards are not enough to engage the participant for the long term. In this research, it seems that intrinsic motivation is dominated by extrinsic motivation (Brabham, 2010).

H6, H7, H8, H9, and H10 tested the moderating effect of trust on participating efforts in crowdsourcing. All the hypotheses were accepted, except for H6, suggesting that "trust in crowdsourcing platforms positively moderates the relationship between external motivation and participation efforts in crowdsourcing".

The outcomes suggest that the majority of members in crowdsourcing challenges have a blend of inborn and extraneous motivation, which is consistent with past studies (Krishnamurthy, 2006; Zheng et al., 2011). The motivation to increase monetary prizes was essentially connected with interest exertion, in spite of the fact that this finding negates a few studies (e.g., Laumeister et al., 2009; Zheng et al., 2011).

The main objective of the present study was to evaluate the motivations that are necessary to influence crowdsourcing for open innovation because innovation is important for any organization to compete in the market and maintain competitiveness. It also uses different motivations that are introduced in previous studies of crowdsourcing that play various roles in engaging participants or to attract them to collect innovative ideas without having them within the organization, which helps the manager or organizer to focus on the factor that works as a motivation for individuals in the service sector.

In addition, the study affirms the speculated directing impacts of trust in crowdsourcing on the connections between different motivations and supporting exertions. The help of trust in crowdsourcing may, somewhat, fortify their inherent motivation and outward motivation with a disguised center, which is predictable with motivational affordances hypothesis (Zhang, 2008b) and Ke and Zhang's (2010) work. Curiously this investigation makes a commitment to finding a noteworthy directing impact of trust in crowdsourcing between outer motivation and supporting exertion. This demonstrates that apparent motivational affordances may mostly moderate the swarming out impact.

8.5.1 LIMITATIONS AND FUTURE DIRECTIONS

Few constraints on this research adversely influence the generalizability of the study. To start with, variables were calculated by employees' perceptions, which were collected in one (cross-sectional) time frame. Regarding crowdsourcing it is better to gather objective data about the actual participation. From a philosophical viewpoint, the cross-sectional technique largely relies on the prompt views of the respondents, who may somewhat experience issues in estimating genuine participation. Future research could direct a longitudinal report to improve the findings by giving progressively relevant data on the potential variables.

The present study shows the impact and relationship of external, identified, integrated, introjected, and intrinsic motivation on participation efforts in crowd sourcing. There have been several research works and each researcher's input assists organizations to recognize the significance of the study. The study was conducted in the service sector of Islamabad to analyze the corresponding independent and dependent variables. Therefore, future studies should be conducted in different sectors to investigate the perceptions of employees working there. Furthermore, other research should be undertaken with unemployed individuals (e.g., students) to know more about their preferences regarding crowdsourcing. These further investigations should be carried out with a larger sample size in order to achieve significant results.

8.5.2 RESEARCH IMPLICATIONS

Our study has some applied and hypothetical implications. In the first place, this investigation depicts motivation as a range, in which the subtypes of extraneous motivation are viewed as falling along a continuum of disguise (Deci and Ryan, 2000; Ryan and Deci, 2000b). In the meantime the discoveries show that different motivations may assume various jobs in identifying with support exertions used in crowdsourcing challenges. This conceptualization advances our comprehension of the differential impacts of various kinds of motivation on interest exertions in crowdsourcing challenges. In any case, given that identified motivation and external motivation have no critical impact on cooperation exertion, we recommend that future research focus on these two motivations, i.e. intrinsic and extrinsic. This examination makes a noteworthy commitment to finding a critical directing impact of trust in crowdsourcing on external motivation.

REFERENCES

Afuah, A., & Tucci, C. L. (2012). Crowdsourcing as a solution to distant search. *Academy of Management Review*, 37(3), 355–375.

Agerfalk, P. J., & Fitzgerald, B. (2008). Outsourcing to an unknown workforce: Exploring opensourcing as a global sourcing strategy. *MIS Quarterly*, 385–409.

Archak, N., & Sundararajan, A. (2009). Optimal design of crowdsourcing contests. *ICIS 2009 Proceedings*, 200.

Bakici, T. (2020). Comparison of crowdsourcing platforms from social-psychological and motivational perspectives. *International Journal of Information Management*, 54, 102121.

Bartol, K. M., & Srivastava, A. (2002). Encouraging knowledge sharing: The role of organizational reward systems. *Journal of Leadership & Organizational Studies*, 9(1), 64–76.

Brabham, D. C. (2008a). Crowdsourcing as a model for problem solving an introduction and cases. *Convergence: The International Journal of Research into New Media Technologies*, 14(1), 75–90.

Brabham, D. C. (2008b). Moving the crowd at iStockphoto: The composition of the crowd and motivations for participation in a crowdsourcing application. *First Monday*, 13(6).

Brabham, D. C. (2010). Moving the crowd at Threadless: Motivations for participation in a crowdsourcing application. *Information, Communication & Society*, 13(8), 1122–1145.

Brabham, D. C. (2012). A model for leveraging online communities. *The participatory cultures handbook*, 120.

Brito, J. (2008). Hack, mash, & peer: Crowdsourcing government transparency. *The Columbia Science and Technology Law Review*, 9, 119–157.

Chan, S. H. (2011). A motivational framework for understanding IS use and decision performance. *Review of Business Information Systems (RBIS)*, 9(4), 102–118.

Choy, K., & Schlagwein, D. (2016). Crowdsourcing for a better world: on the relation between IT affordances and donor motivations in charitable crowdfunding. *Information Technology & People*, 29(1).

Cook, S. (2008). The contribution revolution: Letting volunteers build your business. *Harvard Business Review*, 86(10), 60–69.

Deci, E. L., & Ryan, R. M. (2000). The "what" and "why" of goal pursuits: Human needs and the self-determination of behavior. *Psychological Inquiry*, 11(4), 227–268.

Deci, E. L., & Ryan, R. M. (2008). Self-determination theory: A macrotheory of human motivation, development, and health. *Canadian Psychology/Psychologiecanadienne*, 49(3), 182.

Doan, A., Ramakrishnan, R., & Halevy, A. Y. (2011).Crowdsourcing systems on the worldwide web. *Communications of the ACM*, 54(4), 86–96.

Eagle, N. (2009, July). Txteagle: Mobile crowdsourcing. *International Conference on Internationalization, Design and Global Development* (pp. 447–456). Springer Berlin Heidelberg.

Foss, N. J., Minbaeva, D. B., Pedersen, T., & Reinholt, M. (2009). Encouraging knowledge sharing among employees: How job design matters. *Human Resource Management*, 48(6), 871–893.

Garcia, S. M., Tor, A., & Garcia, S. (2005). Rankings and competition: social comparison in the shadow of standards. *Ann Arbor*, 1001, 48109-1220.

Gassmann, O., & Enkel, E. (2006). Open innovation. *ZeitschriftFührung+ Organisation*, 75(3), 132–138.

Gregg, D. G. (2010). Designing for collective intelligence. *Communications of the ACM*, 53(4), 134–138.

Hau, Y. S., & Kim, Y. G. (2011). Why would online gamers share their innovation-conducive knowledge in the online game user community? Integrating individual motivations and social capital perspectives. *Computers in Human Behavior*, 27(2), 956–970.

Heo, M., & Toomey, N. (2015). Motivating continued knowledge sharing in crowdsourcing: The impact of different types of visual feedback. *Online Information Review*, 39(6), 795–811.

Hetmank, L. (2013). Components and functions of crowdsourcing systems: A systematic literature review. *Wirtschaftsinformatik*, 4.

Hossain, M. (2012a). Performance and potential of open innovation intermediaries. *Procedia-Social and Behavioral Sciences*, 58, 754–764.

Hossain, M. (2012b, May). Users' motivation to participate in online crowdsourcing platforms. *Innovation Management and Technology Research (ICIMTR), 2012 International Conference on* (pp. 310–315). IEEE.

Hossain, M. (2015). Crowdsourcing in business and management disciplines: an integrative literature review. *Journal of Global Entrepreneurship Research*,5(1), 1–19.

Hossain, M., & Kauranen, I. (2015). Crowdsourcing: a comprehensive literature review. *Strategic Outsourcing: An International Journal*, 8(1), 2–22.

Howe, J. (2006a). Crowdsourcing: A definition, crowdsourcing: tracking the rise of the amateur. *Crowdsourcing: Why the power of the crowd is driving the future of business.*

Howe, J. (2006b). The rise of crowdsourcing. *Wired Magazine*, 14(6), 1–4.

Howe, J. (2008). *Crowdsourcing: How the power of the crowd is driving the future of business.* Random House.

Isen, A. M., & Reeve, J. (2005). The influence of positive affect on intrinsic and extrinsic motivation: Facilitating enjoyment of play, responsible work behavior, and self-control. *Motivation and Emotion*, 29(4), 295–323.

Jain, R. (2010, August). Investigation of governance mechanisms for crowdsourcing initiatives. *AMCIS* (p. 557).

Javadi Khasraghi, H., & Aghaie, A. (2014). Crowdsourcing contests: understanding the effect of competitors' participation history on their performance. *Behaviour & Information Technology*, 33(12), 1383–1395.

Jeppesen, L. B., & Lakhani, K. R. (2010). Marginality and problem-solving effectiveness in broadcast search. *Organization Science*, 21(5), 1016–1033.

Kavaliova, M., Virjee, F., Maehle, N., & Kleppe, I. A. (2016). Crowdsourcing innovation and product development: Gamification as a motivational driver. *Cogent Business & Management*, 3(1), 1128132.

Kazman, R., & Chen, H. M. (2009). The metropolis model a new logic for development of crowdsourced systems. *Communications of the ACM*, 52(7), 76–84.

Ke, W., & Zhang, P. (2010).The effects of extrinsic motivations and satisfaction in open source software development. *Journal of the Association for Information Systems*, 11(12), 784.

Kim, H. S., Yong Park, J., & Jin, B. (2008). Dimensions of online community attributes: Examination of online communities hosted by companies in Korea. *International Journal of Retail & Distribution Management*, 36(10), 812–830.

Kim, P., Lee, D., Lee, Y., Huang, C., & Makany, T. (2011). Collective intelligence ratio: Measurement of real-time multimodal interactions in team projects. *Team Performance Management: An International Journal*, 17(1/2), 41–62.

Krishnamurthy, S. (2006). On the intrinsic and extrinsic motivation of free/libre/open source (FLOSS) developers. *Knowledge, Technology & Policy*, 18(4), 17–39.

Lee, C. K. M., Chan, C. Y., Ho, S., Choy, K. L., & Ip, W. H. (2015). Explore the feasibility of adopting crowdsourcing for innovative problem solving. *Industrial Management & Data Systems*, 115(5), 803–832.

Liu, S. B. (2014). Crisis crowdsourcing framework: designing strategic configurations of crowdsourcing for the emergency management domain. *Computer Supported Cooperative Work (CSCW)*, 23(4–6), 389–443.

Luo, S., Xia, H., Yoshida, T., & Wang, Z. (2009). Toward collective intelligence of online communities: A primitive conceptual model. *Journal of Systems Science and Systems Engineering*, 18(2), 203–221.

Luz, N., Silva, N., & Novais, P. (2015).A survey of task-oriented crowdsourcing. *Artificial Intelligence Review*, 44(2), 187–213.

Malhotra, Y., & Galleta, D. F. (2003, January). Role of commitment and motivation in knowledge management systems implementation: Theory, conceptualization, and measurement of antecedents of success. *System Sciences, 2003.Proceedings of the 36th Annual Hawaii International Conference on* (pp. 10). IEEE.

Nam, T. (2012). Suggesting frameworks of citizen-sourcing via Government 2.0. *Government Information Quarterly*, 29(1), 12–20.

Neyer, A. K., Bullinger, A. C., & Moeslein, K. M. (2009).Integrating inside and outside innovators: a sociotechnical systems perspective. *R&D Management*, 39(4), 410–419.

Olson, D. L., & Rosacker, K. (2013). Crowdsourcing and open source software participation. *Service Business*, 7(4), 499–511.

Porter, J. (2010). *Designing for the Social Web, eBook*. Peachpit Press.

Ridge, M. (Ed.). (2014). *Crowdsourcing our Cultural Heritage*. London: Routledge, https://doi.org/10.4324/9781315575162.

Royo, S., & Yetano, A. (2015). "Crowdsourcing" as a tool for e-participation: two experiences regarding CO_2 emissions at municipal level. *Electronic Commerce Research*, 15(3), 323–348.

Ryan, R. M., & Deci, E. L. (2000a). Intrinsic and extrinsic motivations: Classic definitions and new directions. *Contemporary Educational Psychology*, 25(1), 54–67.

Ryan, R. M., & Deci, E. L. (2000b).Self-determination theory and the facilitation of intrinsic motivation, social development, and well-being. *American Psychologist*, 55(1), 68.

Ryan, R. M., & Deci, E.L. (2004). Autonomy is no illusion. *Handbook of experimental existential psychology*, 449–479.

Schweitzer, F. M., Buchinger, W., Gassmann, O., & Obrist, M. (2012). Crowdsourcing: leveraging innovation through online idea competitions. *Research-Technology Management*, 55(3), 32–38.

Stewart, O., Huerta, J. M., & Sader, M. (2009, June). Designing crowdsourcing community for the enterprise. *Proceedings of the ACM SIGKDD Workshop on Human Computation* (pp. 50-53). ACM.

Sun, Y., Fang, Y., & Lim, K. H. (2012). Understanding sustained participation in transactional virtual communities. *Decision Support Systems*, 53(1), 12–22.

Verona, G., Prandelli, E., & Sawhney, M. (2006). Innovation and virtual environments: Towards virtual knowledge brokers. *Organization Studies*, 27(6), 765–788.

Vukovic, M., & Bartolini, C. (2010). Towards a research agenda for enterprise crowdsourcing. *International Symposium on Leveraging Applications of Formal Methods, Verification and Validation* (pp. 425–434). Springer Berlin Heidelberg.

Vukovic, M., Laredo, J., & Rajagopal, S. (2010). Challenges and experiences in deploying enterprise crowdsourcing service. *Web Engineering* (pp. 460–467). Springer Berlin Heidelberg.

Wang, A., Hoang, C. D. V., & Kan, M. Y. (2013).Perspectives on crowdsourcing annotations for natural language processing. *Language Resources and Evaluation*, 47(1), 9–31.

Wang, X., Khasraghi, H. J., & Schneider, H. (2019). Towards an understanding of participants' sustained participation in crowdsourcing contests. *Information Systems Management*, 1–14.

Wasko, M. M., & Faraj, S. (2000). "It is what one does": why people participate and help others in electronic communities of practice. *The Journal of Strategic Information Systems*, 9(2), 155–173.

Whitla, P. (2009). Crowdsourcing and its application in marketing activities. *Contemporary Management Research*, 5(1).

Xia, F., & Zhang, J. (2015). How Social Subsystem and Technical Subsystem Risks Influence Crowdsourcing Performance.

Yang, K. (2019). Research on factors affecting solvers' participation time in online crowdsourcing contests. *Future Internet*, 11(8), 176.

Zhao, Y., & Zhu, Q. (2014). Evaluation on crowdsourcing research: Current status and future direction. *Information Systems Frontiers*, 16(3), 417–434.

Zheng, H., Li, D., & Hou, W. (2011).Task design, motivation, and participation in crowdsourcing contests. *International Journal of Electronic Commerce*, 15(4), 57–88.

Župič, I. (2013). Social media as enabler of crowdsourcing. *Social media in human resources management (Advanced Series in Management)* (12, 243–255). Emerald Group Publishing Limited.

9 Understanding the Relationship between Digital Currencies and Search Engines
An Empirical Analysis

Naveed Ahmad Lone, Yousfi Karima, and Hurmat Sumaiya Binti Bashir

CONTENTS

9.1 INTRODUCTION

Digital currency is a digital representation of value that can be digitally traded and functions as a medium of exchange, a unit of account composed of unique strings of numbers and letters, and/or a store of value but which does not have legal tender status in any jurisdiction (FATF, 2014). Digital currency is an asset that only exists electronically. The most popular cryptocurrencies, such as Bitcoin, were designed for transactional purposes; however, they are often held for speculation in anticipation of a rise in their values (see Bank of England, 2018 for detailed insight into digital currencies). Bitcoin was the first decentralized peer-to-peer payment-convertible global virtual currency and the first cryptocurrency; it is an innovative internet protocol created by the pseudonymous "Satoshi Nakamoto" in 2009 that enables value to be transferred over a communications channel (Satoshi Nakamoto, 2008).

Bitcoins are created (for the purpose of replacing cash, credit cards, and bank wire transactions) as reward for payment processing work in which users offer their computing power to verify and record payments into a public ledger. The activity is called data mining and miners are rewarded with transaction fees and newly created Bitcoins. Transactions are made with no middlemen, so anyone can transfer money anywhere in the world without using any centralized service like a bank or PayPal.

It is based on advancements in peer-to-peer networks and cryptographic protocols for security, and on a distributed register known as "blockchain" to save transactions carried out by users, like any other currency. It is a new form of decentralized electronic currency system; it stands for an IT innovation based on advancement in peer-to-peer networks and cryptographic protocols, and it has low processing fees and is quite trustworthy. The digital currency is based on IT innovation, so has association and dependence on internet access.

So against this backdrop, this chapter attempts to understand the relationship between Bitcoin (BTC) prices and search queries on Google Trends as a measure of the interest in the currency in the world. In this work, Bitcoin is analyzed as a novel digital currency system and the relationship studied that exists between trading volumes of Bitcoin currency and the query volumes of search engines. The frequency of searches of terms about Bitcoin could provide good explanatory power, so it was decided to examine Google, one of the most important search engines. The study reveals whether Web search media activity could be helpful and used by investment professionals analyzing search volumes and the power of forecasting trading volumes of the Bitcoin currency.

9.2   LITERATURE REVIEW

In recent years, some studies have evaluated the ability of keyword analysis to forecast technological factors. For instance, a study by Dotsika and Watkins (2017) used keyword network analysis to identify the potentially disruptive trends in emerging technologies and reported significant influence. Similarly, Dubey et al. (2019) showed that big data and predictive analytics could influence social and environmental sustainability. Some studies have tested the effects of data availability on the internet and in print media on financial asset returns. For illustration, in equity markets, Tetlock (2007) analyzed the role of traditional media, whereas Bollen et al. (2011) used Twitter to forecast equity markets. Likewise, Moat et al. (2013) used Wikipedia as a predictive tool, while Challet and Ayed (2013) showed the importance of keywords in Google for predicting financial market behavior. A study by Preis et al. (2013) analyzed trading behavior using Google Trends.

Mondria et al. (2010) proved that the number of clicks on search results stemming from a given country correlates with the amount of investment in that country. Matta et al. (2015) examined the striking similarity between Bitcoin price and the number of queries regarding Bitcoin recovered on the Google search engine. In their work, Garcia et al. (2014) proved the interdependence between social signals and price in the Bitcoin economy, namely a social feedback cycle based on word-of-mouth effect and a user-driven adoption cycle. They provided evidence that Bitcoin's growing popularity causes an increasing search volume, which in turn results in higher social media activity about Bitcoin. Growing interest inspires the purchase of Bitcoins by users, driving the prices up, which eventually feeds back on the search volumes.

The relationship between Bitcoin price and the interest in the currency as measured by online searches in Wikipedia and Google was examined by Kristoufek (2015). The study finds a strong bi-directional causal relationship between the prices and searched terms and there exists a strong correlation between the price level and the queries in Wikipedia and Google.

Nasir et al. (2019) analyzed the predictability of Bitcoin volumes and returns using Google search values. Using a weekly dataset from 2013 to 2017, the results of the study suggest that the frequency of Google searches leads to positive returns and a surge in Bitcoin trading volumes. Shocks to search values have a positive effect, which persist for at least a week. Our findings contribute to the debate on cryptocurrencies/Bitcoins and have profound implications in terms of understanding their dynamics, which are of special interest to investors and economic policy makers.

9.3   EMPIRICAL UNDERPINNING OF BITCOIN

The study compared USD trade volumes in Bitcoin with those in a media, namely, Google Trends. Being a characteristic of the Google search engine, it demonstrates how often a fixed search term was looked for. Figure 9.1 shows the statistics of the most expensive virtual currencies globally as of October 12, 2018. Bitcoin was the most valuable cryptocurrency at USD6285.99 dollars per unit.

The statistics present the total number of Bitcoins in circulation from the first quarter of 2011 to the third quarter of 2018. The number of Bitcoins has been growing since the creation of this virtual currency in 2009 and reached approximately 17.30 million in September 2018 (Figure 9.2).

The data succinctly depict the total size of the Bitcoin blockchain and the distributed database that contains a continuously growing and tamper-evident list of Bitcoin transactions and records, from the third quarter of 2010 to the latest quarter. The size of the Bitcoin blockchain has been growing since the creation of the Bitcoin virtual currency in 2009, reaching approximately 185 gigabytes in size by the end of September 2018 (Figure 9.3).

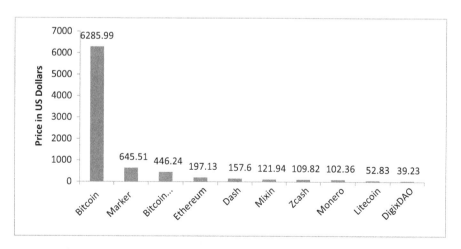

FIGURE 9.1   Most expensive virtual currencies globally as of October 2018

Source: www.statista.com.

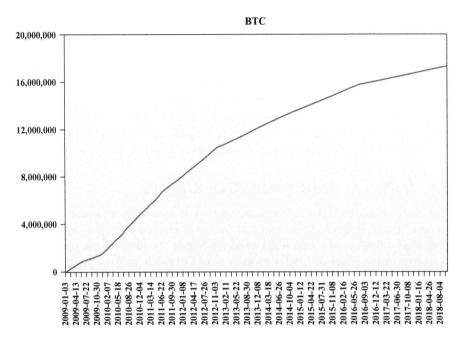

FIGURE 9.2   Number of Bitcoins in circulation worldwide from January 3, 2009 to October 11, 2018 (millions)

Source: www.blockchain.com.

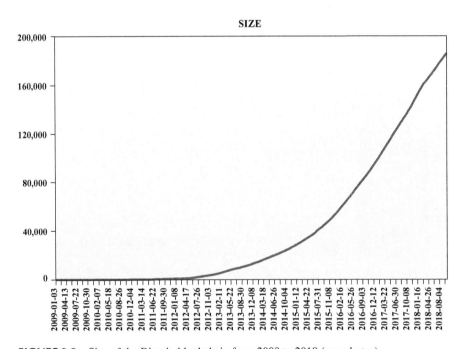

FIGURE 9.3   Size of the Bitcoin blockchain from 2009 to 2018 (megabytes)

Source: https://www.statista.com.

Figure 9.4 presents the evolution of the Bitcoin price index from September 2009 to September 2018. The Bitcoin price index is an average of bitcoin prices across leading global exchanges. The index value for the end of September 2018 amounted to USD6604.97.

Bitcoin came into existence in 2009, and the price of this cryptocurrency remained stable until January 2013, reaching a maximum value of approximately USD20. Thereafter, growth in the monthly price was observed until October 2013 when it reached USD198. The rise was as much as ten times higher, and the value proved to be insignificant in comparison to the price rally in November 2013, when the threshold of USD1100 per coin was broken. After a downfall which followed, the price reached USD1349.19 in April 2017.

The number of Bitcoins in circulation has grown month on month and reached over 17 million in September 2018. The global value of Bitcoin amounted to approximately USD10.1 trillion as of January 2014 and was much higher than the value of other internet currencies, such as Ripple, Litecoin, or Peercoin (Figure 9.5).

Bitcoin has not received nearly as much attention from the media in 2018 as it did in 2012, but Bitcoin-based startups continue to be of great interest to venture capitalists. With two months remaining in 2018, there was already USD1943 million invested in Bitcoin startups, more than in any previous year, with 390 rounds (Figure 9.6).

An ATM (automated teller machine) is a device that enables the holders of debit or credit cards to withdraw cash from their bank accounts. The choice of issuance of statement as a part of the account information is also available to ATM users. The withdrawal of cash from the ATM of the company where the payment card is registered is usually for free, while the owners of cards belonging to other banks have to pay a defined amount of money. There are two main types of Bitcoin ATMs: the basic ones, allowing the users only to purchase Bitcoins, and more complex ones, enabling the users both to buy and sell the virtual money. In the case of complex ATMs, only the members of a particular ATM producer can use the ATM. As of April 2018, the main manufacturers of Bitcoin ATMs were Genesis Coin and General Bytes, with 33.69% and 26.89% of the market share, respectively. As of October 2018, there were 3903 Bitcoin ATMs worldwide. In the same time period, the countries with the highest number of Bitcoin ATMs were the United States (2183), Canada (598), Austria (228), the United Kingdom (200), and the Russian Federation (70) (Table 9.1). In total, approximately 72.10% of global ATMs were concentrated in North America.

A blockchain is a transaction database shared by all nodes participating in a system based on the Bitcoin protocol. A full copy of a currency's blockchain contains every transaction ever executed in the currency. With this information, one can find out how much value belongs to each address at any point in time. A blockchain is vividly one of the most significant and disruptive technologies that has come into existence since the inception of the internet. It is the core technology behind Bitcoin and other cryptocurrencies that have attracted a lot of attention in the last few years. As its core, a blockchain is a distributed database that allows direct transactions between two parties without the need of a central authority. This simple yet powerful concept has great implications for various institutions such as banks, governments, and marketplaces. Any business or organization that relies on a centralized database

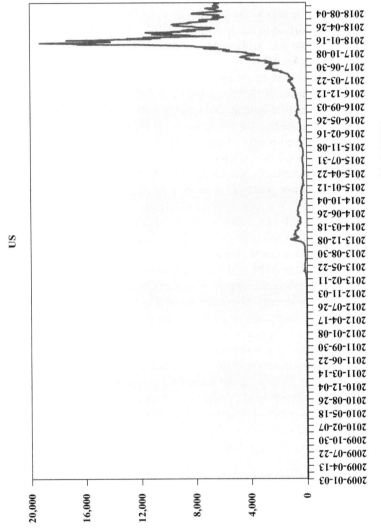

FIGURE 9.4 Bitcoin price index from September 2009 to September 2018 (USD)

Source: www.statista.com.

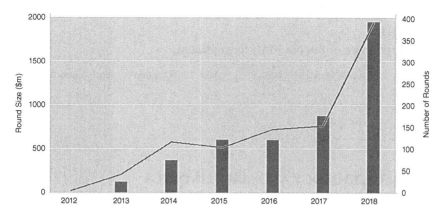

FIGURE 9.5   Venture capital total funding VS number of rounds

Source: coindesk.com.

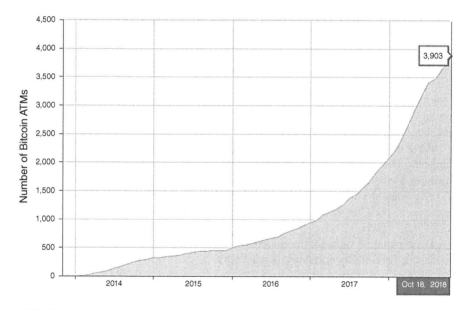

FIGURE 9.6   Number of Bitcoin ATMs installed over time

Source: coinatmradar.com.

as a core competitive advantage can potentially be disrupted by blockchain technology (Kondor et al., 2014).

A blockchain is an online system that provides detailed information about the Bitcoin market. Launched in August 2011, this system shows data on recent transactions, plots on the Bitcoin economy, and several other statistics. It allows users to analyze different Bitcoin aspects: the total Bitcoins in circulation, the number of transactions, the total output volume, the USD exchange trade volume, and the market price (in USD) (Mataa et al., 2015).

TABLE 9.1
Percentage share of Bitcoin ATMs by continent

Country	North America	Europe	Asia	Oceania	South America	Africa
Bitcoin ATMs	72.10	22.98	2.46	1.38	0.90	0.18

Source: coinatmradar.com.

9.4 METHODOLOGY AND MAIN FINDINGS

The data used for the analysis are the Bitcoin price index and Google Trends. The former is an index of the exchange rate between the US dollar (USD) and Bitcoin (BTC). The data is daily and it is formed as a simple average of the covered exchanges. The series is available at Blockchain.com. Google Trends is a feature of the Google Search engine that illustrates how frequently a fixed term is looked for (in our study we downloaded data about how often the term "Bitcoin" was referred to on July 14, 2013, with an end date of October 7, 2018). We obtained the trend data for the "Bitcoin" search keyword at Google Trends at https://trends.google.com. The data are monthly and contain only year and month information, so we converted it to a date data type to join together.

First, we tested the correlation between the variables with the explanatory program. We observed that the prices of Bitcoin were strongly correlated with the search engines (Google Trends) (Figure 9.7).

The R-squared of the model was 0.95 (it should be between 0 and 1, and 1 is the highest). The P-value was 0 (or close to 0), so we can reject the null hypothesis, meaning that the prediction quality of this model is statistically reasonable. The P-value was 0 (or close to 0) and the coefficient was 79.25, meaning that one value increase in the Google Trend score will make the Bitcoin price increase by about USD79.25 (Table 9.2).

The correlation value for the Bitcoin price and Google Trends showed 0.95, which means they are highly correlated (Figure 9.8).

The first step of the analysis stresses the stochastic properties of the series by testing for the presence of unit roots. This allows for the identification of stationary and non-stationary time series, which in turn allows for the specification of a model that should not produce spurious results. Broadly speaking, a stochastic process is said to be stationary if its mean and variance are constant over time and the value of the covariance between two time periods depends only on the distance or lag between them and not on the actual time at which the covariance is measured. Symbolically, letting Y represent a stochastic time series, we say that it is stationary if the following conditions are satisfied (Gujarati & Porter, 2009):

Mean: $E(Y_t) = \mu$

Variance: $\text{Var}(Y_t) = E(Y_t - \mu)^2 = \sigma^2$

Covariance: $\gamma_k = E[(Y_t - \mu)(Y_{t+k} - \mu)]$

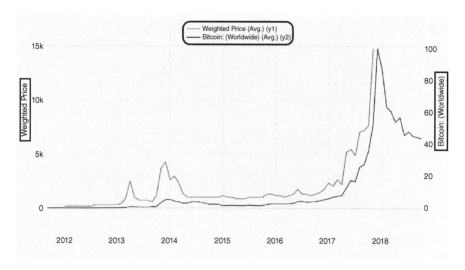

FIGURE 9.7 Bitcoin price index average and Google Trends

Source: coinatmradar.com.

TABLE 9.2

Results of correlation between BTC and Google Trends

R-squared	0.95
Adj R-squared	0.95
RMSE	392.81
F Ratio	20.62
P Value	0.00
Degree of freedom	2
Log-likelihood	−15.97
AIC	31.95
BIC	31.97
Deviance	333.125
Residual DF	2.16

---

where γ_k, the covariance (or autocovariance) at lag k, is the covariance between the values of Y_t and Y_{t+k}, that is, between two values of Y, k periods apart. If k = 0, we obtain γ_0, which is simply the variance of Y (= σ^2); if k = 1, γ_1 is the covariance between two adjacent values of Y.

For all series we tested the null hypothesis of the unit root, using Augmented Dickey-Fuller (ADF), the Phillips-Perron (PP) test, and the Kwiatkowski, Phillips, Schmidt, and Shin (KPSS) unit root test (Maddala & Kim, 1998). Each series was tested for the presence of a unit root. The unit root test statistics suggest the presence of a unit root in the level, while first differencing the series yields the apparent lack of a unit root in the two variables, the Bitcoin price index and Google Trends in the log. From these results, we can conclude that each series has a unit root at levels and it is stationary when the first difference is taken. It can be said that all variables are integrated of order 1, I(1). We then check for the presence of cointegrating relations between these variables (Table 9.3).

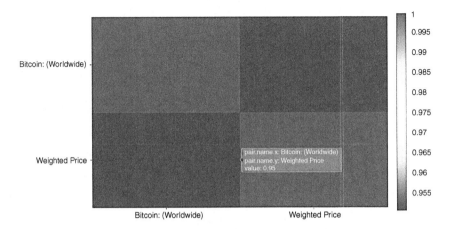

FIGURE 9.8   Correlation between variables

TABLE 9.3
Results of unit root tests

		ADF		KPSS		
		level	1st diff.	level	1st diff.	
BTC	[2]	−0.940262	−15.12464	1.468390	0.106813	I(1)
		[0.7743]	[0.0000]			
	[3]	−1.377811	−15.10049	0.342745	0.114624	
		[0.8655]	[0.0000]			
	[1]	1.936615	−14.91026	//	//	
		[0.9876]	[0.0000]			
Google Trends	[2]	−1.074772	−23.19551	1.839544	0.061067	I(1)
		[0.7263]	[0.0000]			
	[3]	−2.241928	−23.16455	0.192779	0.056207	
		[0.4639]	[0.0000]			
	[1]	1.114131	−22.98511	//	//	
		[0.9313]	[0.0000]			

The cointegration introduced by Granger (1981, 1983), and Engle and Granger (1987), is used to understand the classical long run relationship between the Bitcoin and the search engine. We applied Johansen's trace test and maximum eigenvalue test for cointegration and the results are reported in Table 9.4. The trace test and the maximum eigenvalue test indicate no cointegration, so we can accept the null hypothesis and reject the alternate hypothesis. So, Bitcoin price index series are not cointegrated with Google trends series (no relations in the long run between variables).

We need to turn to the vector auto regression (VAR) methodology applied on the first logarithmic differences with five lags based on the Akaike AIC, Schwarz

TABLE 9.4
Johansen's trace test for cointegration

Unrestricted cointegration rank test (trace)

Hypothesized no. of CE(s)	Eigenvalue	Trace s tatistic	0.05 Critical value	Prob.**
None	0.018945	6.985846	15.49471	0.5793
At most 1	0.006916*	1.859838	3.841466	0.1726

Trace test indicates no cointegration at the 0.05 level
* denotes rejection of the hypothesis at the 0.05 level
** MacKinnon-Haug-Michelis (1999) p-values

Unrestricted cointegration rank test (maximum eigenvalue)

Hypothesized no. of CE(s)	Eigenvalue	Max. eigen statistic	0.05 Critical value	Prob.**
None	0.018945	5.126008	14.26460	0.7258
At most 1	0.006916*	1.859838	3.841466	0.1726

Max. eigenvalue test indicates no cointegration at the 0.05 level
* denotes rejection of the hypothesis at the 0.05 level
** MacKinnon-Haug-Michelis (1999) p-values

SC, Hannan–Quinn HQ, Final Prediction Error FPE, and Likelihood Ratio LR (Table 9.5).

We used an analysis of variance components tool to identify the amount of variation in the prediction of each variable of the model, which is due to an error in the prediction of other variables. In the short run, impulse or shock to BTC accounts for a 91.62% variation of the fluctuation in BTC in period 10 (long run); shock to Google Trends can cause a 8.37% fluctuation in BTC (Tables 9.6 and 9.7). Impulse or shock to Google Trends accounts for a 78.19% variation of the fluctuation in period 10 (long run); and shock to BTC can cause a 21.80% fluctuation in Google Trends.

Figure 9.9 shows the response of a corresponding variable to a shock in the impulse variable. As the study works with logarithmic differences, so the interpretation of these shocks is a proportional reaction to a 1% shock. A one standard deviation shock given by the Bitcoin price index results in a Google Trends change (positive reaction). The study finds that the increased interest in the Bitcoin currency measured by the searched terms increases its price. As the interest in the currency increases, the demand increases as well, causing the prices to increase.

TABLE 9.5
Length of the lag

Lag	LogL	LR	FPE	AIC	SC	HQ
0	−668.9909	NA	0.503096	4.988780	5.015506	4.999513
1	318.7189	1953.389	0.000335	−2.325047	−2.244868	−2.292847
2	345.1271	51.83474	0.000284	−2.491651	−2.358019*	−2.437984*
3	351.6072	12.62290	0.000279	−2.510091	−2.323005	−2.434957
4	353.6213	3.893444	0.000283	−2.495326	−2.254787	−2.398725
5	359.6642	11.59157*	0.000278*	−2.510514*	−2.216523	−2.392447

* Optimum lag length.

TABLE 9.6
Variance decomposition of L trend

Period	S.E.	LTREND	LBTC
1	0.145479	100.0000	0.000000
2	0.176796	98.23755	1.762452
3	0.215347	97.79861	2.201385
4	0.246402	96.76193	3.238069
5	0.266411	96.01423	3.985768
6	0.290332	96.04055	3.959451
7	0.309003	95.94631	4.053686
8	0.327451	96.09216	3.907843
9	0.345109	96.24862	3.751382
10	0.360830	96.38958	3.610421

TABLE 9.7
Variance decomposition of LBTC

Period	S.E.	LTREND	LBTC
1	0.115690	9.262942	90.73706
2	0.167672	16.54287	83.45713
3	0.218675	22.90663	77.09337
4	0.266417	26.14010	73.85990
5	0.300989	27.44252	72.55748
6	0.332006	28.46539	71.53461
7	0.358330	29.06082	70.93918
8	0.381317	29.76560	70.23440
9	0.402682	30.56678	69.43322
10	0.422034	31.30662	68.69338

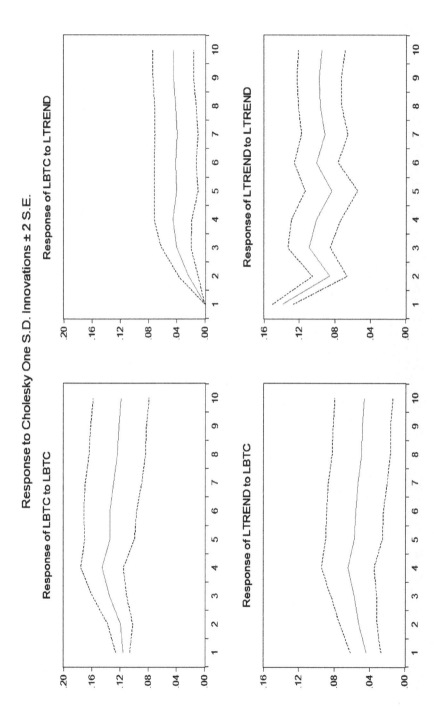

FIGURE 9.9 Impulse response

9.5 CONCLUSION

The internet has been one of the most revolutionary technologies in the last few decades. There was a dramatic change in the majority of the daily activities moving towards a virtual sector, such as Web actions, credit card transactions, electronic currencies, and navigators. Digital currencies have been a hot topic over the last few years, with hundreds of alternative coins in existence, attracting attention from investors. Bitcoin is a prominent platform with its respective currencies. To assess long-term investment potential, investors need to understand the function of cryptocurrencies that serve the underlying technology and their governance structure. Bitcoin prices have struggled in the year being studied, along with the cryptocurrency's ability to capture interest on the internet. Searches for the term "Bitcoin" have dropped more than 75% since the beginning of this year and roughly halved over three months, according to research from Google Trends. Our empirical results confirm that there exists a strong and significant relationship between Bitcoin price movements and the queries in Google Trends by investment professionals in Bitcoins.

REFERENCES

Bank of England. (2018), Digital currencies, available at https://www.bankofengland.co.uk/research/digital-currencies accessed on 10th January 2020.

Bollen, J., Mao, H., & Zeng, X. (2011). Twitter mood predicts the stock market. *Journal of Computational Science*, 2(1), 1–8.

Brooks, C. (2014). *Introductory econometrics for finance*. Cambridge University Press.

Challet, D., & Ayed, A. B. H. (2013). Predicting financial markets with Google Trends and not so random keywords. *arXiv preprint arXiv:1307.4643*.

Dickey, D. A., & Fuller, W. A. (1981). Likelihood ratio statistics for Aautoregressive time series with a unit root. *Econometrica: Journal of the Econometric Society*, 1057–1072

Dotsika, F., & Watkins, A. (2017). Identifying potentially disruptive trends by means of keyword network analysis. *Technological Forecasting and Social Change*, 119, 114–127.

Dubey, R., Gunasekaran, A., Childe, S. J., Papadopoulos, T., Luo, Z., Wamba, S. F., & Roubaud, D. (2019). Can big data and predictive analytics improve social and environmental sustainability?. *Technological Forecasting and Social Change*, 144, 534–545.

Engle, R. F., & Granger, C. W. (1987). Co-integration and error correction: epresentation, estimation, and testing. *Econometrica: Journal of the Econometric Society*, 251–276.

Force, F. A. T. (2014, June). Virtual currencies: key definitions and potential AML/CFT risks. *FATF Report*.

Garcia, D., Tessone, C. J., Mavrodiev, P., & Perony, N. (2014). The digital traces of bubbles: feedback cycles between socio-economic signals in the Bitcoin economy. *Journal of the Royal Society Interface*, 11(99), 20140623.

Granger, C. W. (1969). Investigating causal relations by econometric models and cross-spectral methods. *Econometrica: Journal of the Econometric Society*, 424–438.

Granger, C. W. (1981). Some properties of time series data and their use in econometric model specification. *Journal of econometrics*, 16(1), 121–130.

Granger, C. W., & Weiss, A. A. (1983). Time series analysis of error-correction models. In *Studies in Econometrics, Time Series, and Multivariate Statistics* (pp. 255–278). New York: Academic Press.

Gujarati, D. N. (2009). *Basic econometrics*, 5th edition (pp. 653,654, 744–788). Tata McGraw-Hill Education.

Johansen, S. (1988). Statistical analysis of cointegration vectors. *Journal of Economic Dynamics and Control*, 12(2–3), 231–254.

Johansen, S., & Juselius, K. (1990). Maximum likelihood estimation and inference on cointegration—with applications to the demand for money. *Oxford Bulletin of Economics and Statistics*, 52(2), 169–210.

Kondor, D., Pósfai, M., Csabai, I., &Vattay, G. (2014). Do the rich get richer? An empirical analysis of the Bitcoin transaction network. *PloS One*, 9(2), e86197.

Kristoufek, L. (2015). What are the main drivers of the Bitcoin price? Evidence from wavelet coherence analysis. *PloS One*, 10(4), 1–15.

MacKinnon, J. G., Haug, A. A., & Michelis, L. (1999). Numerical distribution functions of likelihood ratio tests for cointegration. *Journal of Applied Econometrics*, 14(5), 563–577.

Maddala, G. S., & Kim, I. M. (1998). *Unit roots, cointegration, and structural change* (No. 4). Cambridge: Cambridge University Press.

Matta, M., Lunesu, I., &Marchesi, M. (2015, November). The predictor impact of Web search media on Bitcoin trading volumes. *Knowledge Discovery, Knowledge Engineering and Knowledge Management (IC3K), 2015 7th International Joint Conference, Portugal* (Vol. 1, pp. 620–626). IEEE.

Moat, H. S., Curme, C., Avakian, A., Kenett, D. Y., Stanley, H. E., & Preis, T. (2013). Quantifying Wikipedia usage patterns before stock market moves. *Scientific Reports* 3, 1801.

Mondria, J., Wu, T., & Zhang, Y. (2010). The determinants of international investment and attention allocation: Using internet search query data. *Journal of International Economics*, 82(1), 85–95.

Nakamoto, S. (2008). Bitcoin: A peer-to-peer electronic cash system. Available: http://bitcoin. org/bitcoin.pdf

Nasir, M. A., Huynh, T. L. D., Nguyen, S. P., & Duong, D. (2019). Forecasting cryptocurrency returns and volume using search engines. *Financial Innovation*, 5(1), 2.

Preis, T., Moat, H. S., & Stanley, H. E. (2013). Quantifying trading behavior in financial markets using Google Trends. *Scientific reports*, 3, 1684.

Tetlock, P. C. (2007). Giving content to investor sentiment: The role of media in the stock market. *The Journal of Finance*, 62(3), 1139–1168.

10 Water Quality Improvement with a Solar Desalination Station

N. El Moussaoui, I. Atmane, K. Kassmi,
S. Alexopoulos, Z. Mahdi, K. Schwarzer,
P. Schmitz, H. Chayeb, and N. Bachiri

CONTENTS

10.1 INTRODUCTION

Fresh water is a vital substance for humanity and a basic condition for all living organisms (Yadav and Sudhakar 2015). About 97% of the available water in the earth is found in the oceans. Only about 1% of the remaining water is fresh water, meeting international standards, as used by humans, animals, and plants (Vinoth Kumar and Kasturi 2008). The increase in the world population, associated with agricultural and industrial development, has led to an increase in the demand for fresh water; and the supply of humanity with drinking water will become a critical and risky problem in the future (Al-Kharabsheh and Goswami 2003; Abdel-Rehim and Lasheen 2007; Qiblawey and Banat 2008; Shannon et al. 2008). Access to fresh and clean drinking water is one of the major problems in different parts of the world (Yadav and Sudhakar 2015). The best way to provide fresh water to a growing population is by desalination of salt water from the ocean, the sea, rivers, and wells (Vinoth Kumar and Kasturi 2008).

The use of solar desalination techniques, including water with a high salt content in wells (brackish water), to produce drinking water has become an essential alternative to deal with this shortage of drinking water (Chaibi 2000; García-Rodríguez et al. 2002; Abakr and Ismail 2005; Kalogirou 2005; Shatat et al. 2013). Currently, on a regional, national, and continental scale, we observe the absence of drinking water production techniques, in particular, those using solar energy (Ali et al. 2011; Pouyfaucon and García-Rodríguez 2018; Pugsley et al. 2018).

In the literature, the most common techniques are membrane distillation (Eleiwi and Laleg-Kirati 2017), reverse osmosis (Walschot et al. 2020), multi-effect distillation (Sayyaadi and Saffari 2010), flash desalination (Hosseini et al. 2011), vapor compression desalination (Aly and El-Figi 2003), and desalination by dehumidification (Bourouni et al. 2001; Lawal et al. 2020). These techniques are complex and expensive, and the maintenance and cost per cubic meter results in very expensive drinking water (Qiblawey and Banat 2008; Kaushal and Varun 2010). However, distillation equipment, operated by solar energy, is topical, since it requires only a simple mechanism and a low investment to meet the requirement of clean drinking water at low cost (Adhikari et al. 1995; Kaushal and Varun 2010; Schwarzer et al. 2011; Reddy et al. 2012; Tigrine et al. 2014; Feilizadeh et al. 2015; Yadav and Sudhakar 2015; Schwarzer and Bart 2016; Pouyfaucon and García-Rodríguez 2018). This type of solar system is widely adaptable to climatic conditions in countries with a high level of sunshine, particularly on the African continent.

In this context, we propose to realize in collaboration with the Solar Institute of Julich (Germany), in Douar Al Hamri of the province of Berkane (Morocco), within the framework of the Programme of Morocco-German Cooperation for Scientific Research, project PMARS III 2015-64 (2016–2019), a multi-stage solar desalination pilot plant (MSD). This station, equipped with a remote control and supervision system, has the main objective of producing low-cost fresh water for the inhabitants of Douar from solar energy.

In this chapter, we present the structure, operation, numerical modeling, and first results concerning the heating experiment of the MSD multi-stage solar distiller. This system is designed and manufactured in Germany, in collaboration with different partners, and experimented with in Douar Al Hamri.

10.2 STRUCTURE AND FUNCTIONING OF SYSTEM MSD

MSD, which is the subject of our study, is shown in Figure 10.1. The different blocks of this system are:

- A photovoltaic (PV) pumping system for well water and its storage in a salt water storage tank (A).
- A field of four flat-plate solar thermal collectors that heat the distilled water, which circulates between these collectors and the lower basin of the MSD system, using solar energy.

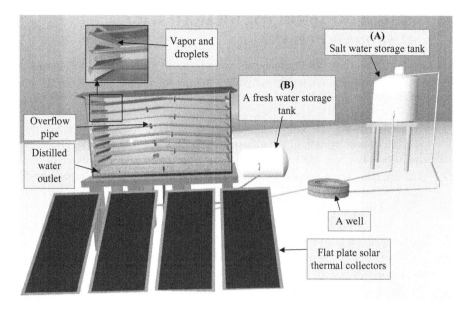

FIGURE 10.1 MSD solar desalination plant, installed at Douar Al Hamri in the Province of Berkane (Morocco)

- A desalination unit formed by a frame containing eight stages and a lower basin. The unit is supplied with well water, stored in reservoir A, from the highest stage (eighth stage). After filling the latter, excess saline water is continuously injected through an overflow pipe to fill and compensate for the lack of water in the lower stages. The same principle is repeated until the first stage is filled.
- The lower basin is supplied with distilled water heated by the four collectors. As a result, each stage is heated by the phase conversion energy released, also known as the enthalpy of evaporation. Then, the saline water, contained in these eight stages, evaporates and condenses on the lower surface of each conical stage. As a result of this condensation, droplets of pure water form and flow into the basin through the drainage gutter, which is welded to the two side walls of the frame. The excess of distilled water in the basin is then drained into the storage tank.
- The characteristics of the stages are:
  - A condensation surface, formed by inclined steel sheets, to form a triangular plateau. The two ends of each stage are placed on supports welded to the sidewalls of the frame.
  - Each stage is manufactured to have an inclination that allows the recovery of the droplets of distilled water produced.
  - An overflow pipe to keep the water level constant in each stage. It returns the excess saline water to the well.
- A fresh water storage tank (B).

10.3 NUMERICAL MODEL FOR THE MULTI-STAGE DESALINATION PLANT

The equations governing the thermal operation of a multi-stage solar MSD are established, for the basin and eight stages, taking into account the following assumptions and conditions (Reddy et al. 2012):

- By reason of the low temperature difference between the adjacent stages, the transfer of heat by radiation, and by natural convection is negligible. The transfer of heat between the hot surface of the saline water and the surface of condensation in each stage is mainly conveyed by the processes of evaporation and condensation.
- The system works at atmospheric pressure.
- The condensation on the walls of the trays is only in the form of a homogeneous and continuous film.
- The losses (leaks) of steam between the stage and the frame inside the distiller are negligible.
- The stages are well placed to ensure a better seal between each stage.
- Fresh water comes out at a temperature equal to the condensation surface temperature.

Assuming the above assumptions and basic relations of thermodynamics (Adhikari et al. 1995; Schwarzer et al. 2011; Reddy et al. 2012; Feilizadeh et al. 2015) and the diagram shown in Figure 10.2, the energy balance at each stage (basin up to the eighth stage) is as follows.

10.3.1 ENERGY BALANCE OF THE BASIN

At this stage the model of heat exchange on the water of temperature T_{sb} is written as a function of ambient temperatures and freshwater (T_{amb}, T_{cb}), the feeding energy $\dot{Q}_{col}$ (the useful energy of the thermal collector), the evacuation energy of the fresh water produced on the stage i ($\dot{m}_{si} \cdot cp_i \cdot T_{ci}$), heat flow lost by evaporation ($\dot{m}_b \cdot h^*_{fgb}$), overall losses $\Delta \dot{Q}_{lossesb}$, and heat loss on stage i ($\dot{m}_{si} \cdot cp_b \cdot T_{cb}$) due to fresh water draining into the tank. Taking into account these different exchanges (Schwarzer et al. 2011; Feilizadeh et al. 2015), the model of heat exchange on the water of the basin is written as the following relation:

$$
\begin{aligned}
M_{sb} \cdot cp_b \cdot \frac{dT_{sb}}{dt} &= \dot{Q}_{col} + \sum_{i=1}^{8} \dot{m}_{si} \cdot cp_i \cdot T_{ci} - \sum_{i=1}^{8} \dot{m}_{si} \cdot cp_b \cdot T_{cb} - \dot{m}_{sb} \cdot h^*_{fgb} - \Delta \dot{Q}_{lossesb} \\
&= A_{col} \dot{E}(t) \eta_0 - \dot{Q}_{losses-col} + \sum_{i=1}^{8} \dot{m}_{si} \cdot cp_i \cdot T_{ci} - \sum_{i=1}^{8} \dot{m}_{si} \cdot cp_b \cdot T_{cb} \\
&\quad - \dot{m}_{sb} \cdot h^*_{fgb} - U_b A_b (T_{sb} - T_{amb})
\end{aligned}
\tag{10.1}
$$

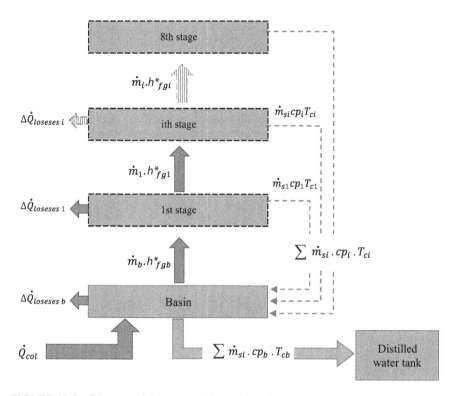

FIGURE 10.2 Diagram of the energy balance of a solar distiller system MSD with a basin and eight stages

where:

i: Stage number (1 to 8),

b: Basin,

M_{sb}: Distilled water mass in the basin (kg),

cp_b: Heat capacity of water in the basin (J/kg·°C),

T_{sb}: Basin water temperature (°C),

A_{col}: Solar collector area (m²),

η_0: Optical collector efficiency,

$\dot{m}_{si}$: The distillate output rate (kg/s) in stages equal to the evaporation rate of the brackish water, which is calculated by the following relation (Abakr and Ismail 2005):

$$\dot{m}_{si} = \frac{(T_{si} - T_{ci})h_{evi}A_{si}}{h_{fgi}}$$ (10.2)

with:

T_{si}: Water temperature in the ith stage (°C),

T_{ci}: Water condensation temperature of the ith stage (°C),

A_{si}: Surface area of the water in the ith stage (m²),

h_{evi}: Evaporative heat transfer coefficient (W/m²·K), which is calculated by the following equation (Kalidasa Murugavel and Srithar 2011; Yadav and Sudhakar 2015):

$$h_{evi} = 16.273 \times 10^{-3} h_{ci} \left(\frac{P_{si} - P_{ci}}{T_{si} - T_{ci}} \right) \tag{10.3}$$

where:

h_{ci}: Convective heat transfer coefficient (W/m²·K), which is calculated by the following relation (Kalidasa Murugavel and Srithar 2011; Yadav and Sudhakar 2015):

$$h_{ci} = 0.884 \times \left((T_{si} - T_{ci}) + \frac{(P_{si} - P_{ci})(T_{si} + 273)}{268.9 \times 10^3 - P_{si}} \right)^{\frac{1}{3}} \tag{10.4}$$

with:

P_{si}: Partial pressure (Pa) at temperature T_{si} calculated by the following relation (Yadav and Sudhakar 2015; Chen et al. 2017):

$$P_{si} = e^{\left(25.317 - \frac{5144}{T_{si} + 273} \right)} \tag{10.5}$$

P_{ci}: Partial pressure (Pa) at temperature T_{ci} calculated by the following relation (Yadav and Sudhakar 2015; Chen et al. 2017):

$$P_{ci} = e^{\left(25.317 - \frac{5144}{T_{ci} + 273} \right)} \tag{10.6}$$

h^*_{fgi}: Defined latent heat of vaporization of water (J/kg), which is given by the following relation proposed by (Feilizadeh et al. 2015):

$$h^*_{fgi} = h_{fgi} + 0.68 \times cp_l (T_{si} - T_{ci}) \tag{10.7}$$

where:

h_{fgi}: Latent heat of vaporization of water (J/kg), which is given by the following relation (Feilizadeh et al. 2015):

$$h_{fgi} = 1000 \times \left(3161.5 - 2.4074 (T_{si} + 273) \right) \tag{10.8}$$

cp_i: Thermal capacity of water (J/ kg·°C), which is defined as a function of water temperature by the following relation (Kalidasa Murugavel and Srithar 2011):

$$cp_i = 1000 \times \left(4.2101 - 0.0022 \times T_{si} + 5 \times 10^{-5} \times T_{si}^2 - 3 \times 10^{-7} \times T_{si}^3\right) \quad (10.9)$$

$\dot{Q}_{losses-col}$: Thermal losses from the collector. They are written according to solar irradiation $\dot{E}(t)$ and ambient temperature T_{amb} by the following relation (Schwarzer et al. 2011):

$$\dot{Q}_{losses-col} = \frac{U_L A_{col}}{\dot{E}(t)} (T_{col} - T_{amb}) \quad (10.10)$$

with:

U_L: Collector overall losses coefficient (W/m²·K).

10.3.2 ENERGY BALANCE FROM THE FIRST TO THE EIGHTH STAGE

At stage i (i varies from 1 to 8), the model of heat exchange on the water of temperature T_{si} – written as a function of ambient temperatures and freshwater (T_{amb}, T_{ci}), heat flow lost by evaporation at the lower stage (for $i = 1$ it is $\dot{m}_b h^*{}_b$, for $i \neq 1$ it is $\dot{m}_{si-1} \cdot h^*{}_{fgi-1}$) due to the condensation of water vapor on the condensation surface, the heat flux lost by evaporation ($\dot{m}_{si} \cdot h^*{}_{fgi}$), overall losses $\Delta \dot{Q}_{lossesi}$, and heat loss ($\dot{m}_{si} \cdot cp_i \cdot T_{ci}$) due to the evacuation of fresh water flowing to the basin – is expressed as:

$$M_{si} \cdot cp_i \cdot \frac{dT_{si}}{dt} = \dot{m}_{i-1} h^*{}_{i-1} - \dot{m}_{si} \cdot h^*{}_{fgi} - \dot{m}_{si} \cdot cp_i \cdot T_{ci} - \Delta \dot{Q}_{lossesi}$$
$$= \dot{m}_{s_{i-1}} h^*{}_{fgi-1} - \dot{m}_{si} h^*{}_{fgi} - \dot{m}_{si} \cdot cp_i \cdot T_{ci} - U_i A_i (T_{si} - T_{amb}) \quad (10.11)$$

with:

M_{si} (kg): Water mass in each stage (first to eighth stage), which is determined from the following conservation relation:

$$\frac{dM_{si}}{dt} = -\dot{m}_{si} \quad (10.12)$$

$\Delta \dot{Q}_{lossesi}$: Heat losses (W) by stage i, which are calculated by the following relation:

$$\Delta \dot{Q}_{lossesi} = U_i A_i (T_{si} - T_{amb}) \quad (10.13)$$

where:

A_i: Surface area of the water in the ith stage (m²),

U_i: Heat loss coefficient (W/m²·°C), defined by:

$$U_i = 1 / \left(\frac{1}{h_{ci}} + \frac{\lambda_i}{\delta_i} + \frac{1}{h_{amb}} \right) \tag{10.14}$$

with:

λ_i: Coefficient of thermal conductivity of the insulating material (W/m·K),

δ_i: Insulating thickness (m),

h_{amb}: Heat transfer convection coefficient exterior (W/m²·K).

10.4 NUMERICAL RESOLUTION

We determined the temperature of the basin (T_{sb}) and that of the stages (T_{si}), as well as the quantities of distilled water produced (m_{si}), numerically solving Equations 10.15 and 10.16 by the Fourth Order Runge-Kutta method (RK4) (El Moussaoui and Kassmi 2019; Talbi et al. 2019). This numerical resolution consists in writing the system of equations MSD (Equations 10.1 and 10.11) in the form of a first-order differential equation, in the following form:

$$\frac{dT_{sb}}{dt} = \frac{1}{M_{sb} \cdot cp_b} (\dot{Q}_{col} + \sum_{i=1}^{8} \dot{m}_{si} \cdot cp_i \cdot T_{ci}$$

$$- \sum_{i=1}^{8} \dot{m}_{si} \cdot cp_b \cdot T_{cb} - \dot{m}_{sb} \cdot h^*_{fgb} - \Delta \dot{Q}_{lossesb}) \tag{10.15}$$

$$\frac{dT_{si}}{dt} = \frac{1}{M_{si} \cdot cp_i} \left(\dot{m}_{i-1} h^*_{i-1} - \dot{m}_{si} \cdot h^*_{fgi} - \dot{m}_{si} \cdot cp_i \cdot T_{ci} - \Delta \dot{Q}_{lossesi} \right) \tag{10.16}$$

where:

i: Stage number (1 to 8),

b: Basin,

T_{sb}: Unknown (basin water temperature),

T_{si}: Unknown (i stage water temperature),

$\dot{m}_{si}$: Unknown (mass of water produced).

The program, developed under the Matlab language, consists in solving numerically the system of first-order differential equations (Equations 10.15 and 10.16) and deducing the temperatures T_{sb} and T_{si} and the quantity of fresh water produced by the MSD system, following the steps in Figure 10.3:

• Initialization of the temperatures and the water mass of the stages and the basin: T_{sb}, T_{s1}, T_{s2}, T_{s3}, M_{sb}, M_{s1}, M_{s2}, M_{s3}.

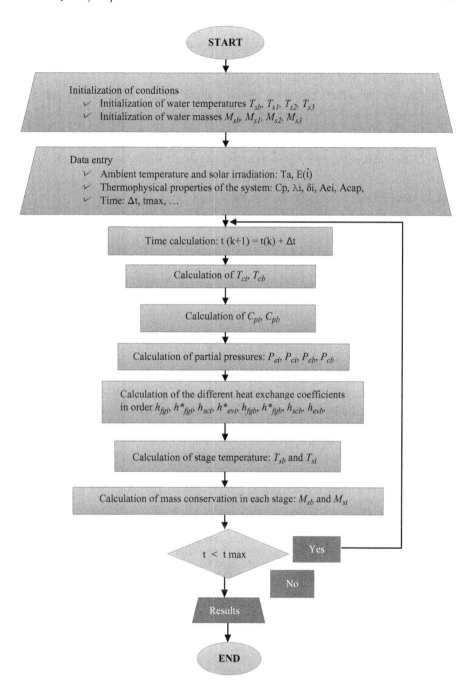

FIGURE 10.3   Flowchart of resolution by the Fourth Order Runge-Kutta method

- Data entry:
  - Thermophysical properties of the system Cp, λ_i, δ_i, A_{ei}, A_{cap} … ;
  - Ambient temperature and solar irradiation T_{amb}, $E(t)$.
- Calculation of T_{cb}, Cp_b, T_{ci}, and Cp_i.
- Calculation of partial pressure vapor P_{si}, P_{ci}, P_{sb}, P_{cb} of water for each stage.
- Calculation of the different heat exchange coefficients in the order h_{fgi}, h^*_{fgi}, h_{sci}, h_{evi}, h_{fgb}, h^*_{fgb}, h_{scb}, h_{evb}.
- Determination of the water temperature of each stage T_{si} and T_{sb}.
- Determination of the body of water M_{si} and M_{sb} in different stages at all times.

10.5 EXPERIMENTATION AND RESULTS

10.5.1 DESCRIPTION OF EXPERIMENTAL PROCEDURE AND MEASUREMENT GUIDELINES

The MSD solar desalination system that is the subject of our study is shown in Figure 10.4. This system is designed and built, in collaboration with the project partners, at the Solar Institute of Julich (Germany), and then installed at Douar Al Hamri in the Berkane region of Morocco, which is located at latitude 35.015029°, longitude −2.386229°, and altitude 408 m.

In accordance with the specification and description in paragraph 2 in Section 10.2, the different blocks of this system are:

- A field of four flat-plate solar thermal collectors of type KS2100 TLP ACR. Each collector has a surface area of 2.1 m², a weight of 35 kg, and is oriented "South" with an inclination of 30°. Their role is to heat, with solar energy, the distilled water circulating in these collectors and the lower basin of the MSD system.
- A frame of eight stages, with surfaces of about 2 m² each, formed by stainless steel sheets, inclined to form a triangular plateau. Each floor, with a filling capacity of 40 l of water, is placed on a support welded to the side walls of the frame.

FIGURE 10.4 Photo taken during the experiment of the MSD solar desalination plant, February 18, 2019

- A basin, which represents the lower floor of the building, with a filling capacity of 60 l of water, is connected to the collectors via the well-insulated pipes.
- A rectangular stainless steel door, length 1.20 cm and width 90 cm. Seals are placed at the ends of this door to prevent heat and vapor leakage to the outside.

The MSD system is tested using these measuring instruments:

- Temperature sensors (type k), with a maximum continuous temperature of approximately 1100°C and an accuracy of approximately ±2.2°C, are available. These sensors allow temperatures to be measured by immersing them directly in the water of the basins, or by holding them in the air.
- A CMP6 pyranometer, with an accuracy of ±5%, connected to the digital multimeter (Keithley model 2700). The set allows us to measure the solar irradiance with an accuracy of ±30 W/m² and a resolution of 0.1 W/m².
- Multimeters of type UNI-T U171C allow the measurement of different temperatures in a range from −40°C to 400°C, with a resolution of 0.1°C–1°C, and an accuracy of ±1°C.

10.5.2 Experimental and Simulation Results

The MSD solar desalination system shown in Figure 10.1 was tested all day in February in 2019. The typical results obtained for the basin and the first three stages are shown in Figure 10.5. For the sake of clarity, the temperatures of stages 4 to 8, with maximum temperatures of 56.4, 54.4, 50.1, 44, and 37.9°C respectively, are not shown in the figure. We have also represented the simulation results taking into account the specific, thermal and geometrical characteristics of the system as presented in Table 10.1 and the meteorological conditions of the day, during the measurements. From these results we can deduce:

- The maximum intensity of the irradiance of 825 W/m² is obtained around 12:00 noon.
- The average ambient temperature during the day is between 22° C and 23°C.
- The temperatures of the stages will only increase significantly from 13:00.
- The temperatures in the basin and on the eight stages decrease with each stages and reach, respectively, the following values: 84.1, 77.1, 69, 62.7, 56.4, 54.4, 50.1, 44, and 37.9°C.
- There are variations in the temperatures of the stages that indicate heat transfer from the basin to the eighth floor. As a result, there occurs condensation of steam and the formation of drops of distilled water below the triangular plates of the eight stages.
- There is very good agreement between simulation and experiment for the basin and each stage.

The overall results show, despite the winter day, that the solar collectors heat the water circulating in the MSD system. The increase could reach the boiling point,

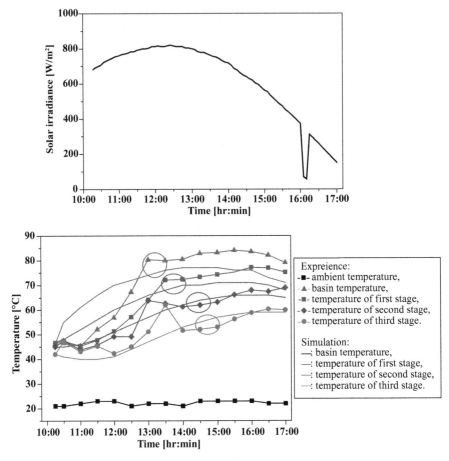

FIGURE 10.5 Experimental plot of solar irradiance, ambient and basin temperature, and temperature of the first three stages of the MSD system. On the same temperature figure, we have shown the simulation of the basin and stage temperatures by setting the parameters of Table 10.1

TABLE 10.1
System parameters and constants

Parameters	Values
Initial temperatures	Tsb = 46.2°C, Ts1 = 46.7°C, Ts2 = 45°C, Ts3 = 42°C, Ts4 = 41.6°C, Ts5 = 39.2°C, Ts6 = 34.5°C, Ts7 = 28.1°C, Ts8 = 19.8°C
Surface area of water in the ith stage	A_{ei} = 2 m²
Coefficient of thermal conductivity of the insulating material	λ_i = 0.044 W/m·°C
Heat transfer convection coefficient with ambient temperature	h_{amb} = 9 W/m²·K
Collector overall losses coefficient	U_L = 6 W/m²·K,
Optical collector efficiency	η_0 = 0.8
Insulating thickness	δ_i = 150 mm
Thermal capacity of water	c_p = 4180 J/kg·°C

which is of the order of 85°C (i.e., an increase by a factor of 4). The significant values of the coefficients set in the simulations, and the good agreement with experience, clearly show the good functioning of each stage of the MSD system, designed and realized during this project.

This work was continued by experimenting with the production of distilled water produced by this MSD system during sunny days in the spring and summer seasons, with significant sunlight favoring the heating of water throughout the day (morning and afternoon).

10.6 CONCLUSION

In this chapter, we have presented the design, the operation, the modelling, and the first experimental results of the MSD system, installed at Dour Al Hamri. The system's operation is based on heating and evaporation of saline water using four solar collectors. The results obtained, during a winter day (February 2019), show:

- Under a maximum solar irradiance of 824 W/m^2, and ambient temperature of 23°C, the maximum temperature value of the basin and the eight stages respectively reach the values of: 84.1, 77.1, 69, 62.7, 56.4, 54.4, 50.1, 44, and 37.9°C. These values are more than sufficient to evaporate salt water, condense it, and form droplets of distilled water.
- There is very good agreement between the experimental and simulated results. The coefficients obtained of global heat loss of the collector and of convective exchange with the outside are respectively of the order of 6 and 9 W/m^2·K.

The analysis of these results and their comparisons with the literature allowed us to conclude the feasibility of this type of MSD system of desalination in eight stages in terms of the heating of the stages by solar energy. This work is being continued, by experimenting with the MSD system to produce drinking water and its consumption, on very sunny days in the spring and summer seasons.

ACKNOWLEDGMENTS

This work falls within the framework of the Program of Cooperation Morocco-Allemand of Scientific Research PMARS III, Project PMARS III 2015-64 (2016–2019), in collaboration with:

- Mohamed Premier University, Laboratory LESPRE (Electromagnetism Signal Processing and Renewable Energies), Team MERE (Materials Electronics and Renewable Energies), Oujda, Morocco.
- Solar-Institut of Jülich, FH Aachen, Germany.
- Engineering Office of Energy and Environmental Technology (IBEU), Jülich, Germany.
- Association "Man and Environment of Berkane", Berkane, Morocco,

REFERENCES

Abakr, Y.A. and A.F. Ismail. "Theoretical and Experimental Investigation of a Novel Multistage Evacuated Solar Still." *Journal of Solar Energy Engineering* 127, no. 3 (July 20, 2005): 381–385. doi:10.1115/1.1866145.

Abdel-Rehim, Z.S. and A. Lasheen. "Experimental and Theoretical Study of a Solar Desalination System Located in Cairo, Egypt." *Desalination* 217, no. 1–3 (November 2007): 52–64. doi:10.1016/j.desal.2007.01.012.

Adhikari, R.S., A. Kumar, and G.D. Sootha. "Simulation Studies on a Multi-Stage Stacked Tray Solar Still." *Solar Energy* 54, no. 5 (May 1995): 317–325. doi:10.1016/0038-092x(95)00001-8.

Ali, M.T., H.E.S. Fath, and P.R. Armstrong. "A Comprehensive Techno-Economical Review of Indirect Solar Desalination." *Renewable and Sustainable Energy Reviews* 15, no. 8 (October 2011): 4187–4199. doi:10.1016/j.rser.2011.05.012.

Al-Kharabsheh, S. and D.Y Goswami. "Analysis of an Innovative Water Desalination System Using Low-Grade Solar Heat." *Desalination* 156, no. 1–3 (August 2003): 323–332. doi:10.1016/s0011-9164(03)00363-1.

Aly, N.H. and A.K. El-Figi. "Mechanical Vapor Compression Desalination Systems—A Case Study." *Desalination* 158, no. 1–3 (August 2003): 143–150. doi:10.1016/s0011-9164(03)00444-2.

Bourouni, K., M.T. Chaibi, and L. Tadrist. "Water Desalination by Humidification and Dehumidification of Air: State of the Art." *Desalination* 137, no. 1–3 (May 2001): 167–176. doi:10.1016/s0011-9164(01)00215-6.

Chaibi, M.T. "An Overview of Solar Desalination for Domestic and Agriculture Water Needs in Remote Arid Areas." *Desalination* 127, no. 2 (February 2000): 119–133. doi:10.1016/s0011-9164(99)00197-6.

Chen, Z., J. Peng, G. Chen, L. Hou, T. Yu, Y. Yao, and H. Zheng. "Analysis of Heat and Mass Transferring Mechanism of Multi-Stage Stacked-Tray Solar Seawater Desalination Still and Experimental Research on Its Performance." *Solar Energy* 142 (January 2017): 278–287. doi:10.1016/j.solener.2016.12.028.

El Moussaoui, N. and K. Kassmi. "Modeling and Simulation Studies on a Multi-Stage Solar Water Desalination System." In *2019 International Conference of Computer Science and Renewable Energies (ICCSRE), Agadir, Morocco* (July 2019). doi:10.1109/iccsre.2019.8807623.

Eleiwi, F. and T.M. Laleg-Kirati. "Observer-Based Perturbation Extremum Seeking Control with Input Constraints for Direct-Contact Membrane Distillation Process." *International Journal of Control* 91, no. 6 (May 22, 2017): 1363–1375. doi:10.1080/00207179.2017. 1314551.

Feilizadeh, M., M.R. Karimi Estahbanati, K. Jafarpur, R. Roostaazad, M. Feilizadeh, and H. Taghvaei. "Year-Round Outdoor Experiments on a Multi-Stage Active Solar Still with Different Numbers of Solar Collectors." *Applied Energy* 152 (August 2015): 39–46. doi:10.1016/j.apenergy.2015.04.084.

García-Rodríguez, L., A.I. Palmero-Marrero, and C. Gómez-Camacho. "Comparison of Solar Thermal Technologies for Applications in Seawater Desalination." *Desalination* 142, no. 2 (February 2002): 135–142. doi:10.1016/s0011-9164(01)00432-5.

Hosseini, S.R., M. Amidpour, and Ali Behbahaninia. "Thermoeconomic Analysis with Reliability Consideration of a Combined Power and Multi Stage Flash Desalination Plant." *Desalination* 278, no. 1–3 (September 2011): 424–433. doi:10.1016/j.desal.2011.05.061.

Kalidasa Murugavel, K. and K. Srithar. "Performance Study on Basin Type Double Slope Solar Still with Different Wick Materials and Minimum Mass of Water." *Renewable Energy* 36, no. 2 (February 2011): 612–620. doi:10.1016/j.renene.2010.08.009.

Kalogirou, S. "Seawater Desalination Using Renewable Energy Sources." *Progress in Energy and Combustion Science* 31, no. 3 (2005): 242–281. doi:10.1016/j.pecs.2005.03.001.

Kaushal, A. and Varun. "Solar Stills: A Review." *Renewable and Sustainable Energy Reviews* 14, no. 1 (January 2010): 446–453. doi:10.1016/j.rser.2009.05.011.

Lawal, D.U., M.A. Antar, A. Khalifa, and S.M. Zubair. "Heat Pump Operated Humidification-Dehumidification Desalination System with Option of Energy Recovery." *Separation Science and Technology* (January 3, 2020) 55: 1–20. doi:10.1080/01496395.2019.170 6576.

Pouyfaucon, A.B. and L. García-Rodríguez. "Solar Thermal-Powered Desalination: A Viable Solution for a Potential Market." *Desalination* 435 (June 2018): 60–69. doi:10.1016/j.desal.2017.12.025.

Pugsley, A., A. Zacharopoulos, J.D. Mondol, and M. Smyth. "Solar Desalination Potential around the World." In *Renewable Energy Powered Desalination Handbook* (2018): 47–90. doi:10.1016/b978-0-12-815244-7.00002-7.

Qiblawey, H.M. and F. Banat. "Solar Thermal Desalination Technologies." *Desalination* 220, no. 1–3 (March 2008): 633–644. doi:10.1016/j.desal.2007.01.059.

Reddy, K.S., K. Ravi Kumar, T.S. O'Donovan, and T.K. Mallick. "Performance Analysis of an Evacuated Multi-Stage Solar Water Desalination System." *Desalination* 288 (March 2012): 80–92. doi:10.1016/j.desal.2011.12.016.

Sayyaadi, H. and A. Saffari. "Thermoeconomic Optimization of Multi Effect Distillation Desalination Systems." *Applied Energy* 87, no. 4 (April 2010): 1122–1133. doi:10.1016/j.apenergy.2009.05.023.

Schwarzer, K., M. Eugênia Vieira da Silva, and T. Schwarzer. "Field Results in Namibia and Brazil of the New Solar Desalination System for Decentralised Drinking Water Production." *Desalination and Water Treatment* 31, no. 1–3 (July 2011): 379–386. doi:10.5004/dwt.2011.2339.

Schwarzer, T. and H.J. Bart. "Mass and Heat Transfer at Different Heat Exchange Surfaces and Their Suitability for Use in Thermal Desalination Plants." *The Open Chemical Engineering Journal* 10, no. 1 (June 3, 2016): 74–86. doi:10.2174/187412310161001 0074.

Shannon, M.A., P.W. Bohn, M. Elimelech, J.G. Georgiadis, B.J. Mariñas, and A.M. Mayes. "Science and Technology for Water Purification in the Coming Decades." *Nature* 452, no. 7185 (March 2008): 301–310. doi:10.1038/nature06599.

Shatat, M., M. Worall, and S. Riffat. "Opportunities for Solar Water Desalination Worldwide: Review." *Sustainable Cities and Society* 9 (December 2013): 67–80. doi:10.1016/j.scs.2013.03.004.

Talbi, S., K. Kassmi, O. Deblecker, and N. Bachiri. "Thermal Heating by Photovoltaic Solar Energy." *Materials Today: Proceedings* 13 (2019): 1125–1133. doi:10.1016/j.matpr.2019.04.080.

Tigrine, Z., H. Aburideh, M. Abbas, D. Zioui, R. Bellatreche, N. Kasbadji Merzouk, S. Hout, and D. Belhout. "Experimental Investigations on a Multi-Stage Water Desalination Prototype." *Desalination and Water Treatment* 56, no. 10 (December 6, 2014): 2612–2617. doi:10.1080/19443994.2014.972735.

Vinoth Kumar, K., and R. Kasturi Bai. "Performance Study on Solar Still with Enhanced Condensation." *Desalination* 230, no. 1–3 (September 2008): 51–61. doi:10.1016/j.desal.2007.11.015.

Walschot, M., P. Luis, and M. Liégeois. "The Challenges of Reverse Osmosis Desalination: Solutions in Jordan." *Water International* 45, no. 2 (February 17, 2020): 112–124. doi:1 0.1080/02508060.2020.1721191.

Yadav, S. and K. Sudhakar. "Different Domestic Designs of Solar Stills: A Review." *Renewable and Sustainable Energy Reviews* 47 (July 2015): 718–731. doi:10.1016/j.rser.2015.03.064.

11 Design and Realization of a Pilot Solar Desalination Plant in Douar El Hamri in the Province of Berkane (Morocco)

I. Atmane, K. Hirech, K. Kassmi, Z. Mahdi, S. Alexopoulos, K. Schwarzer, H. Chayeb, and N. Bachiri

CONTENTS

11.1 INTRODUCTION

Global warming is one of the major problems of climate change, causing the melting of ice, rising water levels, and therefore the degradation of groundwater by excess salt. This increases the number of people suffering from a shortage of drinking water (Nations Unies 2017). This problem has prompted national and international organizations to seek solutions to remedy the problem and provide the population with potable water. The most attractive solution is to find a way to desalinate salt water

(Energy 2012). To do this, we find in the literature several solar desalination systems:

- *Reverse Osmosis System*: Desalination of seawater using the reverse osmosis process currently holds the largest share of global desalination capacity with 60% (Ghaffour et al. 2015). This system has a membrane process. The water is treated by a filter, which retains unwanted accompanying substances in the raw water. We use a membrane permeable to water, but impermeable to dissolved salts and other foreign substances (Sano and Mahidul 2018). In order to generate a volume flow through the raw water membrane toward the pure water side, the application of high pressure, generally from 50 to 200 bar, is necessary on the raw water side (Khalifa et al. 2017). Such pressure requires a large machine, which makes this system too expensive when including maintenance costs for changing membranes and other equipment.
- *MSF system* (*Multi-stage Flash*): In this system, the water to be desalinated is brought under pressure to a temperature of 110°C (maximum temperature). Then, it descends from cell to cell, undergoing a series of detents (each of the cells or stages working at an increasingly low pressure), releasing the energy stored in the form of fogging. This instantaneous vaporization by expansion is called a flash; 28.5% of the desalination systems used worldwide use this technique (Ghaffour et al. 2015). Due to the high-energy requirements, the operation of an MSF system is often only economically viable by coupling it with power plants or other industries with high heat intensity (El-Ghonemy 2018). This makes this type of installation very expensive.
- *MED system* (*Multi-effect distiller*): The distiller is made up of several evaporator modules in series called "effects". The vapor resulting from the first effect condenses at the level of the second effect and the latent heat by condensation is used to evaporate the saline water therein. Then the third effect acts as a condenser for the vapors from the second effect, and so on, the vapor of the last effect being used to heat the feed water of the first (Stärk et al. 2015). This system is characterized by almost zero heat loss but it is less used given its low production of fresh water and its costly installation.

In this context, and within the framework of the Moroccan–German Scientific Research Program "PMARSIII Project 2015–64", we propose the establishment of a multi-stage solar desalination system (MSD) to supply drinking water to Douar Al Hamri, in the rural commune of Boughriba in the province of Berkane (Oujda region), Morocco. This Douar and all regions suffer from the problem of the salinity of the water tables, and therefore from the lack of drinking water. This autonomous system, powered by solar energy (thermal and photovoltaic (PV)), produces 220 L/day and requires little maintenance. In addition, this low-cost system meets the needs of all countries with significant solar deposits (Africa, the Persian Gulf, etc.), in terms of shortages of drinking water. To make this MSD system reliable, the establishment of a system allowing remote monitoring and supervision is essential. In the literature, there are several expensive remote control and supervision systems, but they are for domestic use or for specific applications. The main systems are:

- X. Xiaoli and others (Xiaoli and Daoe 2011) have proposed remote monitoring and control of PV systems based on Zigbee technology, but despite the effectiveness of this method, it cannot cope with huge distances.
- Ahmed and Kim (2013) presented remote monitoring and control of the PV system based on Wi-Fi technology for household applications. This solution requires a more advantageous speed (54 Mbps) than Zigbee (250 kbps), but it is only valid in local network architectures.
- The systems proposed in Ilias et al. (2016), Do et al. (2015), and Mehdi et al. (2016) make it possible to acquire the quantities (of current, voltage, lighting, power, etc.) remotely using the GSM module. These methods are very effective but they become expensive, especially if the supervision is carried out on an international scale.
- The control and supervision system proposed in Feilizadeh et al. (2015) allows the acquisition of the temperatures of each stage of the MSD solar desalination system. This system is useful for local acquisitions, but does not allow supervision or remote control.

We propose the design, production, and operation of a control system for the MSD multi-stage desalination system operated via the Internet. This, on the one hand, guarantees the reliability of the MSD system; on the other hand, promotes employment by creating startups for young people throughout the country and continent.

In this chapter, after having described the structure of the MSD system, we present the structure and typical functioning of a system of remote acquisition, control, and supervision – designed and realized during this work.

11.2 GLOBAL ARCHITECTURE

The proposed MSD desalination system (Figure 11.1) is based on the use of renewable solar energy (thermal and photovoltaic) and the electricity network. Solar thermal energy is used to heat water, which has a high salt content, during the day (Schwarzer et al. 2011). The role of photovoltaic energy is to produce electrical energy to activate the various pumps (borehole water, etc.), light the station, and heat the water in the desalination system. The electrical network will be used overnight to heat the water in the system if necessary.

As shown in Figure 11.1, this MSD desalination system consists of:

- *Bloc A*: Represents the multi-stage desalination system (nine stages) allowing the production, by evaporation, of 220 L of distilled water per day. It was designed, during this work, at the Solar Institute in Julich (Germany).
- *Bloc B*: This block concerns the energy sources used for a hybrid and decentralized fall use, and centralized when needed. The sources used are :
  - Thermal energy produced by four solar panels of 1.82 m^2 to heat the water of the nine floors,
  - Photovoltaic energy, produced by 19 photovoltaic panels of 2.2 kW, to pump 5 m^3 of water per day, from a well 20 m deep. In addition, this energy

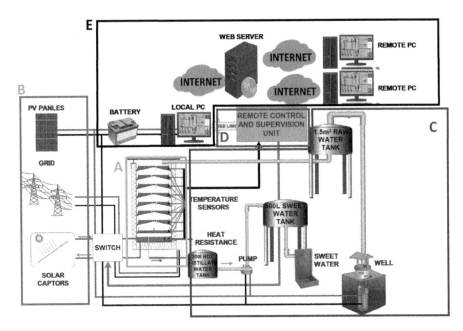

FIGURE 11.1   Diagram of the global architecture of the solar desalination plant equipped with the remote control and supervision system

provides the electrical energy necessary to light the station, to heat the water of the nine stages of the desalination system, and to store the surplus in 540 Ah batteries.

- Power from the electricity network to heat the water in the nine stages of the desalination system during the night or during the day when there is a high-energy demand.
- *Bloc C*: Represents three basins for water storage (salt water pumped from the well, hot distilled water produced directly by the desalination system, cooled distilled water).
- *Bloc D*: This is formed by an electronic card to control the functioning of the photovoltaic system and acquire the electrical and thermal quantities of the system (voltage, current, power, water flow, lighting, and external temperatures), the quantity of salted, distilled water, and sweet, floor temperatures, water salinity rate, etc.).
- *Bloc E*: Represents the remote supervision system for monitoring the operation of all the blocks of the desalination system via the Internet. This system consists of a local PC, at the level of the desalination station, to configure and locally control the entire system and communicate with a Web server installed in our laboratory at the Faculty of Sciences of Oujda. This server stores the station's data in a database accessible via the Internet. In addition, this server allows you to control, order, and acquire all data remotely via a login and password.

11.3 OPERATION OF THE ACQUISITION AND SUPERVISION SYSTEM

11.3.1 STRUCTURE AND FUNCTIONING

Figure 11.2 shows the electrical diagram of the block allowing the acquisition of electrical quantities, and the control and supervision of the solar desalination system (MSD) according to the algorithm of Figure 11.3. As this figure shows, the microcontroller (PIC 18F4550) manages these different tasks:

- Control of power switches, which activate the solar collectors or connect the heating resistance installed in the desalination system to the PV kit or to the electrical network;
- Acquisition of the electrical (current, voltage, and power) and thermal (temperature of each stage) and meteorological quantities (illumination and ambient temperature);
- Acquisition of the salinity levels of the basins, the water of the basins, and the well.

When the desalination station is switched on, the microcontroller acquires all these quantities, and then executes, depending on the choice of use of the station, three scenarios:

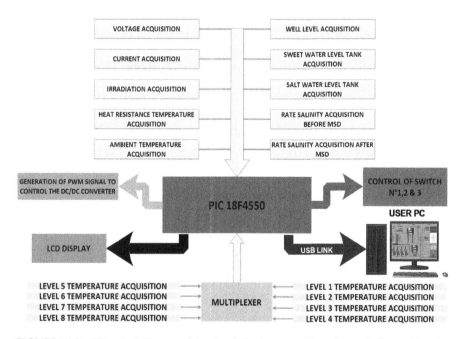

FIGURE 11.2 Electrical diagram of the circuit for the acquisition of electrical quantities, for control and supervision of the solar desalination system

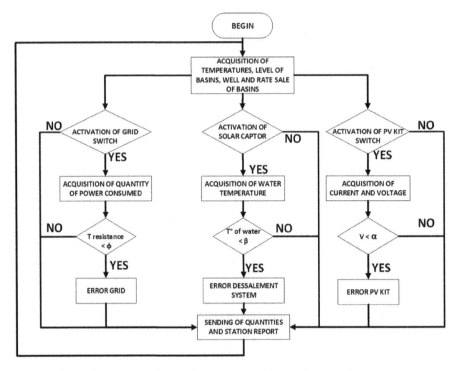

FIGURE 11.3   Overall operation of the remote acquisition, control, and supervision system

- *Activation of switches for solar collectors*: In a very sunny and clear climate, the user activates the solar collectors and the PIC begins the acquisition of the water temperature of the still.
- *Activation of the PV system switches*: The user activates the PV system to supply the heating resistor and the PIC begins the acquisition of electrical quantities relating to the photovoltaic installation (current, voltage, and power).
- *Activation of the electrical network*: If necessary (at night, failure of the solar collectors and PV kit), the user activates the electrical network and the PIC acquires all the electrical and thermal quantities of the network and, as before, of the solar desalination system.

11.3.2   LOCAL AND REMOTE CONTROL AND SUPERVISION OF THE DESALINATION PLANT

To be able to control and monitor the operation of the station, we have created a graphical interface developed in LabView. This interface visualizes all the quantities acquired by the PIC, stores them in a MySQL database, and sends them to the Web server installed in our laboratory (Figure 11.4). This functionality allows control and management of the station by a local computer (Local control & supervision), a laboratory computer (Laboratory Web server: Remote Control & Supervision), and the

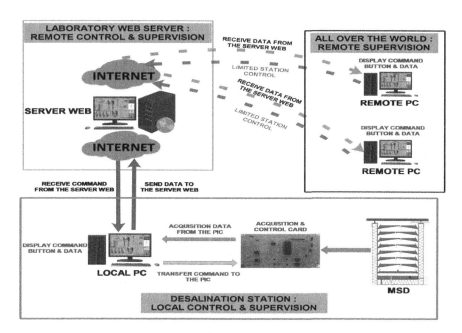

FIGURE 11.4 Synoptic diagram of local and remote control and supervision of the desalination station system

computer of an external user (All Over the Word: Remote Supervision), using this application (Figure 11.4). The interface therefore allows three control modes:

- *Local control & supervision*: This control mode is operated from the interface installed in the desalination station computer according to the algorithm in Figure 11.5. When the interface is launched, it begins to acquire all the data sent by the acquisition and control card and interprets each quantity (current, voltage, temperature, lighting, water level, etc.) in the form of a table and curve, to make it easier for us to read and follow the station. Then, to control the station (by activating the PV panels, the solar collectors, or the electrical network), the interface sends the activation command to the electronic card which automatically activates the corresponding relay, and sends in return the new data acquired as well as the status of the station at the interface. Finally, the interface creates a report containing all the data acquired as well as the status of the station and the transmission to the laboratory Web server.
- *Laboratory Web Server Remote Control & Supervision*: Full remote control of the desalination station is only available from the interface installed in the Web server, located at the laboratory level. This very powerful server is chosen to guarantee the speed of data transfer and processing to all the other remote computers that connect to it. It also has a very powerful protection system against any external intrusion. The goal is to guarantee a smooth flow of data between computers while avoiding any poor control of the station which could cause

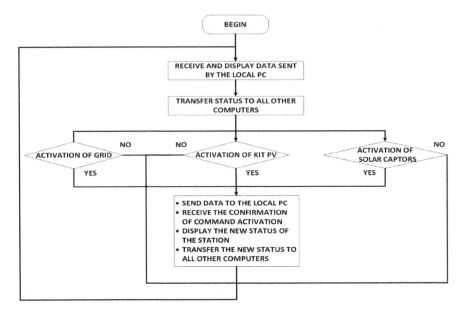

FIGURE 11.5   Operating algorithm of server supervision and remote control of the desalination station

much energy loss and even see material damage. As shown in the algorithm of Figure 11.5, once the interface is launched, it gives us the instantaneous state of the station (on/off, failure) before the Web server transmits it to all the computers connected to it. When we decide to activate the energy sources (PV panels, solar collectors, electrical network), we press the appropriate button and the interface sends our command to the interface of the local PC. The latter will follow the steps mentioned above for local control and await confirmation of its execution. Once confirmation is received, the interface displays the new status of the desalination station and the Web server transmits the update to the other supervisors connected to it. From the computer and interface associated with the Web server, we can follow the instantaneous operation of the station (states of the switches, power of the energy sources, electrical and thermal quantities, water level of the well, quantity of water produced, etc.) and view all the computers connected to it. Following possible malfunctions or improper use, the laboratory computer could intervene and act by performing the appropriate actions, such as shutting down the station or disconnecting users' PCs.

- *Remote control of the station by an external user (All Over the World)*: In order for the control to be carried out only by us (the managers of the station), we have made this mode of operation exclusive and limited to the interface installed in the laboratory's Web server. This will prevent, on the one hand, random control of the station, and on the other hand will guarantee control of the station only by our team (responsible for the station). As shown in Figures 11.6 and 11.7, the launch of the interface is automatically followed by the download of

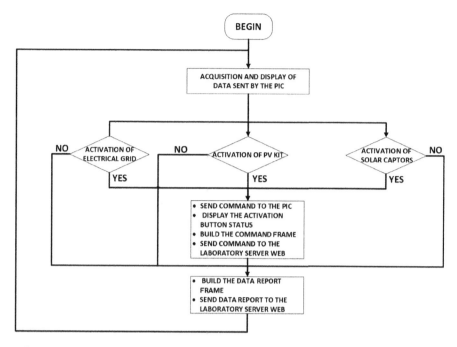

FIGURE 11.6 Operating algorithm of local control of the desalination station

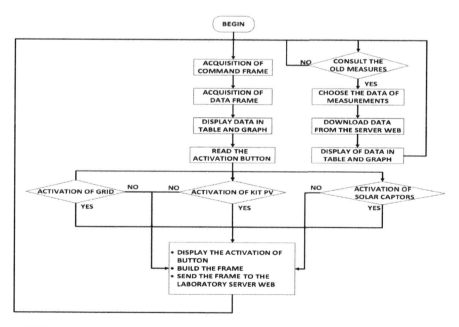

FIGURE 11.7 Operating algorithm of remote acquisition and control of the desalination station

the station state from the SQL database in order to give us an idea of the station state (relays active/deactivated, power on/off of the station, fault detected, etc.). We have also tolerated certain functionalities such as the activation of photovoltaic panels, solar collectors, or the electrical network. For this, the remote user must contact us in order to provide him with the login and password that will allow him to access these features, for a limited time and controlled by us. To do this, the user presses the appropriate relay on the interface, which sends the activation command to the local PC, transmits the information to the electronic card, and executes the command. For supervision, and unlike control, we have given free access to any user with the *guest* version of our interface. Supervision can therefore be done locally from the desalination station, or remotely from the Web server, or from any remote computer connected to the latter. As shown in Figure 11.4, once the interface is launched, it connects directly to the Web server and downloads the status and the latest data from the desalination station. Then it displays the electrical and thermal quantities in table and graph form as well as the state of the desalination system (start, stop, or fail).

In addition to instant supervision, we have also allowed free access to the old measurements of the desalination plant. To do this, the user must enter the desired measurement date and the interface downloads all the data corresponding to this date to display it in the form of curves and tables.

11.3.3   Data Sharing System via the Internet

To be able to connect the laboratory PC to that of the PV station, we have set up an Apache Web server with a fixed IP address. This server is responsible for sending and receiving command and data requests to the various PCs connected to it using the HTTP protocol (Hyper Text Transfer Protocol). This is a client–server protocol whose station plays the role of client and the machine located in the laboratory plays the role of server. The station sends an HTTP request containing the measurements to the laboratory at a regular interval and receives an HTTP response containing the acknowledgment (acknowledgment of receipt) and also the commands if they exist. These are retrieved to configure the station (relay to activate/deactivate). HTTP requests use Web pages created in PHP language and installed on the Web server. These pages have the role of reading/inserting data from/into a MySQL database. Data recording and access to it are done in two different scenarios:

- *Data recording*: When the desalination system interface is launched, it starts by sending an HTTP request containing the address of the Web page to be used and the data frame that we want to save. At the server level, it loads the corresponding Web page (Figure 11.8B) which accesses the MySQL database (Figure 11.8A) to store the data frame sent.
- *Access to data*: In order to access the data, the interface connected to the server begins by sending a request containing the information on the data sought. The Web server then calls on the Web page, responsible for finding data, which will then access the database to load the desired data. Then the Web server sends an

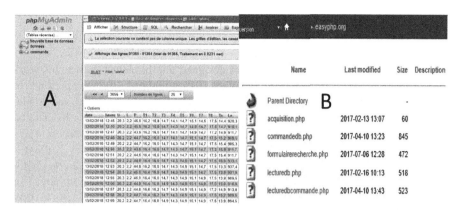

FIGURE 11.8 (A) Web server database, (B) PHP form designed for sending and acquiring data

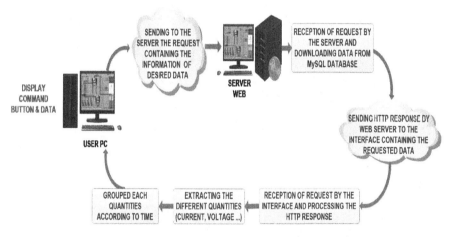

FIGURE 11.9 Example of Internet communication between the user interface and Web server

HTTP response which will contain the document requested by the remote interface. The latter will interpret them in the form of a curve and table. Figure 11.9 shows an example of communication between the Web server and the different interfaces connected to it. When the user activates the remote interface, the latter must retrieve the data in order to show the user the status of the desalination station. To do this, it sends an HTTP request to the Web server containing the Web page responsible for retrieving data from the database as well as the information concerning the desired data. At the server level, it loads the Web page that will access the database and loads the data sought. The Web server then sends an HTTP response in the form of a request containing the desired data. The user interface will extract the data, cut out the frame, and interpret the quantities in the form of a graph and table.

11.4   RESULTS AND DISCUSSIONS

11.4.1   EXPERIMENTAL PROCEDURE

The prototype of the desalination system designed and produced during this work was installed at Douar Al Hamri in December 2019 (Figure 11.10) in the presence of all the project partners. The acquisition and control system designed and implemented during this work was first tested in the laboratory (Figure 11.11). Its installation at Douar is in progress. The different blocks of this (Figure 11.11) are:

- *Bloc A*: Represents a PV generator, formed by monocrystalline silicon photovoltaic panels delivering under the standard test condition (CST) a power of 500 W, to produce the electrical energy necessary for the station. Also, a 10 W

FIGURE 11.10   Solar desalination system installed at Douar Al Hamri

FIGURE 11.11   Remote control and supervision system for the solar desalination system designed and produced in the laboratory

photovoltaic panel, analog amplifiers, and a thermistor with a negative temperature coefficient are installed to determine the intensity of the illumination and the ambient temperature.
- *Bloc B* which is formed by:
  - DC/DC boost converter (operating at 10 kHz switching frequency) whose role is to ensure the adaptation between the PV generator and the load.
  - Data acquisition and power relay control card, which consists of a PIC 18F4550 and a USB link module.
  - An electronic card comprising the control circuits for the power switches, voltage acquisition, current, lighting, ambient temperature, and temperature of the stages of the desalination system. The floors are represented in our test by water basins.
  - Three power switches to activate/deactivate respectively the load, the PV kit, and the electrical network.
- *Bloc C*: Trained by a PC to locally control and supervise our desalination system. This PC is connected to the Internet via a 3G modem to allow the sending and receiving of frames (Tuttlebee 2006).
- *Bloc D*: Represents the Web server installed at the laboratory level to remotely share the data in our acquisition, control, and supervision system.

11.4.2 Typical Operation of a Control and Supervision System

The typical operation of the interface and the remote acquisition, control, and supervision system (Sections 11.2 and 11.3) is as follows:

- When launching the remote monitoring and control interface, the user authenticates by entering the correct login and password (Figure 11.12). Then the main window (Figure 11.13) is displayed. This module displays a real-time animation of the operation of the PV system, and the different electrical and thermal quantities of each block (PV panels, solar collectors, electrical network, temperatures of the MSD system stages, etc.), including the signal cyclic ratio. The pulse width modulation controls the power switches of the DC/DC converters, and the efficiency as well as the level of the pools. It also allows:
  - Using two buttons, to start and stop the software:
    - "Start" button: Starts the acquisition of electrical quantities, the real-time animation of the PV system as well as the recording of data;
    - "Stop" button: Enables us to stop the acquisition and thus the recording of data.
  - Controlling the energy flow or isolating the PV system by closing and opening the various switches, or displaying error messages in the event of a breakdown or malfunction in the system.
  - Accessing other windows by the appropriate buttons on the interface (Measurements and History).
- The measurement window (Figures 11.14 and 11.15): This window displays the evolution of all the electrical and thermal quantities of the desalination station in real time graphically. The numerical values are displayed on the left,

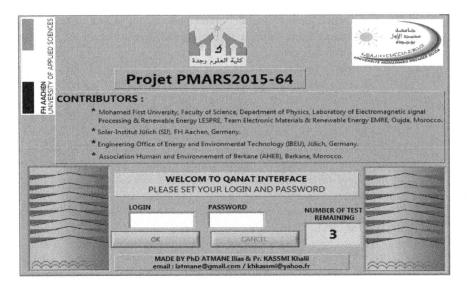

FIGURE 11.12   Startup and authentication window

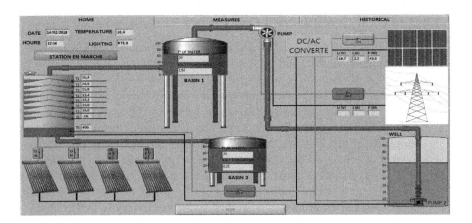

FIGURE 11.13   Home tab of the solar desalination station interface

and the real-time graphical evolution of the voltages, currents, powers, energy efficiency, duty cycles, the different temperatures, and levels of the basins are presented on the right.

• History: The storage and consultation of the operating data of the MSD desalination station are very important functions. Thanks to the data acquired, we can study the reliability and performance of the desalination station designed to improve its performance. This allows us on the one hand to reduce significantly the cost of a malfunction and on the other hand to maximize the production of fresh water.

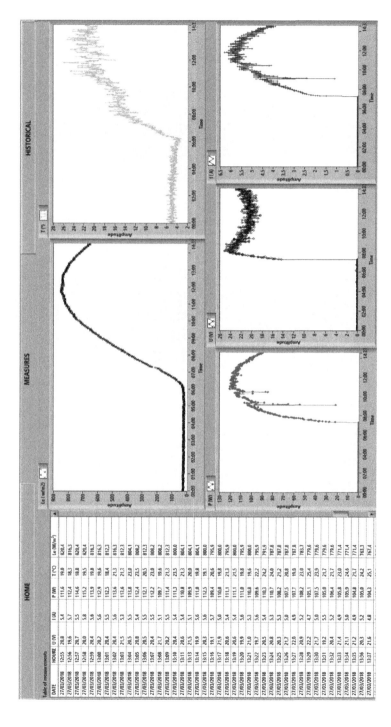

FIGURE 11.14 Display interface for instantaneous measurements of quantities with their plots

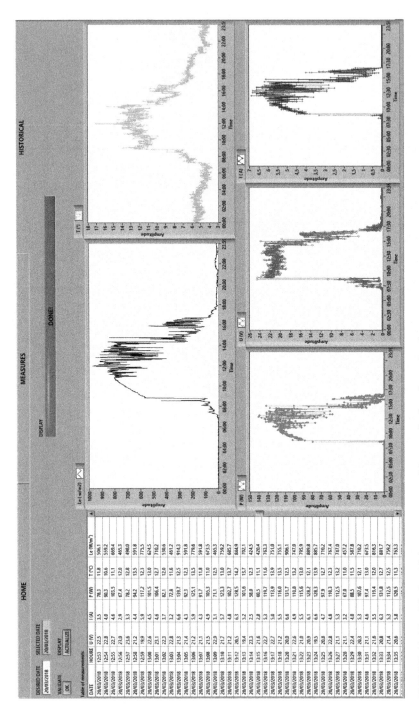

FIGURE 11.15   Display of the measurement history of a day chosen by the user

Comparing with today's multistage desalination systems, which do not have all these functionalities, we can clearly conclude, in terms of operation, that they are neither efficient nor reliable. Consequently, the remote acquisition and control card designed and produced during this work is a contribution to the field of drinking water production by multi-stage desalination systems. It allows monitoring in real time, remotely, its operation, and ensuring the production of drinking water to the population of Douar Al Hamri.

11.5 CONCLUSION

In this chapter, we have presented the MSD solar desalination system equipped with the system of remote acquisition, control, and supervision via the Internet. This system is installed at Douar Al Hamri in the province of Berkane (Morocco), as part of a Moroccan–German collaboration. Particular attention is attached to the system, remote acquisition, and control, which affords reliability and profitability to the MSD system in terms of operation and production of drinking water to the inhabitants of Douar.

After describing the structure and functioning of the MSD system and that of acquisition, control, and supervision, we presented the different algorithms and techniques used as well as the typical operating results obtained. All the results obtained show:

- The proper functioning of the digital and analog circuits of the system;
- The proper functioning of the interface set up in terms of acknowledgments, display, and local control;
- Good Internet communication between the local installation, the Web server, and the user;
- Good mastery of all the tasks carried out by the system so designed and produced;
- The user could follow the operation of the station remotely via the Internet, intervene as needed, download and store the results instantly or store on the Web server.

All of the results obtained and presented in this chapter show the proper functioning of the Internet acquisition, control, and supervision system, designed and implemented during this work. This allows instant monitoring of the functioning of the MSD system, guaranteeing the production of drinking water for the inhabitants, and consequently, a know-how in the field of renewable energy applications which opens up new horizons for young people by creating startups.

ACKNOWLEDGMENTS

This research is supported by Project Qanat n° PMARSIII 2015–64 (2016–2019), Moroccan–German Scientific Research Cooperation Program PMARSIII, Water quality improvement with a solar desalination, in collaboration with:

- Mohamed Premier University, Laboratory LESPRE (Electromagnetism Signal Processing and Renewable Energies), Team MERE (Materials Electronics and Renewable Energies), Oujda, Morocco.
- Solar-Institut of Jülich, FH Aachen, Germany.
- Engineering Office of Energy and Environmental Technology (IBEU), Jülich, Germany.
- Man and Environment Association of Berkane, Berkane, Morocco.

REFERENCES

Ahmed, M. A., & Kim, Y. C. (2013, January). Communication networks of domestic small-scale renewable energy systems. In *2013 4th International Conference on Intelligent Systems, Modelling and Simulation* (pp. 513–518), Bangkok. IEEE.

Do, H. N., Vo, M. T., Tran, V. S., Tan, P. V., & Trinh, C. V. (2015, October). An early flood detection system using mobile networks. In *2015 International Conference on Advanced Technologies for Communications (ATC)* (pp. 599–603), Ho Chi Minh City, Vietnam. IEEE.

El-Ghonemy, A. M. K. (2018). Performance test of a sea water multi-stage flash distillation plant: Case study. *Alexandria Engineering Journal*, 57(4), 2401–2413.

Feilizadeh, M., Estahbanati, M. K., Jafarpur, K., Roostaazad, R., Feilizadeh, M., & Taghvaei, H. (2015). Year-round outdoor experiments on a multi-stage active solar still with different numbers of solar collectors. *Applied Energy*, 152, 39–46.

Ghaffour, N., Bundschuh, J., Mahmoudi, H., & Goosen, M. F. (2015). Renewable energy-driven desalination technologies: A comprehensive review on challenges and potential applications of integrated systems. *Desalination*, 356, 94–114.

Ilias, A., Mustapha, M., Khalil, K., & Kamal, H. (2016, October). Remote control and monitoring of photovoltaic installations equipped with MPPT control. In *2016 International Conference on Electrical Sciences and Technologies in Maghreb (CISTEM)* (pp. 1–6), Marrakech. IEEE.

Khalifa, A., Ahmad, H., Antar, M., Laoui, T., & Khayet, M. (2017). Experimental and theoretical investigations on water desalination using direct contact membrane distillation. *Desalination*, 404, 22–34.

Mehdi, D., Ali, H., & Lassaad, S. (2016, December). Remote control and supervision by the GSM network of a photovoltaic pumping station. In *2016 17th International Conference on Sciences and Techniques of Automatic Control and Computer Engineering (STA)* (pp. 540–545). IEEE.

Mirei, I. (2013). *Water desalination using renewable energy* (Policy brief I12). Energy Technology System Analysis Programme.

Nations Unies. 2017. https://news.un.org/fr/story/2017/07/360702-plus-de-2-milliards-de-personnes-nont-pas-deau-potable-domicile-selon-loms-et (accessed May 1, 2020).

Sano, Y., & Mahidul, I. (2018). Optimum operating condition of a hollow fiber reverse osmosis desalination system. *Cogent Engineering*, 5(1), 1463898.

Schwarzer, K., da Silva, M. E. V., & Schwarzer, T. (2011). Field results in Namibia and Brazil of the new solar desalination system for decentralised drinking water production. *Desalination and Water Treatment*, 31(1–3), 379–386.

Stärk, A., Loisel, K., Odiot, K. et al. (2015). Wetting behaviour of different tube materials and its influence on scale formation in multiple-effect distillers. *Desalination and Water Treatment*, 55(9), 2502–2514.

Tuttlebee, W. H. (Ed.). (2006). *Software Defined Radio: Baseband Technologies for 3G Handsets and Basestations*. John Wiley & Sons.

Xiaoli, X., & Daoe, Q. (2011, April). Remote monitoring and control of photovoltaic system using wireless sensor network. In *2011 International Conference on Electric Information and Control Engineering* (pp. 633–638), Wuhan, China. IEEE.

12 Cooking with a Parabolic Trough Solar Thermal Cooker in a Moroccan Climate

N. El Moussaoui, S. Talbi, K. Kassmi, K. Schwarzer,
H. Chayeb, and N. Bachiri

CONTENTS

12.1 INTRODUCTION

In Africa, wood is the main source of energy used by households (Arenas 2007; Toonen 2009; Kumar et al. 2018). It represents up to 90% of the energy used in certain developing countries for cooking (Schwarzer and da Silva 2003), following shortages or the excessive price of other fuels. However, the use of wood creates serious environmental and ecological problems such as deforestation and the emission of greenhouse gases into the atmosphere (Kumar et al. 2018; Muthusivagami et al. 2010; Schwarzer and da Silva 2003; Yettou et al. 2014).

Currently, solar cooking is considered to be a clean cooking technology, using only the sun to produce the necessary temperature for cooking food (above 100°C). Therefore, this technology is considered as an interesting alternative to solve the problem of the effects caused by the use of conventional sources (e.g., gas and oil) (Otte 2013; Schwarzer and da Silva 2008; Sekhon and Sethi 2018), and to meet the

229

energy needs for cooking in homes. Governments of developing countries are encouraging and strengthening this new cooking method to remedy the problems of deforestation (Toonen 2009), minimize CO_2 emissions into the atmosphere (Hager and Morawicki 2013; Muthusivagami et al. 2010; Sekhon and Sethi 2018; Yettou et al. 2014), and adapt to changing climatic conditions (Arenas 2007).

In the literature, the main solar thermal cookers made and used are:

- *Cookers supplied by a vacuum tube collector* (Esen 2004). This cooking system is mainly composed of three main components: the collector, the heat pipes, and the oven section. The collector is made up of six double wall (concentric) evacuated glass tubes, mounted on parabolic reflectors with a chromium-nickel concentration. Freon 22, Freon 134a, and Freon 407C have been used as working fluids in heat pipes. The cooking time on this type of cooker, under lighting of 1000 W/m^2, is between 27 and 70 min, and the maximum temperature reached is around 175°C (Esen 2004). The disadvantage of this type of cooker is that it is expensive and of rather complex structure.
- *Parabolic trough concentrator (PTC) cookers* (Asmelash et al. 2014; El Moussaoui et al. 2020; Dianda et al. 2019). This cooker has been constructed in a way to concentrate the radiation in a relatively small absorption area to reach a high temperature. A heat transfer fluid is then used to collect and transport the thermal energy to the cooking area. Tests carried out show a maximum cooking stove temperature of 119°C. The efficiency of the system was found to be 6%. The maximum temperature of 126°C was reached in the kettle at about 14:30 under a radiation of 920 W/m^2 (Asmelash et al. 2014). The disadvantage of this type of cooker is that it is complex, fragile, and the pot temperature does not exceed 126°C.
- *Cookers powered by a flat plate thermal collector* (Da Silva et al. 2002, 2005; Schwarzer and da Silva 2003). This system consists of flat-plate collectors, a cooking unit, control valves, and a heat storage tank. In its operation, the working medium is heated in the solar collectors and moves in a thermal siphon circuit. Control valves direct the fluid flow to the cooking unit or to the storage tank for further use. The work presented in Da Silva et al. (2002) and Schwarzer and da Silva (2003) on a solar cooker with a 4 m^2 collector surface, shows a significant efficiency ranging from 34 to 38% and a working fluid temperature that could reach 200°C.
- *Cookers powered by a parabolic concentrator* (Arenas 2007; Craig et al. 2017; Da Silva et al. 2005). This uses a parabolic reflector to focus the sun's rays on a receiver (pot) which is placed at its focal point. Experimental measurements carried out (Da Silva et al. 2005) on a parabolic cooker with a surface area of 1.26 m^2 show a significant efficiency of 24%. In addition, several water cooking tests have been carried out. Using four series of water boiling experiments with a mass of 1.5 kg each, the results show an efficiency of 39% (Craig et al. 2017). The disadvantage of this type of cooker lies in the location of the pot, in the center of the concentrator, which may cause burn problems to the hands and other organs (e.g., the eyes), which are in the field of rays concentrated by the reflectors.

As previously mentioned, the proposed thermal cookers have very limited performance, the cooking temperature and efficiency do not exceed 200°C and 39%. In addition, there is a total lack of thermal analysis of the cookers, based on models developed from the operation of these cookers. In this context, we propose in this chapter an innovative parabolic trough solar thermal cooker (PSTC) along with its thermal modelling, for various applications, such as water heating and cooking.

After having described the basic structure, we present the thermal models at each node of the cooker, the response of the cooker under several intensities of solar irradiance, the storage and relaxation of the temperature of the cooker, and its thermal output. A particular focus will be on the maximum values of the temperatures reached, the duration of water heating, and the cooking of typical dishes in Morocco (therefore in Africa).

12.2   STRUCTURE AND DESCRIPTION OF THE COOKING SYSTEM

The solar cooking represented in Figure 12.1, the PSTC, which is the subject of our study, has a length of 200 cm and an overall weight of 35 kg. It is realized by:

- A concentrating mirror with a surface area of 1.30 m^2 which is made of parabolic cylindrical wood, covered with a reflective layer, and characterized by a high reflection rate.
- A double-walled (concentric) vacuum glass tube placed in the focal line of the concentrator. The inner tube is 1.5 m long and has an outer diameter of 47 mm. The outer tube has a diameter of 58 mm.

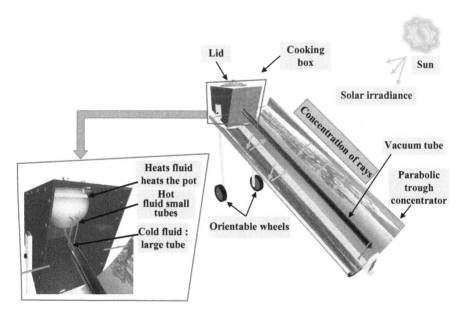

FIGURE 12.1   Schematic diagram of the global system of the PSTC

- A single hermetically closed copper tube, installed slightly inclined in the glass tube. In this tube a thermal oil, considered as a working fluid, circulates. At its upper end, the return of cold oil takes place through this 18 mm diameter tube, and the hot oil through the two small tubes of 9 mm diameter each.
- The wood cooking box, surrounded by thermal insulation made of glass wool, is made up of:
  - A double-jacketed stainless steel pot (volume 5 L), the space between the two jackets filled with thermal oil (the heat transfer liquid);
  - A glass lid.
- Two orientable wheels, equipped with two wooden supports that hold the firing box. They allow the cooker to be moved into the sun, to orientate the concentrators perpendicularly to the sun, and to capture the maximum solar energy.

12.3   FUNCTIONING AND OPERATION

The PSTC solar cooker shown in Figure 12.1, operates on a sunny day in three steps:

- The capture of solar energy by the cylindro-parabolic mirror which then reflects and concentrates solar rays towards the vacuum tube with a double envelope. In this tube, convection losses decrease and the heat is trapped inside by the greenhouse effect. This trapped heat then transfers to the copper tube placed inside.
- In the copper tube, the transfer fluid heats up, moves naturally from the bottom to the top of this tube, through the two secondary tubes, to the cooking box.
- The heat previously produced at the tube level is therefore transferred inside the cooking box. At this stage, we observe in the cooking box:
  - The rising thermal oil, through the two secondary tubes, heats the inner container of the pot by contact;
  - The inner container of the pot immersed in the working fluid heats up by contact, and consequently, the cooking of the food inside the pot;
  - In the cooking chamber, the oil cools and releases a heat, called sensible heat;
  - Then, the cold oil descends from the cooking chamber through the central tube, before the cycle begins again and transfers the heat to the cooking chamber. The rising oil and the falling oil therefore intersect inside the tube (Figure 12.1).

12.4   NUMERICAL MODELS

In this section, we establish the basic equations governing the transient heat transfer balance in our PSTC cooker (Figure 12.1). To do this, our analysis takes into account the following assumptions:

- The temperature is uniform in each component of the concentrator.
- The wind speed is constant.

- The thermo-physical properties of the glass, air, and absorber are considered constant within the range of the operation of the cooker temperatures.
- We neglect the heat transfer by conduction in all the elements of the concentrator.
- Between each node of the cooker (glass (1), absorber (2), transfer fluid (3), pot (4), and interior (5)) (Figure 12.1), the heat exchanges by radiation and convection which are represented by thermal resistances (Figure 12.2). In our approach (Figure 12.2), we represent these by:
  - $Rt_{cv\,(g\,-\,air)}$: convective thermal resistance coefficient between the glass (g) and air (air),
  - $Rt_{r(g\,-\,sky)}$: radiative thermal resistance coefficient between the glass (g) and sky (sky),
  - $Rt_{cv\,(ab\,-\,g)}$: convective thermal resistance coefficient between the absorber (ab) and glass (g),
  - $Rt_{r\,(ab\,-\,g)}$: radiative thermal resistance coefficient between the absorber (ab) and glass (g),
  - $Rt_{cv\,(ab\,-\,F)}$: convective thermal resistance coefficient between the absorber (ab) and fluid (F),
  - $Rt_{cv\,(F\,-\,p)}$: convective thermal resistance coefficient between the fluid (F) and pot (p),
  - $Rt_{cv\,(p\,-\,i)}$: convective thermal resistance coefficient between the pot (p) and interior (i).

These resistances are a function of the coefficient of heat transfer by convection or radiation and the exchange surface. They are written in the form (Srivastva et al. 2019):

$$R_t = \frac{1}{h \cdot A} \tag{12.1}$$

with,

- h: convective or radiative heat transfer coefficient,
- A: exchange surface.

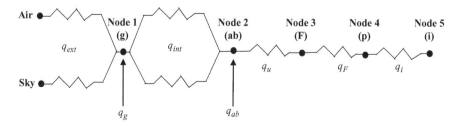

FIGURE 12.2   Thermal resistance network of the PSTC

Taking into account the hypotheses below and the heat exchanges by convection and radiation represented in Figure 12.2, the models of heat exchange at each node are written as follows:

- **Node 1** (glass tube): At this node, the heat exchange model on the glass temperature T_g is written as a function of the solar energy absorbed q_g, the inside heat transfer flow q_{int}, and the outside heat transfer flow q_{ext}:

$$\rho_g C_g A_g \frac{\partial T_g}{\partial t} = q_g + q_{int} - q_{ext} \tag{12.2}$$

where:

q_{int}: is written as a function of the radiation exchange flux $q_{r \cdot (g-ab)}$ and convection $q_{c \cdot (g-ab)}$, between the absorber and the glass, by the following relation:

$$q_{int} = q_{cv \cdot (g-ab)} + q_{r \cdot (g-ab)} = \left(h_{cv \cdot (g-ab)} + h_{r \cdot (g-ab)} \right) A_{ab-e} \left(T_{ab} - T_g \right) \tag{12.3}$$

with:

T_g: glass temperature,

T_{ab}: absorber temperature,

$h_{cv \cdot (g-ab)}$: convective heat transfer coefficient between glass and absorber,

$h_{r \cdot (g-ab)}$: radiative heat transfer coefficient between glass and absorber,

A_{ab-e}: external surface of the absorber,

q_{ext}: outside heat transfer flow.

They are written as a function of the radiation exchange flow $q_{r \cdot (g-amb)}$ and by convection $q_{c \cdot (g-amb)}$, between the glass and the exterior, by the following relation:

$$q_{ext} = q_{cv \cdot (g-air)} + q_{r \cdot (g-sky)} = \left(h_{cv \cdot (g-amb)} + h_{r \cdot (g-amb)} \right) A_{g-e} \left(T_g - T_{amb} \right) \tag{12.4}$$

with:

T_{amb}: ambient temperature,

$h_{cv \cdot (g-air)}$: convective heat transfer coefficient between glass and air,

$h_{r \cdot (g-sky)}$: radiative heat transfer coefficient between glass and sky,

A_{g-e}: external surface of the glass.

- **Node 2** (absorber tube): At this node, the heat exchange model on the absorber temperature T_{ab} is written as a function of the solar energy absorbed q_{ab}, the inside heat transfer flow q_{int}, and the useful heat q_u:

$$\rho_{ab} C_{ab} A_{ab} \frac{\partial T_{ab}}{\partial t} = q_{ab} - q_{int} - q_u \tag{12.5}$$

where:

q_u: useful heat, or exchange flow by convection between the absorber and the fluid $q_{c.\,(F-ab)}$, written as a function of temperatures T_{ab} and T_F (fluid temperature):

$$q_u = q_{cv \cdot (ab-F)} = h_{cv \cdot (ab-F)} A_{ab-i} \left(T_{ab} - T_F \right) \tag{12.6}$$

with:

$h_{cv \cdot (ab-F)}$: coefficient of convection exchange between the absorber and fluid,

A_{ab-i}: internal surface of the absorber.

- **Node 3** (fluid): At this node, the heat exchange model on the fluid temperature T_F is written as a function of the heat converted by fluid movement and useful heat q_u:

$$\rho_F C_F A_F \frac{\partial T_F}{\partial t} = q_u \tag{12.7}$$

where:

ρ_F: density of the heat transfer fluid,

C_F: specific heat of heat transfer fluid,

A_F: internal surface of the absorber.

- **Node 4** (pot) : At this node, the heat exchange model on the pot temperature T_p is written as a function of the convection flow with the thermal fluid q_F and the convection flow with the inside q_i:

$$\rho_p C_p A_p \frac{\partial T_p}{\partial t} = q_F - q_i \tag{12.8}$$

where:

$$q_F = h_{cv \cdot (F-p)} A_{p-i} \left(T_F - T_i \right) \tag{12.9}$$

with:

T_i: interior temperature,

$h_{cv.\,(F-p)}$: convective heat transfer coefficient between fluid and pot,

A_{p-i}: internal surface of the pot.

$$q_i = h_{cv \cdot (p-i)} A_{p-e} \left(T_i - T_p \right) \tag{12.10}$$

with:

$h_{cv \cdot (p-i)}$: convective heat transfer coefficient between pot and interior,

A_{p-e}: external surface of the pot.

- **Node 5** (Interior): At this node, the heat exchange model on the interior temperature T_i is written as a function of convection flow q_i as follows:

$$\rho_i C_i A_p \frac{\partial T_i}{\partial t} = q_i \qquad (12.11)$$

with:

ρ_i: Air density,

C_i: air specific heat,

A_i: internal contact surface of pot with air.

12.5 EFFICIENCY AND OUTPUT POWER

The thermal efficiency η of a cooker is expressed as a function of the sensible output power $\dot{Q}$ (W) of the heating of water of mass m_w, of temperature variation ΔT from 30°C to 95°C in the pots (Schwarzer and da Silva 2003, 2008) for a period of Δt, and the input power of solar radiation I (W/m²). They are given by the expressions (Arenas 2007; Craig et al. 2017; Panwar et al. 2012; Schwarzer and da Silva 2003, 2008):

$$\eta = \frac{\dot{Q}}{I \cdot A} \qquad (12.12)$$

$$\dot{Q} = \frac{m_w \cdot c_p \cdot \Delta T}{\Delta t} \qquad (12.13)$$

where:

A: aperture area of the concentrator,

c_p: specific heat capacity of water at constant pressure.

Taking into account expressions 12.11 and 12.12, the performance is written in the form of:

$$\eta = \frac{m_w \cdot c_p \cdot \Delta T}{I \cdot A \cdot \Delta t} \qquad (12.14)$$

12.6 EXPERIMENTAL RESULTS

12.6.1 EXPERIMENTAL PROCEDURE AND MEASUREMENT GUIDELINES

We experimented with the PSTC cooker of Figure 12.1 during a period of October of the 2019 fall season, on the roof of the LETSER Laboratory of the Faculty of Sciences of the Mohamed Premier University of Oujda (Morocco). The complete measuring tank is formed by (Figure 12.3):

FIGURE 12.3 PSTC solar cooker experimented with at the laboratory, October 2019

- A CMP6 pyranometer, with an accuracy of ±5%, to measure solar irradiance.
- A digital multimeter (Keithley model 2700), connected to the CMP6 pyranometer. The unit, connected to a computer, has an accuracy of ±30 W/m^2 and a resolution of 0.1 W/m^2.
- Sensors (type k), with a maximum continuous temperature of 1100°C and an accuracy of ± 2.2°C, are connected to the large and small pipes, to measure the temperatures of rising and falling fluids. Also, using these sensors, we measured the temperature values: ambient, base, and center of the pot, and food (oil and water, food).
- Multimeters of type UNI-T U171C with a resolution of 0.1°C and an accuracy of ±1°–2°C for measuring temperatures from −40°C to 400°C.

12.6.2 NO-LOAD OPERATION (EMPTY)

We experimented with the cooker of Figure 12.3 without load (empty) by measuring, during entire days in autumn, the intensity of the solar irradiance and the ambient temperatures, of small and large tubes, and of the center and the upper part of the pot. Typical results obtained, on a sunny day, are shown in Figure 12.4:

- The maximum solar irradiance intensity of 700 W/m^2 is obtained at around 1 p.m.
- The ambient temperature is relatively low and varies from 18 to 25°C.
- After 30 min of heating, the intensity of solar irradiance increases from 306 to 408 W/m^2 (an increase of 33%), the temperature of the small tube varies from 13 to 237°C (an increase of a factor of 18.23), and that of the large tube from 13 to 89°C (an increase of a factor of 6.84).

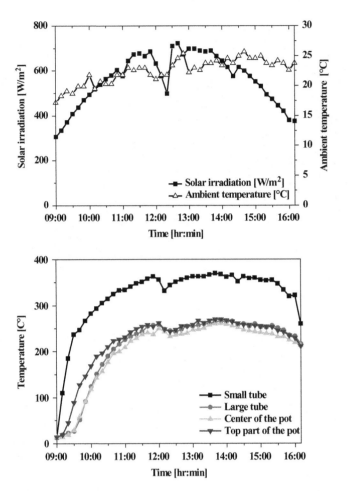

FIGURE 12.4   Typical variation in solar irradiance, ambient temperatures, and temperatures of large and small cooker tubes on a sunny day, October 2019

- After 1.5 h of heating, the temperatures of the large tube and the pot (center and upper part) are practically identical.
- After 2 h of heating, the temperature of the cooker is at the maximum. The small tube reaches the value of 334°C, the large tube and the pot (center and upper part) reach the values 226°C and 231°C respectively.
- After 1 p.m., the intensity of the solar irradiance decreases but the temperatures of the small and large tube and the pot remain constant, around their maximum values (363°C and 267°C).
- The average orientation of the cooker over the sun is about 2.66° per 10 min (or 16° per hour).

The vacuum results clearly show the response of the cooker to the concentration of solar radiation on the heat transfer fluid. By the exchanger, formed by the small tubes, the pot heats up with remarkable performance. Despite the fall season, which has relatively low values and temperatures, the fluid temperature reaches 370°C and the pot 250°C. The loss of 120°C is attributed to heat exchange by convection between the fluid and the pot. Also, we can note the conservation of temperatures after 1 p.m., despite the decrease in solar radiation. This is attributed to the storage of the temperature in the fluid flowing in the tubes.

Comparing these results with those in the literature (Asmelash et al. 2014; Esen 2004), we deduce an improvement of 58% over the time that allows the temperature of 100°C to be reached, and 49% over the maximum value of the temperature of the pot (250°C). All these performances clearly show the good functioning of the PSTC cooker, thus its use to heat water and cook meals during a whole sunny day.

12.6.3 Operation with Charge (Boiling Water)

In order to test the performance of the cooker, water boiling tests were carried out on a day with variable lighting and ambient temperature (Figure 12.5). These variations make it possible to follow the heating time Δt of 1 L of water and note the variation in temperature ΔT and the value of the efficiency η according to Equation (12.14). As shown in Figure 12.5, these efficiency values are determined mainly from 1 p.m. As in the case of the previous section, we measured the temperatures of the small and large tubes, and those of the water. Also, following the rapid boiling of the water every 30 min, we added 1 L of water (Figure 12.5). Typical results obtained show:

- The solar irradiance and the ambient temperature reach respectively the values 800 W/m^2 and 23°C.
- The temperature of the heat transfer oil (temperature of the small tube) could reach the value of 250°C. However, following the boiling of the water, the temperatures of the pot, and therefore of the large tube, do not exceed the boiling temperature which is 98°C.
- The average time and the boiling temperature of a liter of water are around 26 min and 98°C in the afternoon (from 1 p.m.).
- The thermal efficiency of the cooker reaches the value of 52%. In terms of heating capacity and thermal storage, the results obtained in a vacuum confirm those obtained with water. When water is added, the temperature of the heat transfer fluid decreases from 250°C to 160°C, then increases to 250°C after 20 min. Moreover, comparing with the best thermal efficiencies of cookers in the literature in the order of 39% (Asmelash et al. 2014; Craig et al. 2017; Da Silva et al. 2005), using a parabolic reflector, we obtain a better performance with our cookers (PSTC), where the efficiency reaches the value of 52% (an improvement of 23%).

12.6.4 Operation with Charge (Cooking)

This part concerns the cooking of 1.5 kg of vegetables and meat (tagine) on the PSTC cooker in Figure 12.3, during a day with variable solar irradiation (Figure 12.6). As

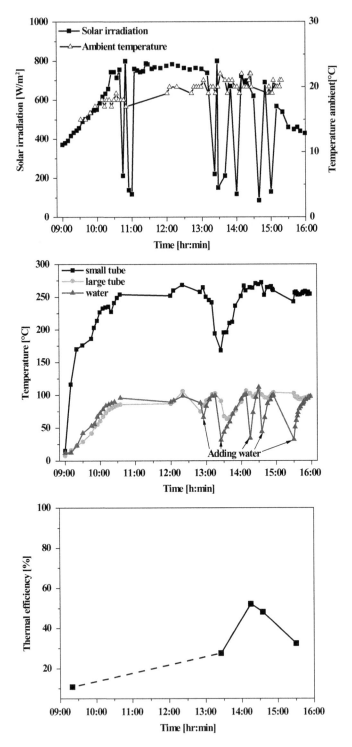

FIGURE 12.5 Variation in solar irradiation, ambient temperature, temperatures (small and large tube, water), and thermal efficiency of the cooker in Figure 12.3 during the day, October 2019

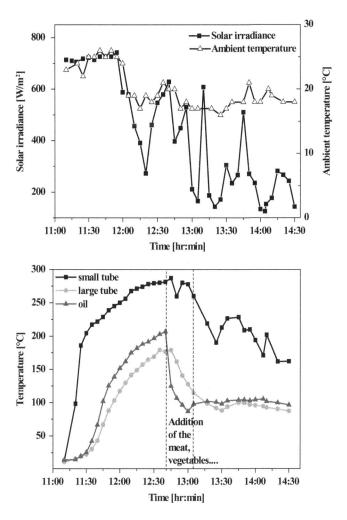

FIGURE 12.6 Variation of the solar irradiance, ambient temperatures, and temperatures of the cooker (small and large tubes, pot) during a changing day, December 2019

before, we recorded the temperatures of the heat-transfer oil (small tubes) and the pot (large tube). The measurements obtained show (Figure 12.6):

- The 0 of the solar radiation and the ambient temperature reach the values 700 W/m^2 and 25°C in the morning, and in the afternoon present rapid variations from 600 to 150 W/m^2, and 18 to 22°C.
- After 1.5 h of heating, the temperatures of the heat transfer oil (small tubes), the large tube, and the pot reach the values of 281.1, 206.7, and 171°C respectively.
- When adding vegetables and meat, the temperature of the pot decreases to 86.8°C, then increases to the stable value of 104°C. However, the temperature of the heat transfer oil and the pot remains stable around 250°C.
- Despite the cloudy afternoon, the temperature of the heat transfer oil drops relatively low by 90°C (from 250 to 160°C). The cooking takes place at a temperature of 100°C for a time of 1.5 h.

The results obtained in this section show that the temperature of the heat transfer fluid reaches its maximum value of 250°C, following the presence of stable solar irradiance in the morning. On the other hand, despite the absence of solar irradiance in the afternoon, the cooking is carried out at a temperature of 100°C, following the storage of the temperature in this heat transfer fluid, where its temperature undergoes a decrease from 250 to 160°C. Consequently, all these results show the good functioning of the cooker, designed and realized during this work, and therefore its use under a changing Moroccan climate.

12.7 CONCLUSION

In this chapter, we have presented the results concerning the operation of a solar cooker of the parabolic cylindrical PSTC type, using a heat transfer fluid. All the results obtained allowed us to conclude:

- Under an irradiation of 720 W/m^2, the rise in temperature of the heat transfer fluid is 2°C/mn under no-load conditions and 2.6°C/mn in the presence of a load (water, food);
- The maximum temperatures reached for the heat transfer fluid and the kettle are of the order of 370°C and 250°C;
- The maximum thermal efficiency of the PSTC cooker is estimated at 52%;
- The cooker has a thermal storage of the order of 3 h, necessary for the cooking of 1.5 kg of food for 1.5 h, without irradiation.

Comparison of all these results with the literature shows an improvement in the maximum pot temperature achieved of 49%, efficiency of 23%, and cooking time of 33%.

ACKNOWLEDGMENT

This research was supported by project INDH 2017/29, in collaboration with:

- Association "Man and Environment of Berkane", Berkane, Morocco.
- National Initiative for Human Development INDH, Province of Berkane, Morocco.
- Engineering Office of Energy and Environmental Technology (IBEU), Jülich, Germany.

REFERENCES

Arenas, J.M. "Design, Development and Testing of a Portable Parabolic Solar Kitchen." *Renewable Energy* 32, no. 2 (February 2007): 257–266. doi:10.1016/j.renene.2006.01.013.
Asmelash, H, M. Bayray, C.Z.M. Kimambo, P. Gebray, and A.M. Sebbit. "Performance Test of Parabolic Trough Solar Cooker for Indoor Cooking." *Momona Ethiopian Journal of Science* 6, no. 2 (November 10, 2014): 39. doi:10.4314/mejs.v6i2.109621.
Craig, O.O., R.T. Dobson, and W. van Niekerk. "A Novel Indirect Parabolic Solar Cooker." *Journal of Electrical Engineering* 5, no. 3 (June 28, 2017): 137–142. doi:10.17265/2328-2223/2017.03.003.

Da Silva, M.E.V., L.L. Santana, A. Pinheiro et al. "Comparative Study of Two Solar Cookers: Parabolic Reflector and Flate Plate Collector Indirect Heating." *World Climate and Energy Event*, Rio de Janeiro, Brazil, 2005, 5.

Da Silva, M.S. et al. "Experimental Results of a Solar Cooker with Heat Storage." in *Proceedings of Rio 02 World Climate and Energy Event*, , Rio de Janeiro, Brazil, 2002, 89–93.

Dianda, B., S.M.T. Ky, I. Ouedraogo et al. "Experimental Investigation of a Solar Parabolic Trough Cooker". *International Journal of Current Research* 11 2019: 1869–1874.

El Moussaoui, N., S. Talbi, I. Atmane, K. Kassmi, K. Schwarzer, H. Chayeb, and N. Bachiri. "Feasibility of a New Design of a Parabolic Trough Solar Thermal Cooker (PSTC)." *Solar Energy* 201 (May 2020): 866–871. doi:10.1016/j.solener.2020.03.079.

Esen, M. "Thermal Performance of a Solar Cooker Integrated Vacuum-Tube Collector with Heat Pipes Containing Different Refrigerants." *Solar Energy* 76, no. 6 (2004): 751–757. doi:10.1016/j.solener.2003.12.009.

Hager, T.J. and R. Morawicki. "Energy Consumption During Cooking in the Residential Sector of Developed Nations: A Review." *Food Policy* 40 (June 2013): 54–63. doi:10.1016/j.foodpol.2013.02.003.

Kumar, A., S.K. Shukla, and A. Kumar. "Heat Loss Analysis: An Approach toward the Revival of Parabolic Dish Type Solar Cooker." *International Journal of Green Energy* 15, no. 2 (January 15, 2018): 96–105. doi:10.1080/15435075.2018.1423978.

Muthusivagami, R.M., R. Velraj, and R. Sethumadhavan. "Solar Cookers with and without Thermal Storage—A Review." *Renewable and Sustainable Energy Reviews* 14, no. 2 (February 2010): 691–701. doi:10.1016/j.rser.2008.08.018.

Otte, P.P. "Solar Cookers in Developing countries—What Is Their Key to Success?" *Energy Policy* 63 (December 2013): 375–381. doi:10.1016/j.enpol.2013.08.075.

Panwar, N.L., S.C. Kaushik, and S. Kothari. "State of the Art of Solar Cooking: An Overview." *Renewable and Sustainable Energy Reviews* 16, no. 6 (August 2012): 3776–3785. doi:10.1016/j.rser.2012.03.026.

Schwarzer, K. and M.E.V. da Silva. "Solar Cooking System with or without Heat Storage for Families and Institutions." *Solar Energy* 75, no. 1 (July 2003): 35–41. doi:10.1016/s0038-092x(03)00197-x.

Schwarzer, K. and M.E.V. da Silva. "Characterisation and Design Methods of Solar Cookers." *Solar Energy* 82, no. 2 (February 2008): 157–163. doi:10.1016/j.solener.2006.06.021.

Sekhon, M. and V. P. Sethi. "Thermal Modeling and Analysis of Novel Twin-Chamber Community Solar Cooker as a Replacement of Biomass-Based Cooking." *International Journal of Green Energy* 16, no. 2 (November 28, 2018): 167–184. doi:10.1080/15435 075.2018.1551221.

Srivastva, U., R.K. Malhotra, K. Ravi Kumar, and S.C. Kaushik. "Comparative Assessment of Classical Heat Resistance and Wilson Plot Test Techniques Used for Determining Convective Heat Transfer Coefficient of Thermo Fluids." *Thermal Science and Engineering Progress* 11 (June 2019): 111–124. doi:10.1016/j.tsep.2019.03.021.

Toonen, H.M. "Adapting to an Innovation: Solar Cooking in the Urban Households of Ouagadougou (Burkina Faso)." *Physics and Chemistry of the Earth, Parts A/B/C* 34, no. 1–2 (January 2009): 65–71. doi:10.1016/j.pce.2008.03.006.

Yettou, F., B. Azoui, A. Malek, A. Gama, and N.L. Panwar. "Solar Cooker Realizations in Actual Use: An Overview." *Renewable and Sustainable Energy Reviews* 37 (September 2014): 288–306. doi:10.1016/j.rser.2014.05.018.

13 Thermal Modeling of Photovoltaic Ovens and Hotplates

S. Talbi, I. Atmane, N. El Moussaoui, K. Kassmi,
O. Deblecker, and N. Bachiri

CONTENTS

13.1 INTRODUCTION

Biomass is used as a fuel for cooking and heating in many countries around the world (Padilla-Barrera et al. 2019; Oyewo et al. 2020). According to World Health Organization (WHO) statistics, approximately 3 billion people still consume solid fuel for cooking and heating in outdoor homes and leaky stoves. This caused 4.3 million deaths worldwide in 2012 (Padilla-Barrera et al. 2019). Similarly, 40% of the world's population uses wood for cooking (Janssen et al. 2017). This leads to forest degradation, and consequently to an increase in greenhouse gases (Aramesh et al. 2019). It is predicted that without a change in the current state of cooking technology, greenhouse gas emissions from cooking will be in the order of 2.5 billion tons (Mohan 2018).

From this perspective, solar cooking is an ecological and non-polluting technology that does not consume any type of fuel (Nkhonjera et al. 2017; Akayleh et al. 2018; El Moussaoui et al. 2020). Solar cookers can have different classifications depending on how they are studied (Farooqui 2014; Ramesh et al. 2019; El Moussaoui et al. 2020; Thirugnanam et al. 2020; Herez et al. 2018). Depending on the mechanism by which the thermal energy from the sun is transferred to the cooker, they can be divided into two main categories: direct and indirect (Farooqui 2014; Herez et al. 2018; Motwani and Patel 2019; El Moussaoui et al. 2020; Thirugnanam et al. 2020). Solar cookers of the "Box" type, reflective panels, and concentrating type solar cookers are the direct type (Herez et al. 2018), as they use solar radiation directly in the cooking process. On the other hand, solar cookers with a solar collector and thermal storage are the indirect type, since they use a heat transfer fluid to move heat from the collector to the cooking unit (Omotoyosi et al. 2017; Panchal et al. 2017; Hosseinzadeh et al. 2020).

The majority of these cookers only work with solar thermal energy. This results in several disadvantages and performance limitations (Rathore and Warkhedkar 2015; Yettou 2015; Harish et al. 2016; Kimambo 2017; Murnawianto et al. 2019; Talbi et al. 2019a): the cooking temperature does not exceed 150°C and their cooking time is twice as long as a conventional cooker (more than 120 min). Recently (Rathore and Warkhedkar 2015; Yettou 2015, 2018), parabolic cookers have been introduced in the summer and experimented with under the sun outside the home. It has been shown that their temperatures could exceed 250°C and burn the food if the contents are not stirred regularly enough. This shows, on the one hand, that the cooking temperature is not controllable, and on the other hand, the difficulty for users to cook with this type of cooker.

Currently, designers favor direct box-type cookers since they offer many advantages, such as: simple to implement, easy to operate, very stable, less expensive, no risk of burns (Yettou 2015). In addition, to improve the performance of this type of cooker, work in the literature focuses on the integration of photovoltaic energy (PV) (Joshi and Jani 2013; Joshi and Jani 2015; Emmanuel and Amo-Aidoo 2018; Talbi et al. 2019a; Talbi et al. 2019b). The operation of these cookers is based on the heating of the thermal resistors by means of the electrical energy stored in the solar batteries, which are charged by the PV panels (Joshi and Jani, 2013; Joshi and Jani, 2015; Emmanuel and Amo-Aidoo, 2018). In the case of a cooker formed by a battery with a capacity of 60 Ah, charged by the 200 W PV panels through a charge regulator, Emmanuel and Amo-Aidoo (2018) show that, under a lighting and ambient temperature of 570 W/m^2 and 35.7°C, the boiling time of a liter of water is very long (about 1 h 45 min). It should be noted that for this prototype, the use of batteries increases the cost of the cooker and the maintenance costs.

In order to contribute in the field and improve the performance of solar cookers, we propose in this work – within the framework of the Cooperation Programme Morocco-Wallonia Brussels (Belgium) (2018–2022) (Project n°4. 2) and socio-economic partners (civil society: Association "Man and Environment of Berkane", Berkane, (Morocco), National Initiative for Human Development INDH (Province of Berkane, Morocco)) – the design and production of innovative solar PV cookers and hotplates, adaptable to the needs of inhabitants, in or outside of their homes in the rural or urban world. The main objective is to promote these new cookers, powered by PV energy, through the establishment of income-generating activists, the creation of startups for young people, adaptation to climate change, and environmental protection.

In this chapter, we present the feasibility and thermal modeling of prototypes of innovative ovens and hotplates, operating on PV solar energy. After describing the structure and operation of the cookers, we present the thermoelectric model of the resistance temperature in transient mode, as well as the results of operation and their validation.

13.2 PHOTOVOLTAIC COOKERS

13.2.1 STRUCTURE AND FUNCTIONING

Figure 13.1 shows a synoptic diagram of the box oven and hotplate powered by PV solar energy. The different blocks of these cookers are:

- PV generator formed by PV panels, made of monocrystalline silicon, with a power of 400 W.
- Box-type cooker formed by:
  - A thermally insulated box (wood, cork, chamotte) to minimize heat loss in the solar cooker;
  - Two thermal resistors, glued on the lower and upper sides inside the oven. They are chosen to withstand a current of 10 A, a power of 600 W, and temperatures above 1200°C.
- Plate containing a heating resistor supporting a current of 10 A, a power of 500 W, and a temperature of 1200°C.
- A system-wide energy acquisition and management system. This is formed by power and control circuits allowing the acquisition, display, and control of all electrical, thermal, and meteorological quantities (temperatures in the resistors and in the oven, lighting and ambient temperature, and energy produced by the PV panels and transferred to the cookers).

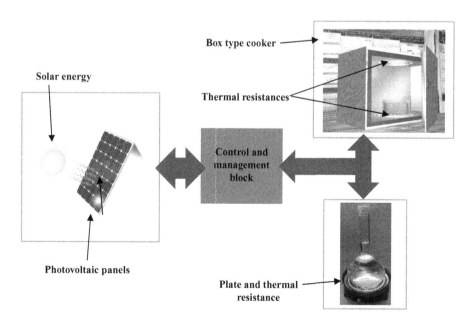

FIGURE 13.1 Synoptic diagram of the box and hot plate cookers powered by PV energy

13.2.2 THEORETICAL MODELS OF THE TEMPERATURE OF THE HEATING RESISTOR POWERED BY PHOTOVOLTAIC ENERGY

The thermal model developed in this work, of the heating of thermal resistors by PV energy, is based on the principle of the thermal energy balance (Terres et al. 2014; Talbi et al. 2019b) of the temperature node of a resistor, taking into account the Pelc electrical power supplied by the PV generator and the heat loss by convection (*Pcv*) and by radiation (PR) (Terres et al. 2014; Talbi et al. 2019b). In our application the temperature thermal resistors (T_r) exchange heat with the interior of the temperature oven (T_{int}) and with the air temperature (T-amb) in the case of the hot plate. Taking these considerations into account, the thermal equations, governing the heating of the thermal resistors, can be written as:

In the case of the oven:

$$m_r \cdot Cp_r \frac{dT_r}{dt} = P_{elc} - P_{cv} - P_R = Vs \cdot Is - h_{r-int} S_r \left(T_r - T_{int}\right) - S_r \sigma \varepsilon_r \left(T_r^4 - T_{int}^4\right) \quad (13.1)$$

and in the case of the hot plate:

$$m_r \cdot Cp_r \frac{dT_r}{dt} P_{elc} - P_{cv} - P_R = Vs \cdot Is - h_{r-amb} A_r \left(T_r - T_{amb}\right) - S_r \sigma \varepsilon_r \left(T_r^4 - T_{amb}^4\right) \quad (13.2)$$

where,

m_r: Mass of heating resistance ($m_r = 0.01634$ kg),

Cp_r: Specific heat of the heating resistance ($Cp_r = 890$ J/kg-K),

V_s et I_s: Voltage and current of the heating resistor,

h_{r-amb}: Coefficient of convective exchange between the heating resistance and its surroundings,

h_{r-int}: Coefficient of convective exchange between the heating resistor and the interior of the box furnace,

S_r: Heating resistance areas ($S_r = 0.0025$ m^2),

σ: Boltzmann constant ($\sigma = 5.669*10^{-5}$ w/m^2·k^4),

ε: Emissivity of heating resistance ($\varepsilon = 0.9$).

13.2.3 THERMAL EFFICIENCY OF HEATING

In the literature, the methods used to calculate the efficiency of solar cookers are based on an energy analysis related to the first law of thermodynamics (Funk 2000; Yetto et al. 2014; Yetto 2015). They provide information for appropriate management in applications where the efficiency depends only on the amount of energy. In our case, when heating ovens and hotplates with PV energy, an electric current I flows through the thermal resistance and a voltage V is supported by it. Thus, the electrical power P_{el} and the energy E_i supplied to the thermal resistance for a time t are written as follows:

$$P_{el} = V * I \qquad (13.3)$$

$$E_i = V * I * t \qquad (13.4)$$

Yetto et al. (2014) and Yettou (2015) establish, by the Funk method, the thermal energy E_o and the thermal power P_o, after a heating time t of a quantity M_e of the water until boiling. They show that these can be written as a function of the variation of the water heating temperature ΔT, in the form:

$$E_o = Me * Cp * (\Delta T) \qquad (13.5)$$

$$P_o = \frac{Me * Cp * (\Delta T)}{t} \qquad (13.6)$$

where, C_p: Specific heat of water ($C_p = 4\ 180$ J/kg-K).
From Equations 13.3–13.6, and according to the first law of thermodynamics, we can deduce the energy yield (η) in the form of (Joshi and Jani 2015; Talbi et al. 2019b):

$$\eta = \frac{E0}{Ei} = \frac{P0}{Pel} = \frac{Me * Cp * (\Delta T)}{V * I * t} \qquad (13.7)$$

13.2.4 Flow Chart of the Numerical Code

The numerical solution of Equations 13.1 and 13.2 can be solved by the Fourth Order Runge–Kutta method (Talbi et al. 2019b) in the C++ language. We have adopted this method since it is remarkably accurate and stable (Talbi et al. 2019b). As shown in the flowchart in Figure 13.2, the main steps of the calculations are as follows:

- Declaration of dimensions and thermal resistance data:
  - Mass of heating resistance ($m_r = 0.01634$ kg).
  - Specific heat of the heating resistance ($Cp_r = 890$ J/kg-K).
  - Heating resistance surfaces ($S_r = 0.0025$ m^2).
  - Emissivity of heating resistance ($\varepsilon = 0.9$).
  - Boltzmann constant ($\sigma = 5.669*10^{-5}$w/m^2·k^4).
  - Coefficient changes convective h_{r-amb} between the heating resistor and its ambient. In our case we have simulated the situations for heating oil, cooking chips, and heating water.
  - Coefficient of convective exchange h_{r-int} between the heating resistor and the inside of the box furnace. In our case, we have simulated the situation of water heating.
- Declaring maximum time (t_{max}) and the calculation step (h).
- Entry of parameters:
  - Hot plate: ambient temperature (T_{amb}) and electrical power by PV solar energy (P_{el});

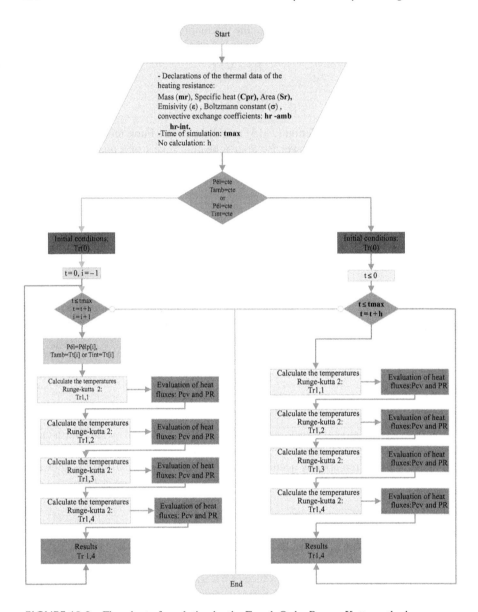

FIGURE 13.2   Flowchart of resolution by the Fourth Order Runge–Kutta method

- Box oven: internal temperature (Tint) and electrical power by PV solar energy (P_{el});
- Time interval (10 min or 15 min) during operation.
- Initial conditions:
  Initial temperatures of the heating resistor ($T_r(0)$).
- Calculate:
  When $t \leq t_{max}$, by incrementing the time t by one step h = 15 min or h = 10 min, the program calculates the temperature of the furnace resistance node or hot

plate resistance optimally by the Runge-Kutta methods. The calculations are carried out as follows:

Hot plate: ambient temperature (T_{amb}) and electric heating power (P_{el}) are constant or variable over time;

Oven box: internal temperature (T_{int}) and electric heating power (P_{el}) are constant or variable over time;

By integration of the differential equations: once at the starting point (RK1), twice in the middle of the interval (RK2 and Rk3), and once at an estimated end point (RK4).

- Stop calculations: when t = 480 min, the program stops execution and displays the results of the temperatures of the thermal resistance T_r with fourth-order accuracy (RK4).

13.3 RESULTS AND DISCUSSIONS

13.3.1 EXPERIMENTAL PROCEDURE

The complete system allowing for the experimentation of solar cookers, of the box oven, and hot plate type, powered by photovoltaic energy, is shown in Figure 13.3. This is completely automated in the laboratory and is formed by:

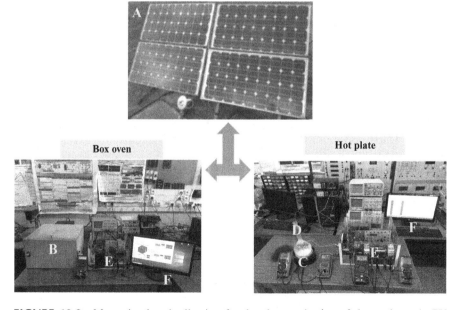

FIGURE 13.3 Measuring bench allowing for the characterization of the cookers. A: PV generator, B: box oven, C: plate and heat resistance, D: water to be heated, E: control, acknowledgment, and system management block, F: PC acquisition and display

- A PV generator consisting of PV panels of optimum voltage, current, and power of 15 V, 4.5 A, and 400 W respectively.
- A CM6 pyrometer, connected to a digital multimeter (Keithley Model 2700), to measure illumination intensity throughout the day.
- A box type furnace formed of:
  - A thermally insulated box (wood, cork, chamotte) to minimize heat loss in the solar cooker.
  - Two thermal resistors, glued on the lower and upper sides inside the oven. They are chosen to support a current of 10 A, a power of 1000 W, and a temperature above 1200°C.
- A heating plate containing a thermal resistor, supporting a current of 10 A, a power of 1000 W, and a temperature of 1200°C. This resistor is placed in a circular ceramic, 20 cm in diameter.
- A block of cards formed by analog and digital electronic components and circuits, connected to a computer. Its role is:
  - Acquisition of electrical, meteorological, and thermal quantities;
  - Management of the energy produced by the PV panels and transferred to the cookers.

13.3.2 Hot Plate

We experimented, depending on the weather conditions (illuminance and ambient temperature), with the operation of the heating plate by PV energy by following the rise of the temperatures of the thermal resistance and the cooking. We analyzed three cases of different cooking applications: heating a container of 0.33 L of oil, cooking 500 g of food (chips), and heating 1 L of water.

For each application, we recorded, during whole days, the intensity of the illumination, the ambient temperature, the electrical power of the heater Pel, the temperature of the heating resistors $T_{r\text{-}exp}$, for oil and water. From these results we simulated and deduced the heating resistor temperature ($T_{r\text{-}sim}$), the water power P0, and the energy efficiency (η) of the heating. The set of results obtained is shown in Figure 13.4:

- In the case of heating a container of 0.33 L of oil:
  - During operation from 9 a.m. to 11 a.m, the illumination intensity varies from 300 W/m^2 to 640.8 W/m^2 (an increase of 53.18% and 170.4 W/(m^2.h)). The temperature of the resistor varies between 15°C and 795°C (or an increase by a factor of 53 and 390°C/h). The oil temperature varies from 16°C to 253°C (or an increase by a factor of 15.81 and 118.5°C/h).
  - The maximum temperature of the resistor (around 821°C) and oil (around 290°C) is reached after 3.5 h of heating.
  - When the illuminance reaches the maximum intensity of 763.3W/m^2 and the ambient temperature reaches 17.5°C, the electrical power, heating resistor, and oil temperature reach the maximum values of 332.64 W, 820°C, and 290°C.
  - There is good agreement between experiment and simulation concerning the temperature of the resistor and setting the convective exchange coefficient between the heating resistor and its ambient $h_{r\text{-}amb}= 87$ w/m^2·k^4.

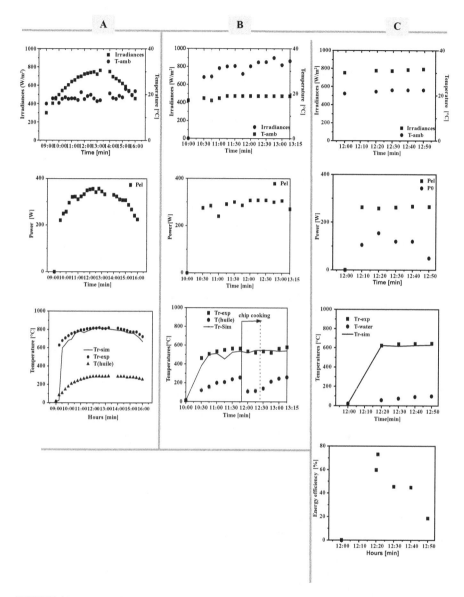

FIGURE 13.4 Experimental plot of the meteorological conditions (irradiances and ambient temperature T_{amb}), experimental and simulated Pel electrical power experimental temperatures T_{r-exp} and simulated Tr-sim of the heating resistor, and thermal powers Po and energy efficiency η. A: heating of 0.3 L of oil, B: cooking of 500 g chips, C: heating 1 L of water

- In the case of cooking the 500 g of chips:
  - The maximum temperature of the resistor, of the order of 580°C, and of the oil, of the order of 265°C, are reached after 2 h of heating.
  - When adding chips, the temperature of the oil decreases to 108°C, then increases to 261°C after 15 min of cooking.

- Cooking took place after 45 min under the electrical power of the 285.94 W heating element. The temperature of the oil and the heating resistors reach 257°C and 564°C respectively.
- In the case of heating 1 L of water:
  - Between 12 and 13 h, the lighting intensity and the ambient temperature reach the values of 775W/m² and 21.98°C. The electrical Pel and thermal power P_0, the temperature of the thermal resistors, and the water temperature reach the values 108.94 W and 262 W, 645°C, and 99°C.
- The boiling temperature of the water (90°C) is reached after 40 min of heating.
- The energy efficiency η reaches the maximum value of 59.56% after 20 min of heating. Beyond this time, the water temperature stabilizes at 98°C and the water evaporates.

In the literature (Joshi and Jani 2015; Joshi and Jani 2013), heating of the thermal resistance is carried out using batteries charged by 65–80W PV panels. The results obtained show that the temperature of the resistance does not exceed 122°C, and for a quantity of water m = 0.385 g, the thermal power and efficiency are of the order 32.2W and 43.6%. Consequently, our studies on cookers, which exploit the energy produced by PV panels through a regulation system, show better results. The improvement also applies to the maximum values of the temperature of the thermal and cooking resistors, of the order of 86% and 43%, and to the thermal power and efficiency, of the order of 79% and 28%.

13.3.3 OVEN

We experimented for a full day with the operation of the solar box oven using the PV energy shown in Figure 13.1. We recorded the intensity of the ambient illumination and temperature, the electrical power (Pel) of the thermal resistor, and the temperature of the heating resistor (Tr-exp) and inside the oven (Tint). From the time (t) of the heating of 1 L, we deduced (according to Equations 13.3, 13.4, 13.5, and 13.6) the power P0 and the thermal efficiency relative to the heating of the furnace. In the case of the thermal resistance, the results of the simulation of the temperature (Tr-sim) during heating are shown in Figure 13.5:

- The maximum illuminance and the ambient temperature reached are about 772 W/m² and 19.8°C, respectively.
- The electrical power of the heating resistor reaches the maximum value of 344 W under an illumination of 771.4 W/m².
- The temperatures in the heating element and inside the cooker reach the values 668°C and 200°C respectively.
- After 20 min of operation:
  - Illuminance and ambient temperature ranging from 334.7 W/m² to 395.9 W/m² (18% increase) and from 14°C to 15.2°C (9% increase) respectively; the Pel electrical powers of the heating and thermal resistor Po, vary respectively from 0W to 226.89 W and from 0 W to 125.7 W; the temperature of the heating resistor Tr ranges from 21°C to 492°C (an increase by a factor

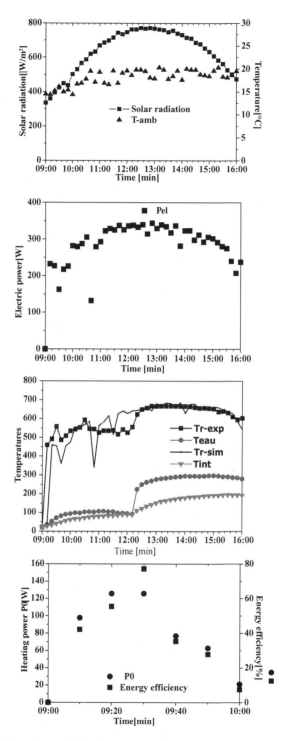

FIGURE 13.5 Experimental plot of meteorological conditions (lighting and external temperature), Pel electrical power, experimental Tr-exp and simulated Tr-sim temperatures of the heating resistor, temperature inside the Tint cooker and boiling point Tint of 1 L of water, heating power P_0, and thermal efficiency of the solar cooker

of 24); the temperature inside the cooker varies from 21°C to 33°C (a 57% increase).

- There is very good agreement between simulated and experimental results concerning the temperature of the thermal resistance. From these results we have deduced and shown that the convective exchange coefficients between the heating resistor and its interior is of the order of $h\_(r\text{-}int) = 180$ w/m²·k⁴).
- The maximum energy efficiency (η) of the heating reaches 77% after 30 min of heating. Beyond this, the water temperature stabilizes at around 98°C and the water evaporates. As a result, the efficiency decreases. From 12 noon (i.e., 3 h of heating), the water is completely evaporated. The temperature measured is therefore that of the water vapor, which attains the value of 300°C.

Comparing these results with those carried out on thermal ovens (Mukaro and Tinarwo 2008) and hybrid "thermal/photovoltaic" ovens (Joshi and Jani 2013; Emmanuel and Amo-Aidoo 2018), we can conclude, on the one hand, the feasibility of box cookers with PV energy is a remarkable performance (an improvement of 78.26% and 31.5% on the shown temperature and its maximum value inside the oven), and on the other hand, the modeling of the heating of the resistors with a coefficient of convective exchange is $h_{r-int} = 180$ w/m²·k⁴.

13.4   CONCLUSION

We have designed, produced, and tested hot plate and box cookers powered by PV energy. To validate the operation of the two cookers, we simulated the heating temperature of the thermal resistors, based on the development of a numerical calculation code of the equations of the heat transfer heat balance in a transient regime. The results obtained allow us to conclude:

- In the case of the hotplate, we showed that the temperature and cooking time improved by 86% and 43%, a good agreement between the results of simulations and experiments, concerning the temperature of the thermal resistance. During one day of operation, the heating of the thermal resistance is modeled by the convective exchange coefficients hr-amb, between the heating resistor and its ambient, varying from 87 W/m²·k to 190 W/m²·k.
- In the case of the can cooker, we have shown that the temperature inside the cooker reaches 200°C after 30 min of heating and the thermal efficiency is around 77%. Also, there is a good agreement between simulation and experimental results concerning the temperature of the thermal resistance. During one day of operation, the heating of the thermal resistance is modeled by the convective exchange coefficients h_{r-int}, between the heating resistance and the interior of the oven, of the order of 180 w/m²·k⁴.

All the results obtained in this work show the feasibility of the two types of cookers, oven and hotplate, running on PV solar energy. The performances obtained in terms of cooking temperatures and times, and the ease of use, are very encouraging. This is a testimony to their use inside and outside the home in rural and urban areas.

ACKNOWLEDGMENTS

This research was carried out in collaboration with Mohamed Premier University (Morocco), Polytechnic Faculty of Mons (Belgium), and Association Man and Environment of Berkane (Morocco), within the framework of projects:

- Morocco-Wallonie Cooperation Program Brussels (2018–2022), Wallonie-Bruxelles-International, project 4, n°2;
- National Initiative for Human Development INDH, Berkane Province, Morocco, project 2017//29.

REFERENCES

Akayleh, Ali S., Mohammed S. Al-Soud, and Saleh A. Al-Jufout. "Design and Development of a Solar-Based Cooker with a Mechanical Sun Tracking System." *International Journal of Ambient Energy* 41, no. 7 (July 13, 2018): 808–812. doi:10.1080/01430750.2018.1472644

Aramesh, Mohamad, Mehdi Ghalebani, Alibakhsh Kasaeian, Hosein Zamani, Giulio Lorenzini, Omid Mahian, and Somchai Wongwises. "A Review of Recent Advances in Solar Cooking Technology." *Renewable Energy* 140 (September 2019): 419–435. doi:10.1016/j.renene.2019.03.021.

El Moussaoui, Noureddine, Sofian Talbi, Ilyas Atmane, Khalil Kassmi, Klemens Schwarzer, Hamid Chayeb, and Najib Bachiri. "Feasibility of a New Design of a Parabolic Trough Solar Thermal Cooker (PSTC)." *Solar Energy* 201 (May 2020): 866–871. doi:10.1016/j.solener.2020.03.079.

Emmanuel, Yeboah Osei and Araba Amo-Aidoo. "Design, Fabrication, and Testing of a Solar Photovoltaic Cooker in Ghana." *2018 International Conference on Applied Sciences and Technology (ICAST)*, Kumasi Technical University, 2018, ISSN 2467-902x:182-188

Farooqui, Suhail Zaki. "A Review of Vacuum Tube Based Solar Cookers with the Experimental Determination of Energy and Exergy Efficiencies of a Single Vacuum Tube Based Prototype." *Renewable and Sustainable Energy Reviews* 31 (March 2014): 439–445. doi:10.1016/j.rser.2013.12.010.

Funk, Paul A. "Evaluating the International Standard Procedure for Testing Solar Cookers and Reporting Performance." *Solar Energy* 68, no. 1 (January 2000): 1–7. doi:10.1016/s0038-092x(99)00059-6.

Harish, Ronge, V. Yenkat Niture and D.S. Ghodake. "A Review Paper on Utilization of Solar Energy for Cooking." *Imperial International Journal of Eco-Friendly Technologies (IIJET)* 1 (2016): 34–37. https://imperialjournal.com.

Herez, Amal, Mohamad Ramadan, and Mahmoud Khaled. "Review on Solar Cooker Systems: Economic and Environmental Study for Different Lebanese Scenarios." *Renewable and Sustainable Energy Reviews* 81 (January 2018): 421–432. doi:10.1016/j.rser.2017.08.021.

Hosseinzadeh, Mohammad, Ali Faezian, Seyyed Mahdi Mirzababaee, and Hosein Zamani. "Parametric Analysis and Optimization of a Portable Evacuated Tube Solar Cooker." *Energy* 194 (March 2020): 116816. doi:10.1016/j.energy.2019.116816.

Janssen, Ellen, Jeroen Poblome, Johan Claeys, Vincent Kint, Patrick Degryse, Elena Marinova, and Bart Muys. "Fuel for Debating Ancient Economies. Calculating Wood Consumption at Urban Scale in Roman Imperial Times." *Journal of Archaeological Science: Reports* 11 (February 2017): 592–599. doi:10.1016/j.jasrep.2016.12.029.

Joshi, S.B., and A.R. Jani. "Design, Development and Testing of a Small Scale Hybrid Solar Cooker." *Solar Energy* 122 (December 2015): 148–155. doi:10.1016/j.solener.2015.08.025.

Joshi, Smita B, and A R Jani. "Certain Analysis of a Solar Cooker with Dual Axis Sun Tracker." *2013 Nirma University International Conference on Engineering (NUiCONE)*, Ahmedbad, India (November 2013). doi:10.1109/nuicone.2013.6780150.

Kimambo, C.Z.M. "Development and Performance Testing of Solar Cookers." *Journal of Energy in Southern Africa* 18, no. 3 (October 23, 2017): 41–51. doi:10.17159/2413-3051/2007/ v18i3a3384.

Mohan, Riya Rachel. "Time Series GHG Emission Estimates for Residential, Commercial, Agriculture and Fisheries Sectors in India." *Atmospheric Environment* 178 (April 2018): 73–79. doi:10.1016/j.atmosenv.2018.01.029.

Motwani, Karan, and Jatin Patel. "Cost Analysis of Solar Parabolic Trough Collector for Cooking in Indian Hostel – a Case Study." *International Journal of Ambient Energy* 40 (August 25, 2019): 1–7. doi:10.1080/01430750.2019.1653968.

Mukaro, R., and D. Tinarwo. "Performance Evaluation of a Hot-Box Reflector Solar Cooker Using a Microcontroller-Based Measurement System." *International Journal of Energy Research* 32, no. 14 (November 2008): 1339–1348. doi:10.1002/er.1441.

Murnawianto, S. S. and S. B. Rahardjo. "Solar Cookers as a STEM-Based Learning Media of Heat Transfer Topic at Middle School Level." *Journal of Physics: Conference Series* 1153 (February 2019): 012130. doi:10.1088/1742-6596/1153/1/012130.

Nkhonjera, Lameck, Tunde Bello-Ochende, Geoffrey John, and Cecil K. King'ondu. "A Review of Thermal Energy Storage Designs, Heat Storage Materials and Cooking Performance of Solar Cookers with Heat Storage." *Renewable and Sustainable Energy Reviews* 75 (August 2017): 157–167. doi:10.1016/j.rser.2016.10.059.

Omotoyosi O. Craig, Robert T. Dobson, and Wikus van Niekerk. "A Novel Indirect Parabolic Solar Cooker." *Journal of Electrical Engineering* 5, no. 3 (June 28, 2017). doi:10.17265/2328-2223/2017.03.003.

Oyewo, Ayobami Solomon, Arman Aghahosseini, Manish Ram, and Christian Breyer. "Transition Towards Decarbonised Power Systems and Its Socio-Economic Impacts in West Africa." *Renewable Energy* 154 (July 2020): 1092–1112. doi:10.1016/j. renene.2020.03.085.

Padilla-Barrera, Zuhelen, Ricardo Torres-Jardón, Luis Gerardo Ruiz-Suarez, Telma Castro, Oscar Peralta, María Isabel Saavedra, Omar Masera, Luisa Tan Molina, and Miguel Zavala. "Determination of Emission Factors for Climate Forcers and Air Pollutants from Improved Wood-Burning Cookstoves in Mexico." *Energy for Sustainable Development* 50 (June 2019): 61–68. doi:10.1016/j.esd.2019.02.004.

Panchal, Hitesh, Jay Patel, and Sudhir Chaudhary. "A Comprehensive Review of Solar Cooker with Sensible and Latent Heat Storage Materials." *International Journal of Ambient Energy* 40, no. 3 (November 3, 2017): 329–334. doi:10.1080/01430750.2017.1392357.

Rathore, Mahesh M. and Ravi M. Warkhedkar. "A Review of Solar Cookers." *International Journal of Modern Trends in Engineering and Research* 2, no. 7 (2015): 1997–2004. www.ijmter.com.

Talbi, S., K. Kassmi, O. Deblecker, and N. Bachiri. "Thermal Heating by Photovoltaic Solar Energy." *Materials Today: Proceedings* 13 (2019a): 1125–1133. doi:10.1016/j. matpr.2019.04.080.

Talbi, Sofian, Khalil Kassmi, Ilias Atmane, Olivier Deblecker, and Noureddine Elmoussaoui. "Feasibility of a Box-Type Solar Cooker Powered by Photovoltaic Energy." *2019 7th International Renewable and Sustainable Energy Conference (IRSEC)*, Agadir, Morocco (November 2019b). doi:10.1109/irsec48032.2019.9078275.

Terres, Hilario, Arturo Lizardi, Raymundo López, Mabel Vaca, and Sandra Chávez. "Mathematical Model to Study Solar Cookers Box-Type with Internal Reflectors." *Energy Procedia* 57 (2014): 1583–1592. doi:10.1016/j.egypro.2014.10.150.

Thirugnanam, C., S. Karthikeyan, and K. Kalaimurugan. "Study of Phase Change Materials and Its Application in Solar Cooker." *Materials Today: Proceedings* (March 2020). doi:10.1016/j.matpr.2020.02.780.

Yettou, F., B. Azoui, A. Malek, A. Gama, and N.L. Panwar. "Solar Cooker Realizations in Actual Use: An Overview." *Renewable and Sustainable Energy Reviews* 37 (September 2014): 288–306. doi:10.1016/j.rser.2014.05.018.

Yettou, F., A. Gama, B. Azoui, A. Malek, and N. L. Panwar. "Experimental Investigation and Thermal Modelling of Box and Parabolic Type Solar Cookers for Temperature Mapping." *Journal of Thermal Analysis and Calorimetry* 136, no. 3 (October 13, 2018): 1347–1364. doi:10.1007/s10973-018-7811-9.

Yettou, Fatiha. 2015. "Conception et réalisation d'un système de cuisson solaire destine au site saharien (Ghardaia,Algerie)," Doctorat en sciences, Département d'Electrotechnique, Universite HADJ LAKHDAR Batna, Faculte de technologie.

14 A Hot Plate Operated by Photovoltaic Solar Energy

I. Atmane, S. Talbi, K. Hirech, M. Melhaoui,
K. Kassmi, O. Deblecker, and N. Bachiri

CONTENTS

14.1 INTRODUCTION

Currently, millions of people still have limited access to cooking fuels. In developing countries, particularly in the rural world, cooking is done by using wood from forests. This causes much deforestation, and therefore degrades the environment. To remedy this, several alternative solutions have been proposed based on renewable energy sources, such as solar (Planète Urgence 2006). Solar ovens have been proposed and used in the countries of the South and Latin America, which have a significant solar deposit all the year (Colla 2017). In this context, the different types and performance of solar ovens offered are:

- Box-type solar ovens operating outside of homes with thermal energy (Panchal et al. 2019). These types of ovens were able to reach, in 4 h of use, under an irradiance of 858.11 W/m² and an ambient temperature of 37.9°C, a maximum baking temperature of the order of 140°C, with maximum thermal efficiency not exceeding 54% (Kumar et al. 2010). This type of oven requires, during its use, orientation and displacement according to the position of the sun.

- Parabolic solar ovens are based on the concentration and focusing of solar rays by the parabolic reflectors on the bottom of the container (the pot) (Indora and Kandpal 2018). This type of cooker has, when optimally used, high cooking temperatures between 200°C and 300°C (Yettou 2015) and thermal yields which vary from 43.45% to 77% (Grupp et al. 2009). The disadvantages of this type of system are: it is bulky, orientation of the reflectors with the wire of the sun, it is expensive, and it presents risks of burns for users.
- Concentrated solar ovens, based on the heating of a fluid, have recently been proposed (El Moussaoui et al. 2020). This type of oven, operating on thermal energy, attains a temperature of 200°C after 30 min, with an irradiance of 720 W/m² and an ambient temperature of 26°C. Despite the interesting performance of this type of oven, its use remains limited outside of homes under the sun.
- Currently, designers of solar cookers are moving towards improving the performance of boxed solar ovens by integrating photovoltaic (PV) energy. The work carried out concerns the heating of the elements by solar batteries of 24 V and a capacity of 45 Ah, charged by PV panels (Osei and Amo-Aidoo 2018). The use of batteries increases the purchase cost of the stove and the maintenance costs. Despite the use of PV energy, performance is very limited. Under an irradiance of 950 W/m² and an ambient temperature of 20°C, and a battery power of 76.9 W, the temperatures inside the oven and the heating element do not exceed 124°C (Joshi and Jani 2013). In addition, for a quantity of water of 0.385 g, the thermal efficiency does not exceed 43.6% (Joshi and Jani 2015).

In this context, we propose the design and production of innovative solar cookers, operating on PV energy outside and inside homes. This work was carried out in collaboration with the Polytechnic Faculty of MONS (Belgium) (project: Wallonie-Bruxelles International WBI, 2018–2022, N°4.2), and the socio-economic sectors and civil society (National Initiative for Human Development INDH from the province of Berkane, project N°29/2017, Association AHEB). The main objective of the work is the realization of ovens, which are reliable, of good performance, and adaptable to user needs, in terms of cooking at temperatures adjustable up to 300°C.

In this chapter, we propose the feasibility of a solar heating plate, powered by photovoltaic panels of power 280 W/Crete. After describing the structure of the hot plate, we present the first results obtained in the laboratory, concerning water heating and cooking temperatures, and the estimation of the thermal efficiency throughout the day.

14.2 PHOTOVOLTAIC HOT PLATE

14.2.1 SYSTEM STRUCTURE

Figure 14.1 shows a diagram of the PV energy heating plate, proposed in this work. The different blocks of our equipment are:

- *Block A*: formed by PV panels, with a peak power 280 W, and producing electrical energy according to the intensity of the illumination throughout the sunny

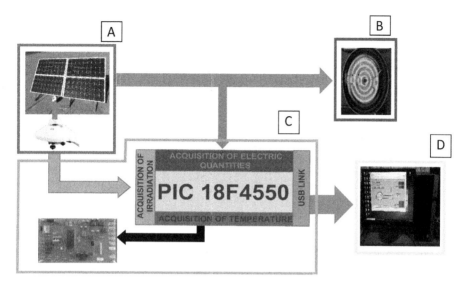

FIGURE 14.1   Synoptic diagram of the solar hot plate system with PV energy, equipped with a system of acquisition measurement

day. They are placed on the roof of the laboratory, oriented south and exposed to the sun at a fixed inclination of 40°. In order to accurately determine the intensity of the illumination and the ambient temperature, a pyranometer and CTN sensor fitted with an adequate circuit are calibrated and installed with the PV panels.

- *Block B*: represents the heating plate, which is constituted by a steel support, placed in a ceramic with a radius of 16 cm, supporting a temperature higher than 1200°C. This ceramic is engraved, in the form of a coil, to house a thermal resistance. This resistance, 16 cm long, is sized and chosen to withstand a temperature above 900°C and produce sufficient thermal power for cooking.
- *Block C*: is a system for acquiring electrical (current, voltage, and power), thermal (temperature of thermal resistance and cooking), and meteorological (irradiance and ambient temperature) quantities. This is achieved by means of a card, put in place during this work, and a microcontroller to monitor the instantaneous operation of the equipment.
- *Block D*: represents the display interface developed within the laboratory. The latter, connected to the electronic card via a USB connection, displays electrical, meteorological, and thermal quantities in real time in the form of digital data and graphs. This interface also allows the storage of all data and remote monitoring of operation.

14.2.2   THERMAL MODELS

We simulated our thermal model based on the energy balance (Mahdavi and Delavar 2018) of a temperature node of the heating resistor, taking into account the electric power P_{elc} supplied by the PV generator and the heat loss by convection P_{CV} (Mahdavi and Delavar 2018) and by P_R radiation (Mahdavi and Delavar 2018). At this node, at a given irradiance and ambient temperature T_{Amb}, the temperature gradient of the resistance T_r, in a transient regime, is written in the following form:

$$m_r \cdot Cp_r \frac{dT_r}{dt} = P_{elc} - P_{cv} - P_R$$

$$= Vs \cdot Is - h_{r-amb}A_r\left(T_r - T_{Amb}\right) - A_r\sigma\varepsilon_r\left(T_r^4 - T_{Amb}^4\right) \tag{14.1}$$

where

m_r: Mass of the heating element ($m_r = 0.01634$ kg),

Cp_r: Specific heat of the heating element ($Cp_r = 890$ J/kg-K)

Vs and Is: Electric voltage and current of the heating resistance,

h_{r-Amb}: Coefficient of convective exchange between the heating resistance and its exterior,

Ar: Heating resistance surfaces ($S = 0.0025$ m²),

σ: Boltzmann constant ($\sigma = 5.669*10^{-5}$ w/m²·k⁴),

ε: Heating resistance emissivity ($\varepsilon = 0.9$).

The numerical resolution of Equation 14.1 is carried out by the Fourth Order Runge–Kutta method (Talbi et al. 2019) in the C ++ language. We have adopted this method since it presents remarkable accuracy and stability of calculation (Talbi et al. 2019).

The estimate of the power Po and the thermal efficiency η of the plate is obtained from the rise in the water temperature ΔT during time intervals $\Delta t = 10$ min, using the Funk model (Talbi et al. 2019), according to the expressions:

$$Po = \frac{m \cdot Cp \cdot \Delta T}{\Delta t} \tag{14.2}$$

$$\eta = \frac{P_o}{P_{pv}} \tag{14.3}$$

where

m: Mass of heated water (m = 1 kg),

Cp: Specific heat of water (Cp = 4190 J/kg-K),

P_{pv}: Electrical power of photovoltaic panels.

14.3   RESULTS AND DISCUSSION

14.3.1   EXPERIMENTAL PROCEDURE

The solar PV cooker, which is the subject of our work, designed and produced in our laboratory, is shown in Figure 14.2. This cooker is composed of:

A. PV panels generating 280 W/peak power. This block also has a weather station formed by a pyrometer to measure the total irradiance and a CTN thermistor sensor to measure the ambient temperature. This station is linked to our acquisition card, which transfers all the data to our acquisition interface.
B. A heating resistance of 12 Ω and which can provide a temperature of 1000°C.
C. An electronic system formed by:
   • An acquisition card based on the use of a microcontroller, which ensures the acquisition and transfer of data to the graphical interface.
   • An LCD display in real time of the system power, the lighting, the ambient temperature, and the temperature of the heating element.
   • A power supply card that generates the voltages (+12 V, −12 V, +5 V, GND) necessary for the proper functioning of the system. A small battery (50 Ah) which is charged by a 10 W panel, via a charge/discharge regulator, powers this card. This card makes our system autonomous and independent of any other non-PV energy source.

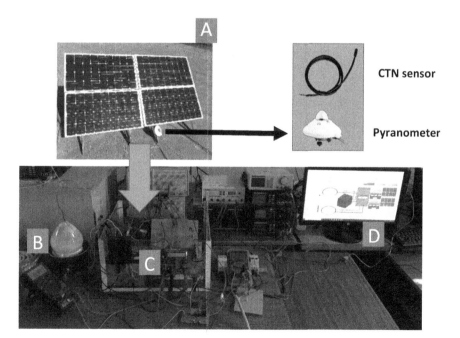

FIGURE 14.2   Solar cooker made in the laboratory

D. A supervision computer, linked to our system via a USB laision, for the acquisition of data sent by the acquisition card. This computer has a graphical interface created in LabVIEW, which processes and displays the data acquired in the form of a graph and table.

14.3.2 PHOTOVOLTAIC GENERATOR

The dimensioning of PV panels, which are the source of energy for our hot plate, requires knowledge of the electrical characteristics of the thermal resistance and of the PV panels used (Mambrini 2014). To do this, we noted the current–voltage and power–voltage characteristics of the PV panels used. The typical results obtained under an irradiance intensity of 250, 500, and 900 W/m² and an ambient temperature of 25°C are shown in Figure 14.3. From these characteristics, we have deduced and

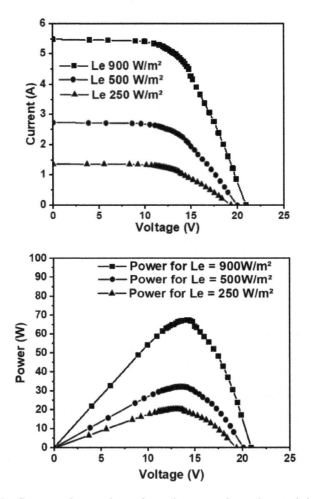

FIGURE 14.3 Current–voltage and experimental power–voltage characteristics of a panel under the irradiance of 250, 500, and 900 W/m² and an ambient temperature of 25°C

TABLE 14.1
Optimal electrical quantities of a PV panel

Le (W/m²)	Vopt (V)	I(opt) (A)	Ropt (Ω)	Popt (W)
250	14.8	1.3	11.38	19.24
500	14.9	2.5	6	37.25
900	15	5	3	75

represented in Table 14.1, the optimal electrical quantities of the PV panel (voltage, current, power, and resistance).

From the value of the resistance used (Rthermal = 12 Ω) and the optimal value of the resistance of a panel, we connected four PV panels in series. This minimizes power losses (Dang 2014) and produces an electrical power adaptable to thermal resistance.

14.3.3 OPERATION OF THE PHOTOVOLTAIC COOKER

We experimented with the solar cooker shown in Figure 14.2 on an entirely sunny day by heating 0.3 L of oil. We noted the intensity of the irradiance, the ambient temperature, the electric power supplied by the PV panels, and the temperature of the thermal resistance and of the oil. We have plotted the temperature of the resistance simulated by Equation 14.1. The results obtained, shown in Figure 14.4, show:

- During the measurement day, around noon, the intensity of the irradiance reaches 730 W/m² and the ambient temperature 25°C.
- The best performance is obtained around noon when the power of the PV panels reaches the value of 230 W.
- When the intensity of the irradiance is 500 W/m² and the power supplied by the PV panels is around 150 W, the resistance temperature varies from 24°C to 500°C after 30 s (i.e., an increase of 16°C/s).
- The oil temperature exceeds 100°C, after 30 min of operation.
- The oil temperature reaches 200°C, after 120 min of heating, under an irradiance varying from 500 to 660 W/m² and an electrical power of the PV panels varying from 142 to 225 W.
- Optimal performance is obtained around noon. The intensity of the irradiance is 730 W/m², the power of the PV panels is 230 W, the temperatures of the thermal resistance and of the oil is 680°C and 250°C.
- There is good agreement between the experience and the simulation of the temperature of the thermal resistance by taking the convective exchange coefficients between the heating resistance and its ambient varying from 87 to 190 W/m²·k.

Comparing these results with thermal ovens, in particular PV, with the literature (Joshi and Jani 2015), we obtain better performances on our cooker. The rise in

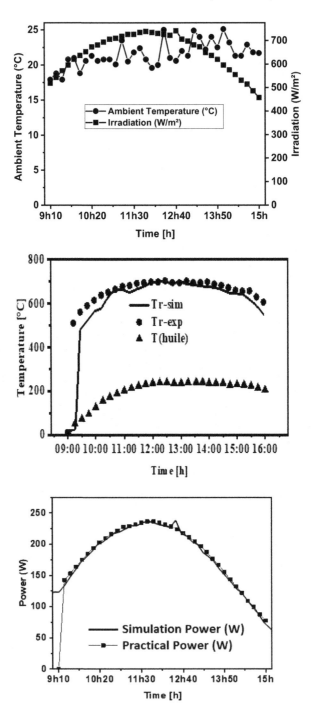

FIGURE 14.4 Intensity of irradiance and temperature, electric power supplied by the experimental and simulated PV panels, and experimental and simulated temperature of the oil of the heating resistance, obtained during characterization of the heating plate of Figure 14.1

temperature of the thermal resistance and of the oil is improved by 86% and their maximum values reach 43%. This testifies to the proper functioning of the cooker designed, produced, and tested during this work.

14.3.4 SOLAR COOKER EFFICIENCY

We have estimated the thermal efficiency of our solar cooker based on the funk method using Equations 14.2 and 14.3. The measurements are carried out under the same meteorological conditions as those in the previous section (Figure 14.4). Typical results are shown in Figure 14.5. We can deduce that:

- The intensity of the irradiance and the ambient temperature are around 700 W/m^2 and 22°C.
- After 10 min of heating, the water temperature increases from 20°C to 40°C, i.e., 2°C/min.
- Between 12 and 1 p.m., the electrical powers Pel and thermal Po, the temperatures of the thermal resistances and of the water, reach the values 257.2 W, 153.6 W, 645°C, and 99°C, respectively.
- The boiling point of water (98°C) is reached after 40 min of heating.
- The energy efficiency η reaches the maximum value of 77.5% after 20 min of heating. Beyond that, the water temperature stabilizes around 98°C and the water evaporates.

Comparing these results to those found on conventional ovens, we can deduce a better performance by our equipment: heating speed, maximum heating temperature of 250°C, thermal efficiency of around 77%. In addition, its use in homes does not require the orientation of the cooker to the sun. All of these results clearly show the feasibility of the operation of the hot plate, designed and produced during this work, with PV energy.

14.4 CONCLUSION

In this chapter, we have shown the feasibility of a new method of cooking with renewable PV energy. We have sized, depending on the nature of the PV panels and the thermal resistance used, a hot plate, supplied with a power of 230 W, to heat 0.33 L of oil and 1 L of water. The results obtained show:

- The heating resistance temperature reaches 500°C after 30 s of heating;
- A rapid rise in oil temperature of 7.7°C/min and 2°C/min in the case of water;
- For a power of 230 W, after 10 min of heating, the cooking temperature reaches 240°C;
- A very good agreement between the simulation and experimental results concerning the temperature of the heating resistor;
- A thermal efficiency of around 77%.

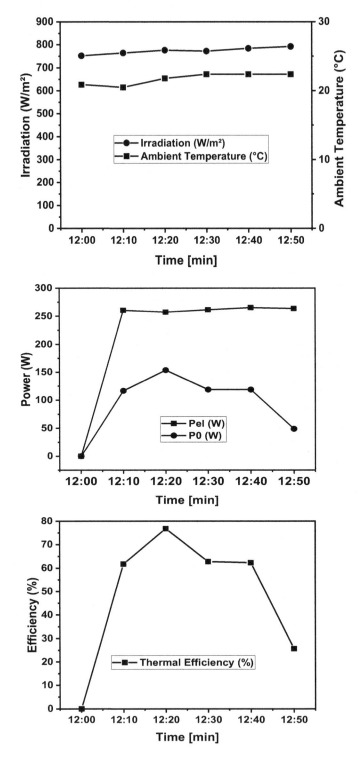

FIGURE 14.5 Variation of the intensity of the irradiance and the temperature, electric and thermal power, and thermal efficiency of the cooker of Figure 14.2

The comparison of these results with those of the literature on conventional box ovens shows:

- Improvement of the maximum value of the oil heating temperature of 73%;
- Improvement of the rise in water heating by 73%;
- Improvement of thermal efficiency compared to conventional thermal ovens of 20%.

All the results and performance obtained in this work clearly show the good functioning and the feasibility of the PV energy heating plate, and hence its use in homes in the rural and urban world.

ACKNOWLEDGMENTS

This research was carried out in collaboration with Mohamed Premier University (Morocco), Polytechnic Faculty of Mons (Belgium) and Association Man and Environment of Berkane (Morocco), within the framework of the projects:

- Morocco-Wallonie Cooperation Program Brussels (2018–2022), Wallonie-Bruxelles-International, project 4, n°2;
- National Initiative for Human Development INDH, Berkane Province, Morocco, project 2017/29.

REFERENCES

Colla, S. 2017. https://www.wedemain.fr/Ce-four-solaire-peut-etre-utile-en-Amerique-latine-en-Afrique-Et-dans-votre-jardin_a3047.html (accessed May 1, 2020).

Dang, X. L. 2014. *Contribution à l'étude des système Photovoltaïque/Stockage distribués. Impact de leur intégration à un réseau fragile* (Doctoral dissertation).

El Moussaoui, N., Talbi, S., Atmane, I. et al. 2020. Feasibility of a new design of a Parabolic Trough Solar Thermal Cooker (PSTC). *Solar Energy*, 201, 866–871.

Grupp, M., Balmer, M., Beall, B. et al. 2009. On-line recording of solar cooker use rate by a novel metering device: Prototype description and experimental verification of output data. *Solar Energy*, 83(2), 276–279.

Indora, S., & Kandpal, T. C. 2018. Institutional and community solar cooking in India using SK-23 and Scheffler solar cookers: A financial appraisal. *Renewable Energy*, 120, 501–511.

Joshi, S. B., & Jani, A. R. 2013. Certain analysis of a solar cooker with dual axis sun tracker. In *2013 Nirma University International Conference on Engineering (NUiCONE)* (pp. 1–5). IEEE.

Joshi, S. B., & Jani, A. R. 2015. Design, development and testing of a small scale hybrid solar cooker. *Solar Energy*, 122, 148–155.

Kumar, N., Chavda, T., & Mistry, H. N. 2010. A truncated pyramid non-tracking type multi-purpose domestic solar cooker/hot water system. *Applied Energy*, 87(2), 471–477.

Mahdavi, A., & Delavar, M. A. 2018. Thermal performance of wavy fin in a compact heat exchanger duct using Galerkin method. *Applied Thermal Engineering*, 130, 1290–1298.

Mambrini, T. 2014. *Caractérisation de panneaux solaires photovoltaïques en conditions réelles d'implantation et en fonction des différentes technologies* (Doctoral dissertation, Paris 11).

Osei, E. Y., & Amo-Aidoo, A. 2018. Design, fabrication, and testing of a solar photovoltaic cooker in Ghana. In *International Conference on Applied Science and Technology Conference Proceedings* (Vol. 4, No. 1, pp. 182–188).

Panchal, H., Patel, J., & Chaudhary, S. 2019. A comprehensive review of solar cooker with sensible and latent heat storage materials. *International Journal of Ambient Energy*, 40(3), 329–334.

Planète Urgence. 2006. Les principales causes et solutions à la déforestation. https://www.notre-planete.info/actualites/891-causes_solutions_deforestation (accessed May 1, 2020).

Talbi, S., Kassmi, K., Deblecker, O., & Bachiri, N. 2019. Thermal heating by photovoltaic solar energy. *Materials Today: Proceedings*, 13, 1125–1133.

Yettou, F. 2015. *Conception et réalisation d'un système de cuisson solaire destiné au site saharien (Ghardaïa, Algérie)* (Doctoral dissertation, Thèse de doctorat unique).

15 Autonomous Photovoltaic Installations and Injection into the Electricity Network at the Technopole of Oujda (Morocco)

A. Lamkaddem and K. Kassmi

CONTENTS

15.1 INTRODUCTION

The production of electrical energy is a real challenge from an ecological and economic point of view. The energy needs of industrialized societies as well as developing countries continues to grow. Today, most of the world's energy production comes from non-renewable fossil sources. This type of energy gives rise to greenhouse gas emissions and therefore an increase in pollution (Al-Badi, Abdullah 2019; Mensah, Yamoah, and Adaramola 2019). In addition, the excessive consumption of natural resource stocks dangerously reduces the reserves of this type of energy (Al-Otaibi 2015; Attri et al. 2016; Shukla et al. 2016; Benalouache 2017). At the national level, the promotion of solar energy in Morocco shows a greater interest in renewable energy sources, since this type of resource is more profitable in terms of cost and environmental protection (Shiva Kumar and Sudhakar 2015; Nallapaneni et al. 2017; Herbazi et al. 2019). In addition, the average annual value of solar radiation in Morocco, which is 2,600 kWh/m^2/year, qualifies the country to develop the photovoltaic (PV) sector in terms of electrical energy. Currently, several solar PV power plant projects are underway: Noor Argana for 100 MW, Noor Atlas 1 for 200 MW, and Noor Atlas 2 in a later longer-term phase for an additional 800 MW. Also, the Moroccan Agency for Sustainable Energy (MASEN) is developing four other large solar power plants in Midelt (400 MW), Tarfaya (500 MW), Laayoune (400 MW), and Boujdour (100 MW) (Attari 2016; Laura 2016; Salimi, Hossein et al. 2020). All of these achievements show the strategy of the Moroccan government to develop, innovate, and encourage autonomy and injection into the electricity network PV installations. Moreover, Morocco sees the African continent as a source of economic and diplomatic relations. It is a major partner facing a continent which has recorded overall growth of 5% per year since the 2000s. The country's regional vision is also turning toward Europe. Already electrically interconnected with Spain, Morocco does not despair of exporting electricity from renewable sources to the member countries of the EU. In the absence of a horizontal regional integration (Maghreb) for which Morocco militates openly, the latter intends to constitute an essential hub between Europe and sub-Saharan Africa, thus creating a corridor along a vertical axis (Lamkaddem and Khalil 2016; Al-Badi, A. 2018). Allowing Morocco to organize the United Nations Climate Conference-COP22 is also recognizing the kingdom as a great country of ecology. According to the independent organization Climate Action Tracker (CAT), Morocco is one of the four greenest countries in the world alongside Costa Rica, Bhutan, and Ethiopia. This Moroccan ecological shift has materialized in a very ambitious energy plan since the kingdom aims to bring the contribution of renewable energies to electricity production to 52% by 2030 (Amegroud 2015; Attari 2016, Lamkaddem et al. 2016). Many projects are already underway, such as the Ouarzazate solar power plant. The first Noor I unit has been in service since February 4, 2016, and in the long term the program aims to produce 2000 MW in 2020. Noor will be the largest solar project in the world in the long term (El-Katiri 2016). In this energy context, Mohammed Premier University of Oujda (Morocco) aligns with the government strategy, in terms of energy protection, production of electric energy by renewable sources (PV), and reduction of electricity bills. As part of a pilot project, the university is proposing the production of electrical energy entirely by PV renewable energies on the university CAMPUS of the Oujda Technopole. Our MEER research team from the LETSER laboratory is responsible for the formulation, execution, and monitoring of this project. Two pilot installations are then proposed and carried out: an autonomous one with storage in batteries (10 kW, 48 V), to supply single-phase loads, and an injection into the grid (10 kW, 50 Hz, 230 V) to supply three-phase loads; the surplus is injected into the grid. In this chapter, we present the dimensioning, the stages of realization, and the first test results of these two pilot installations.

15.2 SIZING OF THE PHOTOVOLTAIC INSTALLATION

15.2.1 Geographical Location of the Oujda Technopole

The Oujda region benefits from a 200 km long sea front, offering significant opportunities for its economic development. Currently, the region offers the Technopole, with an area of 4 hectares, for investors in various fields (renewable, electronic, agricultural, etc.). In terms of service infrastructure, the Technopole benefits from the proximity advantages of Oujda Angad airport, served by the Oujda–Saidia expressway, and is located 120 km from the port of Nador-BeniNssar. The university campus, attached to Mohammed University of Oujda, is located in the Oriental region. The GPS coordinates of the university campus are:

- Latitude: 34.6805200°;
- Longitude: −1.9076400°.

During discussions with the President of Mohammed Premier University of Oujda in 2015, our team proposed the installation of two pilot stations, each with a power of 10 kW, on the university campus. One installation is autonomous with storage in solar batteries (eight batteries of 150 Ah) and the other is connected to the electrical grid (10 kW, 50 Hz, 230 V). According to the data provided on the consumption of electrical energy, these installations provide the necessary energy during the day and at night, throughout the year.

On the basis of in-depth studies of the energy needs of the campus and simulations in the appropriate software, we proposed two installations (Figure 15.1): grid-connected and stand-alone with storage in batteries. The different equipment required is shown in Table 15.1. The structure and operation of these facilities are:

- The grid-connected system consists of 40 PV panels of 250 W. They are mounted in two parallel strings, of 20 panels in series, and a 10 kW central inverter and a 1000 V protection box. The three-phase inverter is connected to the three-phase grid (230 V, 50 Hz) and to the buildings (three-phase load). When there is sufficient sunlight, this installation feeds the buildings and the surplus is injected into the grid.
- The stand-alone installation with storage is formed by two branches, connected to eight batteries (12 V, 150 Ah), two 5 KVA DC/AC inverters, two 1000V protection boxes and in particular to the buildings (single-phase load). Each branch is formed by 20 PV panels, connected in series of four strings of five panels in parallel. In the presence of sufficient sunlight, this installation charges the batteries, which power the buildings. In the absence of sunlight, or during the night, the batteries supply the buildings.
- When there is no sunlight and the energy stored in the batteries is not available, the buildings are powered directly from the electrical grid.

15.2.2 FUNCTIONING

To ensure the correct dimensioning of the two installations, we analyzed their operation in the PVsyst software, then extracted the information on: irradiance; normalized production and loss factors; normalized production and loss factors; daily energy at the exit of the field from the panels; cos(φ); and the cost of the energy produced. The results obtained are shown in Figures 15.2, 15.3, and 15.4. We can thus conclude:

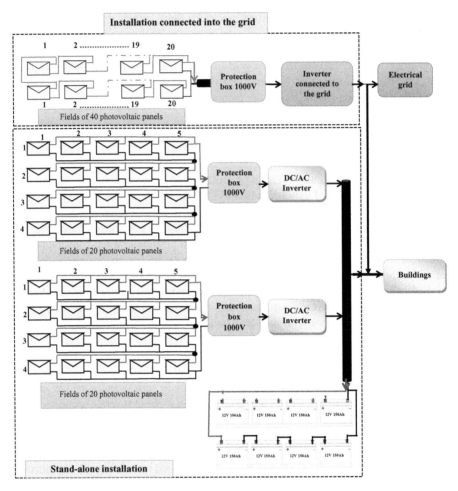

FIGURE 15.1 Installations, grid-connected and stand-alone with storage, realized at the University Campus of the Oujda Technopole

- **Grid-connected PV system:**
  - The average daily irradiance value is 5.519 kW/m²/day (Figure 15.2A).
  - Average annual temperatures vary between 15°C and 20°C. Maximum temperatures can exceed 34°C, while absolute minimum temperatures decrease below 4°C (Figure 15.2B).
  - The average daily energy produced by the PV plant is 5.52 kWh/kWp/day, or 55.2 kWh/day.
  - The energy injected into the grid is of the order of 4.33 kWh/kWp/d, i.e., 43.3 kWh/d and 15.93MWh/year. This energy represents 78.4% of the energy produced by the PV installation, the rest of the energy 21.6% is lost (e.g., in cables) (Figure 15.2C, D).
  - The average value of the voltage of the PV field in June is around 515.3V.

TABLE 15.1
Characteristics of the equipment of the global PV installation in Figure 15.1

Equipment		Characteristics
1 PV panel		P_{max} = 250 W; V_{mp} = 29.7 V; I_{mp} = 7.75 A; V_{oc} = 36.6 V; I_{cc} = 8,36 A
Photovoltaic installation connected to the grid	40 PV panels	P_{syst} = 9.20 kW; V_{syst} = 594 V; Isyst = 15.5A
	Three-phase inverter, 10 kW	Type of grid connection, AC, Ps = 20,000 W, Vs= 400V_{ac}, I_s = 33.0 A
	electrical cables	40 m
	1 protective box	10,000V
Stand-alone photovoltaic installation	40 PV panels	P_{syst} = 9,207 kW, V_{syst} = 594 V, I_{syst} =15.5 A
	two inverters	4KVA 48VDC
	Batteries	8 batteries, 12V, 150 Ah, 100 A
	Electrical cables	80 m
	2 protective boxes	10,000 V

- The energy injected into the power grid increases according to the overall incident irradiance on the active surface of the PV panels. For a global irradiation of approximately 5,519 kWh/m²·day, the energy injected into the grid is approximately 22 kWh/day (Figure 15.2F).

- **Stand-alone PV installation with storage in solar batteries:**
  - The average daily irradiation is around 5,519 kWh/m²·d.
  - The maximum PV energy produced by the PV installation is around 48 kWh/d.
  - The average daily energy produced by the PV installation is 45.80 kWh/d; 74.8% of this energy is supplied to the user, while 25.1% is lost (e.g., in cables, system, battery charge, PV field) (Figure 15.4E).

15.3 FIELD INSTALLATION

15.3.1 INSTALLATION EQUIPMENT

Figure 15.4 shows the equipment of the two stand-alone and grid-connected PV installations used at the University Campus of Oujda, according to the specifications in the previous section. After discussion with the company, which took over this contract, it was agreed to deliver and install the following equipment (Figure 15.5):

- **A Grid-connected photovoltaic system with a power output of 10 kW:**
  - A PV generator formed with 40 PV panels, where the power of one panel is 230 Wp (Figure 15.5A). This is formed with 2 PV strings, with 20 panels in series each, to obtain a voltage of 580 V. These two strings are connected to the ABB 10 kW, 200–850 V inverter (Figure 15.5B).
  - An inverter (ABB 10 kW THREE-PHASE PVI-10.0-TL-OUTD-BWP ON-SITE EXCHANGE) (Figure 15.5B) whose characteristics are shown in

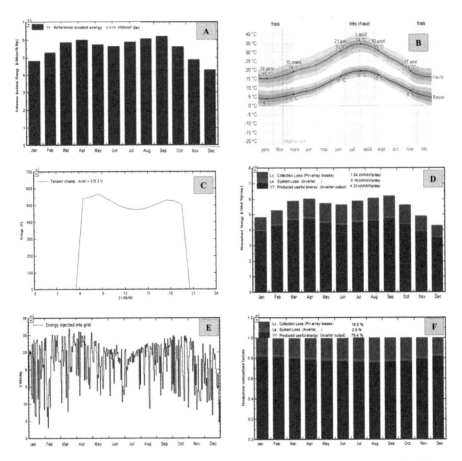

FIGURE 15.2 Operation of the system connected to the grid. A: Variation of the irradiance during the year. B: Evolution of average, maximum, and minimum temperature during the year. C: Average June voltage of the field. D: Normalized production (KWp installed) during the year. E: Daily energy at the exit of the system injected into the grid, during the year. F: Normalized production and loss factors, during the year

Table 15.1. While the inverter is operating, the display shows all the information related to the installation (Figure 15.6): voltage, current, powers, frequencies, and so on.

- A 1,000 V protection box whose role is to protect and control the operation of the installation. It is made up of circuit breakers, visible, contactor, distributor, and so on.
- A public LV distribution network that is three-phase 50 Hz with a distributed neutral. The technical characteristics of this network are:
- Reference voltage: 3×.220/380 V–50 Hz- 4 wires:
- Voltage range: from 0.8 to 1.15 one.
- Current range: 5–60 A:
  - Base current: 5 A minimum;
  - Maximum current: 80 A maximum.

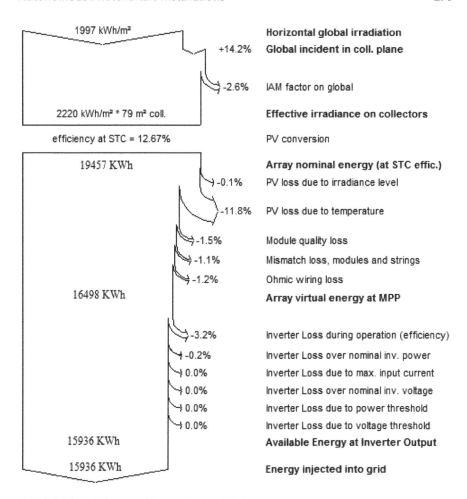

1997 kWh/m²	**Horizontal global irradiation**
+14.2%	**Global incident in coll. plane**
-2.6%	IAM factor on global
2220 kWh/m² * 79 m² coll.	**Effective irradiance on collectors**
efficiency at STC = 12.67%	PV conversion
19457 KWh	**Array nominal energy (at STC effic.)**
-0.1%	PV loss due to irradiance level
-11.8%	PV loss due to temperature
-1.5%	Module quality loss
-1.1%	Mismatch loss, modules and strings
-1.2%	Ohmic wiring loss
16498 KWh	**Array virtual energy at MPP**
-3.2%	Inverter Loss during operation (efficiency)
-0.2%	Inverter Loss over nominal inv. power
0.0%	Inverter Loss due to max. input current
0.0%	Inverter Loss over nominal inv. voltage
0.0%	Inverter Loss due to power threshold
0.0%	Inverter Loss due to voltage threshold
15936 KWh	**Available Energy at Inverter Output**
15936 KWh	**Energy injected into grid**

FIGURE 15.3 Diagram of losses for the global system

- Reference frequency: 50 Hz:
- Accuracy class: 1;
- Resistance to alternating voltage for 1 min: 4 kV–50 Hz;
- Resistance to shock voltage: 8 kV.
- **Stand-alone photovoltaic installation of 10 kW with storage in solar batteries:**
  - A PV generator formed by two identical PV fields (Figure 15.5A). Each field is formed by 20 PV panels, where the power of one panel is 230 Wp. The field is connected as five parallel strings of four panels in series. This connection provides 104 V and is adaptable to the Hybrid solar inverter Ph18 MPPT 1KVA~5KVA (Figure 15.4D).
  - Two DC/AC Hybrid Solar PH1800 DC/AC inverters, single-phase, with an efficiency of more than 93%. Their input is 48 V and output is 230V. Each consists of a powerful pure wave current inverter, a charge regulator, and a battery charger (Figure 15.5D). Also, it has a screen that allows us to

configure its parameters and display all the electrical quantities during its operation (Figure 15.7): input voltage, input frequency, PV voltage, battery charge current, battery voltage, output voltage, output frequency, percentage of charge, VA charge, and Watt charge (Table 15.2).

- Eight batteries, where one battery is 12 V voltage and 300 Ah capacity. The batteries are connected by two chains in parallel to form 48 V. Each chain consists of four batteries in series. The role of these batteries is to store the electrical energy produced by the PV panels and to ensure the continuity of the power supply during the night and on non-sunny days.

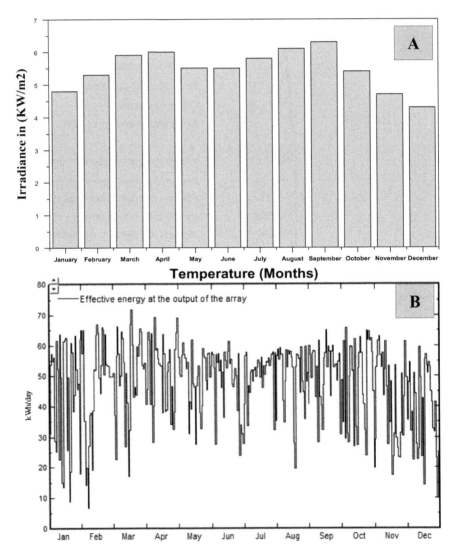

FIGURE 15.4 Operation of the stand-alone installation. A: Irradiance variation during the year. B: Daily energy produced by the PV installation during the year.

(*Continued*)

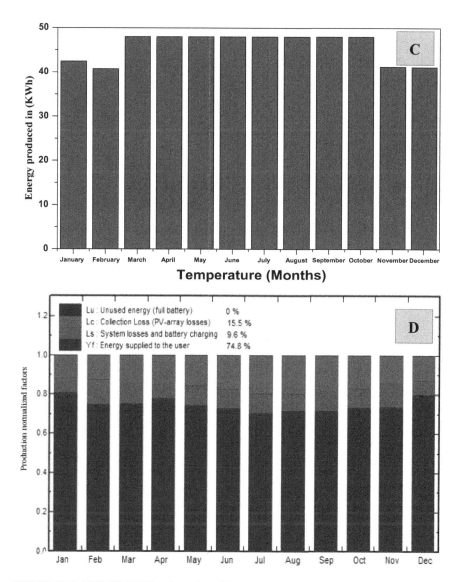

FIGURE 15.4 (CONTINUED) Operation of the stand-alone installation. C: Daily energy at the exit of the PV field during the year. D: Normalized production and loss factors during the year.

- A 1000 V protection box which is the same as the ones in the installation connected to the electrical grid.

15.3.2 RESULTS AND DISCUSSION

15.3.2.1 Grid-Connected Installation

We followed the daily operation of the installation connected to the electrical grid by directly downloading the measurements provided by the acquisition system

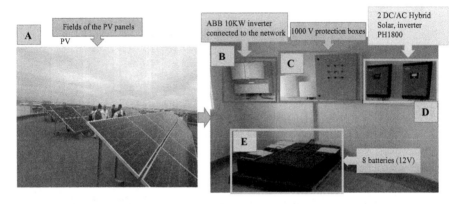

FIGURE 15.5 Stand-alone and grid-connected installation at the Oujda University campus. A: Fields of the PV panels. B: Grid-connected ABB 10KW inverter. C: 1000 V protection boxes. D: Two DC/AC Hybrid Solar inverter PH1800 (E) 8 batteries (12V)

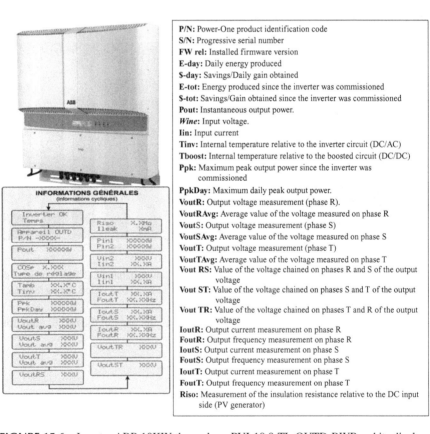

P/N: Power-One product identification code
S/N: Progressive serial number
FW rel: Installed firmware version
E-day: Daily energy produced
$-day: Savings/Daily gain obtained
E-tot: Energy produced since the inverter was commissioned
$-tot: Savings/Gain obtained since the inverter was commissioned
Pout: Instantaneous output power.
***Wine*:** Input voltage.
Iin: Input current
Tinv: Internal temperature relative to the inverter circuit (DC/AC)
Tboost: Internal temperature relative to the boosted circuit (DC/DC)
Ppk: Maximum peak output power since the inverter was commissioned
PpkDay: Maximum daily peak output power.
VoutR: Output voltage measurement (phase R).
VoutRAvg: Average value of the voltage measured on phase R
VoutS: Output voltage measurement (phase S)
VoutSAvg: Average value of the voltage measured on phase S
VoutT: Output voltage measurement (phase T)
VoutTAvg: Average value of the voltage measured on phase T
Vout RS: Value of the voltage chained on phases R and S of the output voltage
Vout ST: Value of the voltage chained on phases S and T of the output voltage
Vout TR: Value of the voltage chained on phases T and R of the output voltage
IoutR: Output current measurement on phase R
FoutR: Output frequency measurement on phase R
IoutS: Output current measurement on phase S
FoutS: Output frequency measurement on phase S
IoutT: Output current measurement on phase T
FoutT: Output frequency measurement on phase T
Riso: Measurement of the insulation resistance relative to the DC input side (PV generator)

FIGURE 15.6 Inverter ABB 10KW three phase PVI-10.0-TL-OUTD-BWP and its display

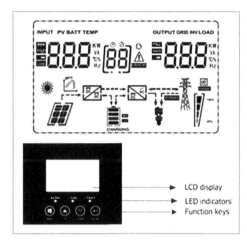

FIGURE 15.7 PH1800 Hybrid Solar inverter and its LCD display

TABLE 15.2
Description of display screen parameters
Selectable information and LCD display

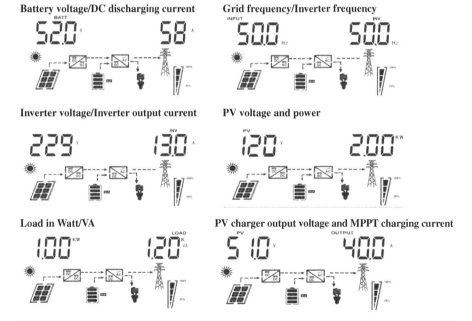

integrated in the inverter. Typical results obtained during spring (Figure 15.7) and summer (Figure 15.8) days show:

- Irradiance ranges from 600 W/m^2 to 1,000 W/m^2.
- Inverter input (PV panels):
  - The average value of the input voltage is 470 V–500 V.
  - The value of the electric current measured at the input varies between 8 A and 17 A.
  - The input power, produced by the PV panels, and the output power of the inverter, supplied to the buildings, is practically the same. Depending on the light intensity, they vary from 4000 W to 8000 W.
- Inverter output (grid feed):
  - The temperature inside the inverter does not exceed 60°C at the boost circuit (DC/DC) and 50°C at the DC/AC converter.
  - The values of the frequencies measured at the output of the three phases (T, R, S) remain fixed during operation. They are of the order of 50 Hz.
  - The average values of the voltages measured on the phases (T, S, R) are 228 V–238V.
  - The values of the voltage chained on the R/S, S/T, T/R phases of the output voltage are of the order of 400 V–418 V.
  - The values of the currents measured between the phases (T, R, S) vary from 6A to 14A.
  - Cos (φ) remains constant and equal to 1, throughout the day. During the whole year no three-phase equipment is installed in the buildings. This shows that the apparent power produced which is the active power and varies from 4 kW to 8 kW is totally applied to the grid (absent of reactive power).
  - The daily energy produced during the day is of the order 50–60 kWh/D.
  - The insulation resistance measured at the DC input side (panel field) remains constant. It is equal to 20 MΩ in spring and 6.5 MΩ in summer. Since this insulation resistance is higher than 1 MΩ, the inverter is well connected to the grid.
  - On this day, following the injection of 7–8 h on the grid, the gain obtained is 10$ to 12$ (i.e. 100–120 DH).

All the results obtained during this day show, on the one hand, the absence of energy consumption by the building (absent of three-phase equipment); and on the other hand, a very good agreement between the simulation results and those indicated by the inverter display. As a result, we see the correct operation of the installation designed and installed at the University Campus of the Oujda Technopole.

15.3.2.2 Stand-Alone Installation

On a sunny day, we followed the daily operation of the stand-alone installation, by reading directly the measurements provided by the inverter's display. The typical results obtained during one day (Figures 15.9 and 15.10) show:

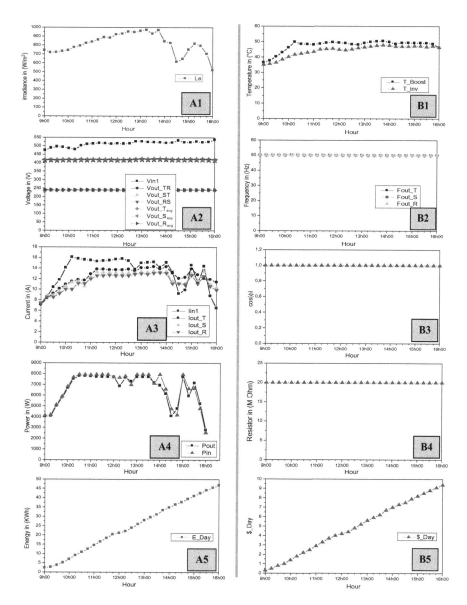

FIGURE 15.8 Typical operation of the grid-connected PV system during one day in spring. A1: Variation of an irradiance. B1: Temperatures of DC/DC and DC/AC converters. A2: Voltage variations. B2: Frequency variation. A3: Current variation. B3: Power factor. A4: Variation of input and output power. B4: Relative insulation resistance. A5: Energy produced. B5: Gain in dollars of energy produced

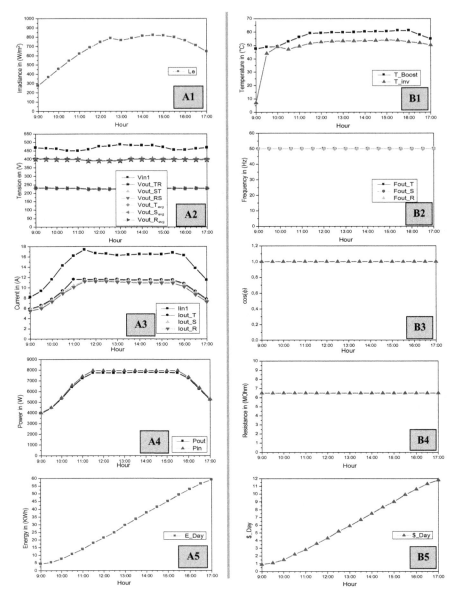

FIGURE 15.9 Typical operation of the grid-connected PV system during one day in summer. A1: Variation of an irradiance. B1: Temperatures of DC/DC and DC/AC converters. A2: Voltage variations. B2: Frequency variation. A3: Current variation. B3: Power factor. A4: Variation of input and output power. B4: Relative insulation resistance. A5: Energy produced. B5: Gain in dollars of energy produced.

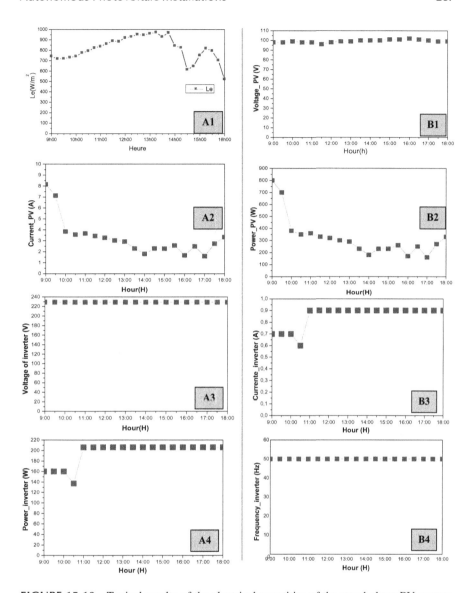

FIGURE 15.10 Typical results of the electrical quantities of the stand-alone PV system. A1: Variation of an irradiance. B1: Voltage of the PV generator during one day. A2: Electrical charging current of the storage unit. B2: Power of the PV panels supplied to the batteries. A3: Output voltage of the inverter during one day. B3: Output current of the inverter. A4: Power consumed by the user. B4: Frequency of the signal at the output of the inverter.

- Irradiation ranges between 700 W/m² and 1,000 W/m².
- Single Phase Inverter Input:
  - The PV field voltage at the inverter input is 98 V–102 V.
  - The electric current measured at the input of the inverter, which is the battery charging current, varies from 2 A to 8 A, and the voltage from 48 V.
  - The power supplied to the batteries varies from 200 W to 800 W. This power, and the charging current, are low compared to the power of the installation, which is 10 kW. This is due to the state of the batteries, which are fully charged, as a result of the low consumption of the buildings (single phase) throughout the year. These buildings are not yet operational due to a lack of staff and researchers.
- Single Phase Inverter Output (Building):
  - The voltage at the output of the inverter is equal to 230 V, during the whole day of operation.
  - The electrical output current varies between 0.7 A and 1 A.
  - The output power is around 210 W. This power is relatively very low, as mentioned above. The University Campus buildings are not yet operational. The recorded consumption is that of the lighting and PCs of two staff (management and secretariat).

All of the results presented in this section are consistent with those obtained in the simulations, and as a result, we observe the proper functioning of the stand-alone facility designed and installed at the University Campus of the Oujda Technopole.

15.4 CONCLUSION

In this chapter, we have presented the structure and operation of two 10 kW PV systems installed at the University campus of the Oujda Technopole. One installation is connected to the grid (10 kW, 50 Hz, 230 V), and the other is stand-alone with storage in eight batteries (10 kW, 48V, 300 Ah/battery). The monitoring of the operation during one day shows:

- In the case of the PV installation connected to the grid, the results obtained show an energy output of 60 kWh/J, a gain of $10 to $12, a three-phase voltage (230 V, 418 V), an output and input power varying with the illumination with an efficiency of 98%, a power factor equal to 1, and an insulation resistance greater than 1 MΩ. All these results show a good functioning of the installed system, and its connection to the grid is adequate to the specifications.
- In the case of the 10 kW stand-alone installation, the results obtained show that the field of panels delivers 100 V, a power output dependent on the state of charge of the batteries, a voltage at the output of the inverter of 230 V, an electric current and an output power dependent on the consumption. During this period, consumption is relatively low, since the buildings of the university campus are not yet operational. The recorded consumption is that of the lighting and PCs of two staff members (management and secretariat). All of these results show that the stand-alone installation is functioning well in accordance with the specifications.

This fieldwork is being continued by the team by monitoring the operation of the two installations throughout the year. Attention will be paid to the energy consumption of the buildings, the injection into the grid, the return on investment, and the cost of the electricity bill of the University Campus.

ACKNOWLEDGMENTS

This work is carried out as part of the project: University Campus Lighting (2016–2019) (06 Installation-technopole/20126): Photovoltaic installations autonomous and connected to the network of University Campus at the Oujda Technopole.

REFERENCES

Adaramola, Muyiwa S., and Emil E.T. Vågnes. "Preliminary Assessment of a Small-Scale Rooftop PV-Grid Tied in Norwegian Climatic Conditions." *Energy Conversion and Management* 90 (January 2015): 458–465. doi:10.1016/j.enconman.2014.11.028.

Al-Badi, A. "Measured Performance Evaluation of a 1.4 kW Grid Connected Desert Type PV in Oman." *Energy for Sustainable Development* 47 (2018): 107–113. doi:10.1016/j.esd.2018.09.007.

Al-Badi, Abdullah. "Performance Assessment of 20.4 kW Eco-House Grid-Connected PV Plant in Oman." *International Journal of Sustainable Engineering* 13, no. 3 (September 6, 2019): 230–241. doi:10.1080/19397038.2019.1658824.

Al-Otaibi, A., A. Al-Qattan, F. Fairouz, and A. Al-Mulla. "Performance Evaluation of Photovoltaic Systems on Kuwaiti Schools' Rooftop." *Energy Conversion and Management* 95 (May 2015): 110–119. doi:10.1016/j.enconman.2015.02.039.

Amegroud, Tayeb. "Morocco's Power Sector Transition: Achievements and Potential". Policy Paper, February 2015. https://media.africaportal.org/documents/OCPPC-PP1509.pdf.

Attari, K., Elyaakoubi, A., and Asselman, A. "Performance Analysis and Investigation of a Grid-Connected Photovoltaic Installation in Morocco". *Energy Reports* 2 (2016): 261–266. doi.org/10.1016/j.egyr.2016.10.004.

Ayompe, L.M., A. Duffy, S.J. McCormack, and M. Conlon. "Measured Performance of a 1.72kW Rooftop Grid Connected Photovoltaic System in Ireland." *Energy Conversion and Management* 52, no. 2 (February 2011): 816–825. doi:10.1016/j.enconman.2010.08.007.

Benalouache, N. 2017. "L'énergie solaire pour la production d'électricité au Maghreb: transition énergétique et jeux d'échelles". Université de Sfax, 30 June 2017.

Chokmaviroja, S., Wattanapong, R., Suchart, Y. "Performance of a 500kWP grid connected photovoltaic system at Mae Hong Son Province, Thailand". *Renew. Energy* 31 (2006): 19–28. doi10.1016 / j.renene.2005.03.004.

El-Katiri, Laura. "Morocco's Green Energy Opportunity," Policy Paper, October 2016. https://media.africaportal.org/documents/OCPPC-PP1614v1.pdf.

Kazem, Hussein A., Tamer Khatib, K. Sopian, and Wilfried Elmenreich. "Performance and Feasibility Assessment of a 1.4kW Roof Top Grid-Connected Photovoltaic Power System Under Desertic Weather Conditions." *Energy and Buildings* 82 (October 2014): 123–129. doi:10.1016/j.enbuild.2014.06.048.

Kumar, Nallapaneni Manoj, M. Rohit Kumar, P. Ruth Rejoice, and Mobi Mathew. "Performance Analysis of 100 kWp Grid Connected Si-Poly Photovoltaic System Using PVsyst Simulation Tool." *Energy Procedia* 117 (June 2017): 180–189. doi:10.1016/j.egypro.2017.05.121.

Lamkaddem, A., S. Talbi, M. Melhaoui, and K. Khalil. "Energy Produced by Photovoltaic Plants Injected on the Electric Network." *JMES* 8, no. 3 (2017): 849–859.

Lamkaddem, Ali, and Kassmi Khalil. "Design and Modeling of a Photovoltaic System Connected to the Electrical Network." *2016 International Conference on Electrical Sciences and Technologies in Maghreb (CISTEM)*, (October 2016). doi:10.1109/cistem.2016.8066785.

Matiyali, Kanchan, and Alaknanda Ashok. "Performance Evaluation of Grid Connected Solar PV Power Plant." *2016 2nd International Conference on Advances in Computing, Communication, & Automation (ICACCA)* (Fall), (September 2016). doi:10.1109/icaccaf.2016.7748989.

Mensah, D., O. Yamoah, and S. Adaramola. "Performance Evaluation of a Utility-scale Grid-tied Solar Photovoltaic (PV) Installation in Ghana." *Energy for Sustainable Development* 48 (2019): 82– 87. doi:10.1016/j.esd.2018.11.003.

Milosavljević, Dragana D.,Tomislav M. Pavlović, and Danica S. Piršl. "Performance Analysis of A Grid-Connected Solar PV Plant in Niš, Republic of Serbia." *Renewable and Sustainable Energy Reviews* 44 (April 2015): 423–435. doi:10.1016/j.rser.2014.12.031.

Rachid, Herbazi, Khalid Amechnoue, Ahmed Khouya, Adil Chahboun, Mustapha Diani, Mohamed Louzazni, and Mohammed Addou. "Performance Evaluation and Analysis of Polycrystalline Photovoltaic Plant Located in Northern Morocco." *International Journal of Ambient Energy* (December 2, 2019): 1–7. doi:10.1080/01430750.2019.1694985.

Salimi, Hossein, Hossein Ahmadi Danesh Ashtiani, Arash Mirabdolah Lavasani, and Reza Fazaeli. "Experimental Analysis and Modeling of Weather Condition Effects on Photovoltaic Systems' Performance: Tehran Case Study." *Energy Sources, Part A: Recovery, Utilization, and Environmental Effects* (June 5, 2020): 1–13. doi:10.1080/15567036.2020.1765902.

Sharma, Vikrant, and S.S. Chandel. "Performance Analysis of a 190 kWp Grid Interactive Solar Photovoltaic Power Plant in India." *Energy* 55 (June 2013): 476–485. doi:10.1016/j.energy.2013.03.075.

Shiva Kumar, B., and K. Sudhakar. "Performance Evaluation of 10 MW Grid Connected Solar Photovoltaic Power Plant in India." *Energy Reports* 1 (November 2015): 184–192. doi:10.1016/j.egyr.2015.10.001.

Shukla, Akash Kumar, K. Sudhakar, and Prashant Baredar. "Simulation and Performance Analysis of 110 kWp Grid-Connected Photovoltaic System for Residential Building in India: A Comparative Analysis of Various PV Technology." *Energy Reports* 2 (November 2016): 82–88. doi:10.1016/j.egyr.2016.04.001.

Spertino, Filippo, and Fabio Corona. "Monitoring and Checking of Performance in Photovoltaic Plants: A Tool for Design, Installation and Maintenance of Grid-Connected Systems." *Renewable Energy* 60 (December 2013): 722–732. doi:10.1016/j.renene.2013.06.011.

Ya'acob, M. Effendy, Hashim Hizam, Tamer Khatib, and M. Amran M. Radzi. "A Comparative Study of Three Types of Grid Connected Photovoltaic Systems Based on Actual Performance." *Energy Conversion and Management* 78 (February 2014): 8–13. doi:10.1016/j.enconman.2013.10.064.

Yadav, S. and K. Sudhaka. "Different Domestic Designs of Solar Stills : A Review." *Renewable Sustainable Energy Review* 47 (2015): 718–731. doi: 10.1016/j.rser.2015.03.064.

16 Migration and Livelihood Sustainability

A Case Study of the Migratory Workers of Lucknow, Uttar Pradesh

Shambhavi Singh and Suhail Ahmad Bhat

CONTENTS

16.1 INTRODUCTION

India is a labor abundant country, where every year millions enter the labor force market. However, due to a lack of adequate jobs a large number of Indians remain unemployed in the country. The Indian economy is also witnessing sectorial changes, due to which the employment scenario of the country is becoming different from sector to sector and from region to region (rural to urban). Nonetheless, the economy is predominantly a rural based one, where two-thirds of the population and 70% of the workforce live in rural areas (Reddy and Dutta, 2018). The agricultural sector occupies a predominant place in rural areas. However, a decline in the share of the sector in the gross domestic product (GDP) of the country and also a decline in land among small and marginal landholders have given much rise to labor migration in India (Chandrasekhar and Sahoo, 2018). Therefore, a massive flow of workers is moving from rural to urban areas in search of better employment or livelihood. The migrated laborers include self-employed persons, regular workers, daily wage workers, rickshaw pullers, street vendors, and shop owners (Cooper, 2012). Therefore, labor migration has been identified as a survival strategy utilized by the people, especially by rural people to combat poverty and unemployment. The assessment of the effects of migration on rural areas has remained relevant, since migration acts as a catalyst in the transformation process of not only the destiny of individual migrants, but also of the conditions of family members left behind, local communities, and the wider regions (Castelli, 2018). Therefore, migrating from one place to another in search of employment and for a better livelihood is a strategic feature of human nature (Sanyal and Maity, 2018).

Nonetheless, migration in India take places with both pull and push factors. The push factor mainly consists in the fact that, after independence, the economy underwent rapid urbanization (Kone et al., 2018) and, on the other side, that the agricultural sector provides only a subsistence wage rate to laborers, who must endure poverty. Poverty and economic misery compels laborers to migrate from rural to urban areas in search of better jobs, as well as for better economic opportunities (Gheasi and Nijkamp, 2017). However, the pull factor also attracts labor migration within India. This is mainly because there are still huge regional and economic disparities in the country. Therefore, due to uneven development state-wise, the developed states or regions attract laborers to migrate for better wages and for a better life (Das and Saha, 2013). The National Commission on Rural Labour (NCRL) has pointed out that migration in India is influenced by the pattern of development along with inter-regional disparities. Nonetheless, the National Survey Sample Organization (NSSO) has published its various quinquennial rounds on labor migration. The NSSO's 64th round considered the economic reasons, such as searching for employment, better jobs/salaries, and movement of parents, for the cause of migration. Historically, Uttar Pradesh and Bihar are the two highest outward-migration states of India and they still remain the highest (Bhagat et al., 2020). According to an International Labor Organization report (2018), a massive flow of laborers from Uttar Pradesh, Bihar, and Madhya Pradesh are migrating to Delhi, Mumbai, and Gujarat for better employment and living, because there is a lot of economic disparity and uneven development in these states.

On the other hand, Uttar Pradesh is one of the most populous state of India which comprises a 19.98 crore population, with an area of 243,286 km^2 (2011 census). But unfortunately it is recognized as a backward state in terms of having a lack of employment opportunities, uneven development, low per-capita income, poverty, and regional disparities. However, the economy is primarily agrarian where the majority of people are largely dependent on it and its allied activities. As per the census of 2011, 65% of the total workforce in the state depends on agriculture, and most of them live below the poverty line. However, with increasing population pressure the per capita land holding size is declining, hence the massive flow of laborers migrating from rural areas to the adjacent cities or states in search of a livelihood. Therefore, declining land holding among farmers, lack of employment opportunities, undesirable work, and low wages are attributed as the main drivers of out-migration in the state. Consequently, Uttar Pradesh has witnessed an enormous increase in out-migration in recent times as a result of the above-mentioned factors. However, Uttar Pradesh has some highly developed cities which include Lucknow, Ghaziabad, and Noida. Hence, a large chunk of rural people from Uttar Pradesh migrate toward these cities in search of better jobs and a livelihood. The present chapter examines this scenario at the macrolevel by using secondary data, and it also examines the impact of migration on the socio-economic conditions of migratory workers in Lucknow, Uttar Pradesh. Nonetheless, migration acts as a sustainable source of livelihood among these migratory workers, who have come from rural areas to Lucknow.

16.2 LITERATURE REVIEW

Bhagat (2017) explored migration and urban transition in India. The author highlights that in India there is a close relationship between urbanization and economic development. About 65% of GDP in India is accumulated in urban areas. Thus, the emerging pattern of urbanization in India is not simply a rural to urban transfer of labor and population, but is a complex process of change in the characteristics of human settlement. The author also points out that the sustainable development goals (SDGs) were followed by the New York Declaration for refugees and migrants. The New Urban Agenda was agreed by world leaders in 2016, which reaffirms the objectives of SDGs in achieving sustainable cities and urbanization and which enables also the positive contribution of migrants to cities and to strengthen urban–rural linkages.

Gheasi and Nijkamp (2017) examined international migration motives and impacts, with specific reference to foreign direct investment (FDI). The authors highlighted that globalization and free movement among developed countries have made a significant impact on cross-border migration, capital mobility, innovation, and economic development. The study tried to observe the long-term impact of immigration on FDI. It discovered that newly arrived migrants in Europe are mostly educated people, and their education costs have usually been financed by the country of origin. This is a huge economic benefit to the receiving countries and this flow may bring long-term economic benefits. Therefore, there is a positive and statistically significant relationship between international migration and FDI.

Bhagat (2018) tested the relationship between migration, urbanization, and development. The author points out that migration has emerged as a strong force, which is shaping cities and urbanization. It is influencing the demand and supply of labor, economic growth, and human wellbeing. Further, the author highlights that, earlier, migration was seen as a development failure and that policy makers were busy in suggesting ways to reduce it, though later it has been observed that urbanization has emerged as a major driver of economic development in recent times. It is the spatial outcome of an economic process which leads to the concentration of population and economic activities in some places and thereby attracts many skilled and unskilled laborers for a livelihood.

Castelli (2018) has examined the drivers of migration. The study reveals that more than 244 million international migrants were estimated who are living in foreign countries in 2015. However, there are many factors which induce labor migration such as macrofactors, mesofactors, political factors, demographic factors, socio-economic conditions, and environmental situations, which are major. Moreover, the author points out that among macrofactors, the inadequate human and economic development of a country, demographic increase, urbanization, dictatorships and wars, social factors, and environmental change are the major contributors to migration, while, mesofactors are linked with the individual and to his or her ethnic group or religious community, land grabbing, communication technology, and diaspora – all of which play an important role for labor migration across world.

Kone et al. (2018) analyzed the internal borders of migration within India. The study reveals that internal mobility is a vital component of economic growth and development, because it enables the reallocation of labor to more productive opportunities, sector-wise and region-wise. Moreover, it finds out that average migration between neighboring districts in the same state is at least 50% larger than in neighboring districts on different sides of a state border, even after accounting for linguistic differences. Therefore, the results of the study suggest that inter-state mobility is inhibited by state-level prerogative schemes, ranging from access to subsidized goods through the public distribution system to the bias of states in favor of their own residents in access to tertiary education and public sector employment.

Mohanakumar (2018) published an edited book on internal migration in contemporary India. This book is the collection of many research papers, which examine the various issues and concerns of interstate migration in India. The authors explain the complexities of life, societal aspirations, and livelihood of migrants of different types, which are defined in terms of purpose of migration, gender, and geographical location. The growing mismatch between the rapid growth of urban population and the basic amenities of life is attributed to a lackadaisical approach to urban planning and the absence of a well-thought-out action plan in urban areas, which leads to internal migration.

Sanyal and Maity (2018) examined labor migration in India, and its trends, causes, and impacts. The authors highlight that uneven development is the predominant factor behind this migration. The results show that disparities in socio-economic conditions, disparities in the development policies, and wage rate differentials induce laborers to migrate. However, the study concludes that migration is having a miserable impact on the living standards of migrants. Both migrant persons and family members are deprived of health, education, and other essential public services, as well as the basic amenities of hygiene, drinking water, and so on.

Malhotra and Devi (2019) explored internal migration in India by using primary data. They highlighted that this migration, particularly interstate and rural urban streams, has increased, because it not only fills the demand and supply gaps, but also acts as an instrument of survival for millions of the poor population in the concentrated states of Uttar Pradesh, Bihar, Rajasthan, and Madhya Pradesh. Moreover, the study finds that better work facilities are the major pull factors. However, internal migration is basically driven by push factors, which include lack of job opportunities for young males, family problems, agriculture not being profitable, unemployment, and landlessness.

16.3 RESEARCH METHODOLOGY

The present study is based on both primary and secondary data. Nonetheless, NSSO is the main source of data for labor migration in India. NSSO has published various quinquennial rounds on labor migration. The last round is the 64th (July 2007–June 2008). However, after 2008 the NSSO has not published any more issues. Therefore, due to the non-availability of the latest secondary data an attempt has been made to study and examine the impact of migration on the socio-economic and livelihood sustainability of migrant workers in Lucknow, Uttar Pradesh by collecting primary data. This has been done from migrant workers through a structured schedule. A purposive random sampling method was used, through which 100 samples were taken. Descriptive statistics were used for the data analysis.

16.4 MAIN RESULTS

16.4.1 THE MIGRATION SCENARIO IN INDIA

A number of economic, social, cultural, and political factors play an important role in the decision to migrate from one place to another place. The change in residence can take place either on a permanent or semi-permanent or temporary basis. Internal migration involves a change of residence within national borders, while international migration involves migrating from one's own country to another (Kone et al., 2018). The Census of India categorizes migration into two types – migration by birth place and migration by place of last residence. In census methodology, when a person is enumerated at a place other than his or her place of birth, they are considered as a migrant by place of birth. Similarly, a person is considered a migrant by place if he or she had last resided at a place enumerated other than his or her home town. Nonetheless, the Indian constitution provides the basic freedom to reside in any part of the country and to earn the livelihood of choice. Thus, migrants are not required to register either at the place of origin or at the place of destination. A proper understanding of the patterns of migration would help in the estimation of future population redistribution.

16.4.2 MIGRATION RATE AMONG MALES AND FEMALES IN RURAL AND URBAN AREAS IN INDIA

This section explores the migration rates in India. The migration rate has been estimated as the number of migrants belonging to the male and female category of both rural and urban areas per 1000 persons. Table 16.1 depicts migration rates at the all-India level in both rural and urban areas, which have been presented by different NSS rounds from the 38th (1983) to the 64th (2007–2008). Table 16.1 highlights that in both rural and urban areas migration rates have been gradually increasing from 1983 to 2007–2008 (i.e., from the 38th NSS round to the 64th). However, a marginal fall has been seen in the migration rate in the NSS 49th round. The migration rates in rural areas have raised approximately from 21% in the 38th round (1983) to 26% in the 64th round (2007–08). In urban areas it has increased from 32% in the 38th round to 36% in the 64th round in urban areas.

TABLE 16.1
Migration rates among males and females obtained by different NSS rounds

Category of persons	38th round (Jan–Dec, 1983)	43rd round (Jul 1987– Jun 1988)	49th round (Jan– Jun, 1993)	55th round (Jul 1999– Jun 2000)	64th round (Jul 2007– Jun 2008)
Rural male	7.2	7.4	6.5	6.9	5.4
Rural female	35.1	39.8	40.1	42.6	47.7
Rural persons	20.9	23.2	22.8	24.4	26.1
Urban male	27	26.8	23.9	25.7	25.9
Urban female	36.6	39.6	38.2	41.8	45.6
Urban persons	31.6	32.9	30.7	33.4	35.4

Source: NSS Report No. 533: Migration in India: July 2007–June 2008.

Table 16.1 also shows that in rural areas the male migration rate has declined by 2 percentage points, i.e., it has declined from 7% to 5%. A similar trend has been observed in urban areas, where the male migration rate has decreased by 1 percentage point from 27% to 25.9%. While female migration shows an increasing trend, the rate of females in rural areas has increased by 13 percentage point change, i.e., it has been increased from 35% to 48%, while in urban areas it has increased by 9 percentage points, i.e., it has increased from 36.6% to 45.6%. The major reason for higher migration rates of females in both rural and urban areas is marriage and employment.

16.4.3 State-Wise In-Migrants and Out-Migrants in India

NSSO estimated in-migrants and out-migrants and, from that, net migrants. The net migrants are obtained by the difference of in-migrants and out-migrants, where in-migrants are those coming from other states of India or from other countries, and out-migrants are those who migrate to other states or abroad. The number of net migrants per 1000 of the population gives the net migration rate. Table 16.2 shows that in the 55th round of 2005, the most in-migrants were in Maharashtra at 60,462 persons, followed by Uttar Pradesh at 38,680 persons, West Bengal at 29,002, Haryana at 23,674 persons, and Madhya Pradesh at 22,817 persons. The least in-migrants were in Assam which comprises 1469 persons, followed by Bihar with 5923 persons, and Orissa with 7352 persons. However, in the case of out-migrants, in the 55th round the highest number were in Uttar Pradesh which comprises 51,320 persons, followed by Bihar with 34,483 persons, Maharashtra with 21,561 persons, and Rajasthan with 18,622 persons. The table also shows that the highest net-migrants were in Maharashtra, which comprises 38,901 persons, followed by West Bengal with 19,684 persons, and Haryana with 15,007 persons.

However, in the case of the 64th round, the scenario has changed. The highest number of in-migrants were at Uttar Pradesh which comprises 51,320 persons, followed by Bihar with 34,483 persons, Maharashtra with 21,561 persons, and Rajasthan with 18,622 persons. However, in the case of out-migrants, in the 55th round then the highest number were at Uttarakhand which comprises 81,405 persons, followed by

TABLE 16.2
State-wise in-migrants and out-migrants in India

Major States	In-migrants (thousands)		Out-migrants (thousands)		Net-migrants (thousands)	
	55th round	55th round	55th round	55th round	55th round	55th round
Andhra Pradesh	11,658	10,153	11,059	12,324	599	−2171
Assam	1469	1070	2578	2282	−1109	−1212
Bihar	5923	5505	34,483	47,077	−28,560	−41,572
Chhattisgarh	NA	9651	NA	3193	NA	6458
Gujarat	18,569	20,778	9995	10,879	8574	9899
Haryana	23,674	22,349	8667	14,175	15,007	8174
Himachal Pradesh	NA	3040	NA	3442	NA	−402
Jammu & Kashmir	NA	824	NA	1791	NA	−967
Jharkhand	NA	3913	NA	8129	NA	−4216
Karnataka	14,329	20,130	18,482	14,173	−4053	5957
Kerala	10,050	10,691	8498	8096	1552	2595
Madhya Pradesh	22,817	13,168	15,249	17,035	7568	−3867
Maharashtra	60,462	56,584	21,561	15,414	38,901	41,170
Orissa	7352	5303	5418	9648	1934	−4345
Punjab	18,138	18,586	12,848	11,697	5290	6889
Rajasthan	15,288	17,582	18,622	20,841	3334	−3259
Tamil Nadu	14,487	9906	15,595	13,675	−1108	−3769
Uttar Pradesh	38,680	10,005	51,320	6619	−12,640	3386
Uttarakhand	NA	32,326	NA	81,405	NA	−49,079
West Bengal	29,002	23,670	9318	12,303	19,684	11,367

Source: NSS Report No. 533: Migration in India: July 2007–June 2008.

Bihar with 34,483 persons, Rajasthan with 20,841 persons, and Maharashtra with 15,414 persons. Table 16.2 also shows that the highest net-migrants were at Maharashtra, which comprises 41,170 persons, followed by West Bengal with 11,367 persons, and Gujarat with 9899 persons.

16.4.4 REASONS FOR MIGRATION

There are many reasons for migration in India, which include economic, social, cultural, and political factors. The NSSO has covered the major reasons in its 49th round (1993), 55th round (1999–2000), and 64th round (2007–2008). Table 16.3 highlights various reasons of migration, which include employment-related reasons, study, marriage, and movement with parents or earning family members. The data on reasons for migration are useful for understanding the motives behind the movement of people. Therefore, it has been observed that employment among males, and marriages among females, is the major reason for migration. Table 16.3 highlights that in the 49th round (1993) the highest reason for migration among rural males was employment, i.e., 47.7%. But this declined to 30.3% in 1999–2000 and has further

TABLE 16.3
Reasons for migration (%)

Reasons for migration	49th round (1993)	55th round (1999–2000)	64th round (2007–2008)
Rural Males			
Employment related reasons	47.7	30.3	28.6
Studies	4.1	5.3	10.7
Marriage	2.3	9.4	9.4
Movement of parents/ earning member	20.8	26	22.1
Other reasons (incl. non-residents)	25.1	29	29.2
All	100	100	100
Urban Males			
Employment related reasons	41.5	51.9	55.7
Studies	18	6.2	6.8
Marriage	9	1.6	1.4
Movement of parents/ earning member	28.3	27	25.2
Other reasons (incl. n.r.)	11.3	13.3	10.9
All	100	100	100

Source: NSS Report No. 533: Migration in India: July 2007–June 2008.

declined to 28.6% in 2007–2008. However, the reason for migration being study has increased from 4.1% in 1993 to 10.7% in 2007–2008. Similarly, migration with marriage has increased from 2.3 to 9.4% in the same period. Migration for movement with parents/earning members has increased from 20.8% in 1993 to 26% in 1999–2000, but it has declined to 22.1% in 2007–2008. The other reasons for migration in rural areas have enlarged from 25.1% to 29.2% in the same period.

Similarly in the case of urban males, Table 16.3 also highlights that in the 49th round (1993) the highest reason for migration was employment, i.e., 41.5%, which increased to 51.9% in 1999–2000 and has further increased to 55.7% in 2007–2008. However, the reason of migration as study among urban males has declined from 18% in 1993 to 6.8% in 2007–2008. Similarly, migration with marriage has declined from 9 to 1.4% in the same period. Migration for movement with parents/earning members has also diminished from 28.3% in 1993 to 25.2% in 2007–2008. The other reasons for migration in urban males have shrunk from 11.3 to 10.9% in the same period.

16.4.5 OCCUPATIONAL STATUS OF MIGRANT HOUSEHOLDS

The occupational distribution always plays a crucial role in the economic development of a country, because the occupational pattern reflects economic activity and the socio-economic conditions of the labor force. Nonetheless, changes in

TABLE 16.4
Occupational status of individual migrant workers (%)

Occupational status of individual migrants	Before migration	After migration
Unemployed	45	0
Casual worker	28	13
Self-employed	20	63
Regular worker	7	24
Total	100	100

Source: Field Survey Data.

occupational structure with decent jobs are highly associated with economic development. Therefore, the present section highlights the economic activities of migrant households. Table 16.4 shows the occupational distribution of households, before and after migration.

This is divided into three parts, i.e., casual workers, self-employed persons, and regular workers. The casual worker includes rickshaw pullers and wage workers, whereas the self-employed worker includes street vendors and shop owners, and the regular worker includes salesmen in shops and persons with a monthly income from a job. The results of the table show that before migration 45% of people were unemployed whereas 28% were casual workers, 20% were self-employed, and 7% were regular workers respectively. However, after migration, unemployed persons find jobs easily; therefore after migration the majority of respondents, which comprises 63%, find self-employed jobs; 24% find regular worker-type jobs; and the remaining 13% find casual-worker jobs. Hence, it has been observed from the occupational pattern of migrants that the proportion of casual workers declines after migration and that there is a substantial increase in regular/self-employed jobs. The overall result of the table indicates that migration provides better employment opportunities for the respondents.

16.4.6 Agricultural Land Holding and Plot Purchasing of Migrant Households

Employment in agriculture has almost stagnated in India. In certain sub-sectors of it, like livestock, forestry, and fishing, employment has declined drastically, due to which laborers are shifting to non-farm sectors. The present section highlights the agricultural landholding size of migrant households before migration and Table 16.5 shows the plots purchased in Lucknow. The table shows that before migration, out of 100 surveyed households, 44%had agricultural land, while the remaining 56% had none. The other part of table shows that, after migration, 42% of household purchased plots in Lucknow. Moreover, 5% bought plots in the range of less than 400 ft^2, while another 5% bought land in the range of 800–1200 ft^2. The maximum proportion of households, which comprises 32%, had bought a plot in the range of 400–800 ft^2. Owning a plot in the capital highlights that, after migration, the economic status of workers has improved significantly.

TABLE 16.5
Agricultural land holdings and plot purchasing of migrant households before and after migration

Agricultural landholding before migration		Plot purchasing after migration			
		Plot purchased		Size of plot (in ft²)	Frequency	Percentage
NO	56.0	NO	58.0	Households did not purchase plot	58	58.0
YES	44.0	YES	42.0	< 400	5	5.0
Total	100.0	Total	100.0	400–800	32	32.0
				800–1200	5	5.0
				Total	100	100.0

Source: Field Survey Data.

TABLE 16.6
Monthly income of individual migrant workers (rupees)

Monthly income	Frequency	Percentage
< 5000	2	2.0
5001–10,000	47	47.0
10,001–15,000	38	38.0
Above 15,000	13	13.0
Total	100	100.0

Source: Field Survey Data.

16.4.7 MONTHLY EARNINGS OF MIGRANT WORKERS

Income always plays a vital role in determining the socio-economic condition of individuals. It is the income which decides who is above and below the poverty line. Higher income is always determined by decent work. The International Labour Organization (ILO) launched the concept of "decent work" in 1999 in its 1987 conference on employment, social protection, workers' rights, and social dialog (Ghai, 2003). Decent work helps to reduce poverty among the masses and to improve the socio-economic conditions of the workers. This section explores the income comparison of individual migrants by analyzing their incomes before and after migration.

From the Table 16.6 it is clear that out of 100 respondents only 2% were earning less than 5000 rupees per month. However, 47%, which is the majority, were earning between 5001 to 10,000 rupees per month. Similarly, 38% were earning between 10,001 to 15,000 rupees per month, whereas 13% were earning 15,000 and above rupees per month. Therefore, the overall results show that migration makes a positive impact on the income of migratory persons, because before migration 45% were unemployed, whereas after no one remained unemployed. After migration the income pattern of respondents has changed significantly.

16.4.8 MONTHLY INCOME MEAN AND STANDARD DEVIATION

The present section highlights the individual and family average monthly income of the household. The results of Table 16.7 show that after migration the minimum earnings of individual migrants is 4500 rupees per month and the maximum earnings are 22,000 rupees per month. The minimum total household income per month of migrants is 4500 rupees and the maximum is 47,000 rupees per month. The mean individual income is 10,720 rupees with a standard deviation of 3600.87, while in the case of total family income it is 14,580.00, with a standard deviation of 6627.81, respectively. Therefore, the table shows that after migration the economic conditions of migratory households have improved significantly.

16.4.9 SAVING BEHAVIOR OF MIGRANTS

Classical economists believed that savings are a necessary and sufficient condition for investment creation. They believed that if savings go up, investment increases, because the interest rate declines and economic growth will be imminent. The neoclassical economist Solow (1956) suggested that savings affected economic growth because higher savings lead to capital increase. So in the context of Indian financial inclusion, this is confined to ensuring minimum access to a saving bank account without any frills, but in the context of the international perspective, financial inclusion is wider than the Indian context. In the international perspective, financial inclusion does not mean only having a current bank account or savings bank account but it encompasses having a minimum access to a wide range of financial services which include short term and long term credit, savings insurance pensions, money transfers, and mortgages (Sujlana and Kiran, 2018).

Hence, it is important to explore and identify the financial services of the surveyed households. The present section highlights the savings behavior of migrant households by analyzing their previous and present bank account status and the savings status of migrant households. Table 16.8 depicts the number of persons having a bank account in both situations (before and after migration). It exhibits that before migration, out of a total of 100 respondents, only 5% had a bank account, while the remaining 95% did not. However, after migration, access to banking services among the migratory respondents had increased: 67% had a bank account and only 33% did not.

Table 16.9 highlights the savings of migrant workers after migration in Lucknow, Uttar Pradesh. It reveals that, before migration, 82% of households had no savings,

TABLE 16.7
Standard deviation of monthly income of migrant workers

Average monthly income	Min.	Max.	Mean	Std. deviation
Individual migrant income	4500	22,000	10,720.00	3600.87
Total household income	4500.00	47,000.00	14,580.0000	6,627.81

Source: Field Survey Data.

TABLE 16.8
Bank account status before and after migration

Having bank account	Before migration		After migration	
	Frequency	**Percent**	**Frequency**	**Percent**
YES	5	5.0	67	67.0
NO	95	95.0	33	33.0
Total	100	100.0	100	100.0

Source: Field Survey Data.

TABLE 16.9
Monthly savings status of migrant households

Savings after migration (in Rs.)	Before migration (%)	After migration (%)
No saving	82	52
< 500	18	14
500–1000	2	17
1001–1500	0	6
1501–2000	0	6
Above 2000	0	5
Total	100	100

Source: Field Survey Data.

whereas 18% had savings of less than 500 rupees per month. Only 2% had savings in between 500 and 1000 rupees. However, after migration, 52% still had no savings, possibly due to the large size of the family and high expenditure, whereas 14%had savings less than 500 rupees and 17% had savings in between 500 and 1000 rupees per month. Moreover, 6% percent of households had monthly savings between 1001 and 1500 rupees per month and 1501 to 2000 rupees per month, and the remaining 5% had savings above 2000 rupees per month. The above analysis thus shows a big difference between the saving behaviors of migrant households and their previous condition.

16.4.10 SOCIAL PARTICIPATION OF MIGRANTS

Social gathering and social participation are quite important for a healthy and peaceful life. Therefore, the present section explores perceptions of street vendor migrants regarding social gathering and social participation in the newly migrated locality. Table 16.10 highlights that 51% of street vendor migrants socially gathered to some extent, whereas 25% did so to a very large extent, and 24% did not socially gather at Lucknow. The table also reveals that 41% of migrants responded that people are cooperative, whereas 17% the street vendors responded that people are less cooperative. However, 42% of vendors responded that there is not any difference in the new locality as far as people's cooperation is concerned, compared with the cooperation of their native places.

TABLE 16.10
Social participation or gathering of migrants

Social participation/gathering after migration (%)		Perception of migrants about mew locality (%)	
Some extent	51	Cooperative	41
Large extent	25	Less cooperative	17
Nil	24	No difference	42
Total	100	Total	100

Source: Field Survey Data.

TABLE 16.11
Problems at the place of residence

Problem at your residence	Frequency	Percent
No problems	30	20.0
Large amount of rent	49	49.0
Intolerance of landlord	15	25.0
Other problems	6	6.0
Total	100	100.0

Source: Field Survey Data.

16.4.11 PROBLEMS FACED BY MIGRANTS AFTER MIGRATION

There are many earlier studies which concluded that migrant workers face problems in their residence, hiring rooms, or renting land; therefore, taking that into consideration, respondents were asked about such problems. Table 16.11 highlights that 30% of migrant households have not faced any kind of problems, whereas 49% have faced problems in hiring rooms and land, and 15% have faced harassment from their land owners. However, the remaining 6% have reported having some kind of issues with their neighbors and locality. The overall results of the table show that migrants encounter difficulties related to hiring rooms and also face some harassment from land owners.

16.5 CONCLUSION

From the above study it is concluded that, as per the various NSSO rounds on migration, after 1993 migration for employment in rural areas has declined, which is mainly because the government of India has implemented various employment generation schemes in rural areas. However, migration for study purposes increased in this period, which is considered a positive indicator for rural literacy. The primary survey data reveal that, although the majority of workers find casual jobs after migration, none of the workers has remained unemployed. Nonetheless, the results of the primary data study also show that before migration a significant number of the migrants were unemployed, but later on they found jobs easily in the capital Lucknow. Besides, this research demonstrates after migration the majority of workers were earning monthly an income between 5000 and

10,000 rupees per month. In terms of social values, the study shows that migration does not affect the social values of migratory persons. The overall results of the study highlight that migration provides a positive impact on the socio-economic conditions of migrant workers in Lucknow, Uttar Pradesh. The current research cas also be used to compare the socio-economic conditions of migratory laborers with the skill composition of existing ones. The second scope of the study is a fuller analysis which could examine why rural people are still migrating, when the government of India is already providing employment opportunities through the Mahatma Gandhi National Rural Employment Guarantee scheme. Better evidence could help to explore whether the MNREGA is successful at the ground level or is only providing a subsistence wage rate due to which people are migrating to other places for better socio-economic conditions. The present study has some limitations. In fact, after 2008 the NSSO, which is the major source of migration data, has not published any further information on labor migration. Therefore, due to the non-availability of the latest secondary data on migration, the present study mainly focuses on labor migration in Lucknow, Uttar Pradesh, which is wholly based on field survey data.

REFERENCES

Bhagat, R. B. (2017, September). Migration and urban transition in India: Implications for development. In *United Nations Expert Group Meeting on Sustainable Cities. Human Mobilities and International Migration*, New York.

Bhagat, R. B. (2018). Development impacts of migration and urbanisation. *Economic and Political Weekly*, 53(48), 15–19.

Bhagat, R. B., Reshmi, R. S., Sahoo, H., Roy, A. K., & Govil, D. (2020). *The COVID-19, migration and livelihood in India* (Working paper No. id: 13054). eSocialSciences.

Castelli, F. (2018). Drivers of migration: Why do people move?. *Journal of Travel Medicine*, 25(1), tay040

Chandrasekhar, S., & Sahoo, S. (2018). *Short-term migration in rural India: The impact of nature and extent of participation in agriculture* (No. 2018-016). Indira Gandhi Institute of Development Research, Mumbai, India.

Cooper, R. (2012). Growth and inclusion: Theoretical and applied perspectives. *Current Opinion in Environmental Sustainability*, 24, 52–57.

Das, K. C., & Saha, S. (2013). Inter-state migration and regional disparities in India. http:// iussp. org/sites/default/files/event_call_for_papers/Inter-state%20migration_ IUSSP13. pdf (accessed March 15, 2015).

Ghai, D. (2003). Decent work: Concept and indicators. *International Labour Review*, 142(2), 113–145.

Gheasi, M., & Nijkamp, P. (2017). A brief overview of international migration motives and impacts, with specific reference to FDI. *Economies*, 5(3), 31.

Kone, Z. L., Liu, M. Y., Mattoo, A., Ozden, C., & Sharma, S. (2018). Internal borders and migration in India. *Journal of Economic Geography*, 18(4), 729–759.

Malhotra, N., & Devi, P. (2019). Determinants of internal migration: A case study of urban informal sector of Punjab. *International Journal of Development and Sustainability*, 8(1), 1–18.

Reddy, T. K., & Dutta, M. (2018). Impact of agricultural inputs on agricultural GDP in Indian economy. *Theoretical Economics Letters*, 8(10), 1840–1853.

Sanyal, T., & Maity, K. (2018). On labour migration in India: Trends, causes and impacts. *Economic Affairs*, 63(1), 57–69.

Solow, R. M. (1956). A contribution to the theory of economic growth. *The Quarterly Journal of Economics*, 70(1), 65–94.

Sujlana, P., & Kiran, C. (2018). A study on status of financial inclusion in India. *International Journal of Management Studies*, 2(3), 96–104.

17 The Urban Poor and Their Financial Behavior

A Case Study of Slum Dwellers in Lucknow (India)

Firdous Ahmad Malik, D.K. Yadav,
Hebatallah Adam, and Amina Omrane

CONTENTS

17.1   INTRODUCTION

The pace of urbanization in India is increasing, and with it, urban poverty and the urban poor are also growing day by day. This phenomenon gives birth to various challenging issues related to urban poverty, such as migration, labor, access to basic services/logistics, and the appalling conditions of India's slums and beggars. Poverty in India is being urbanized and this urbanization is projected to reach 50% (World Bank, 2000). More specifically, urban poverty in India is over 25% and almost 81 million people are living in urban areas on incomes that are below the poverty line. At the national level, rural poverty remains higher than urban poverty; but the gap will be closed by 2030 (Patel, 2014). Poverty estimates in India are contested at 102 million poor people and 20 million in Uttar Pradesh (UP). The urban poor have reduced by 25% overall; but in UP this decline has been only 3% (Rangarajan and Dev, 2014). So, the state continues to stagger under a huge burden of urban poverty. For these reasons, plenty of problems arise and remain a challenge for the growth and development of the state and each of its counties. Poverty mostly affects migrant workers from both rural and urban areas, so that they move from one area to another in order to minimize the harsh conditions of poverty. The nature of urban poverty poses distinct challenges for housing, water, sanitation, health, education, social security, livelihoods, and the special needs of vulnerable groups such as women, children, and the aged. An important manifestation of urban poverty and vulnerability is the increasing slum population.

The main concern of the present study is then to analyze the financial behavior of the poorest of the urban poor with special reference to slums and beggars based specifically in Lucknow. From this perspective, there is simple prior evidence related to the poor's lack access to basic financial services. The distribution of poverty among countries differs. Some of them face challenges related to health, education, and employment. Some others are handling issues attached to illiteracy, malnutrition , and exclusion. Therefore, policymakers across the globe pay special attention to financial exclusion so that people can be made bankable by providing them with financial services. In India, the financial needs and practices of the poor differ greatly across rural and urban areas.

Eradication of poverty through inclusive growth has become the mainstay of development policy in India over the past decades. One of the most important empirical relationships revealed in the last decade has been the establishment of the causal link between financial depth and growth (Honohan, 2004). More specifically, despite the phenomenal spread of microfinance in India, financial inclusion remains the most challenging mechanism.

"Why is financial inclusion so important?" The Committee of Financial Inclusion (2008) defines it as "the process which guarantees the timely services of financial access, like savings, credit, insurance, and other banking facilities to the vulnerable groups such as weaker sections and low-income groups, at an affordable cost" (Report of the Committee for Financial Inclusion, 2008). Another question may arise from the first one : "Can poor people save?" (Schreiner & Sherraden, 2007). We deduce that such a debate for policymakers and academics remains not yet resolved, appealing to them to investigate the grassroot phenomenon of these poor, so that suitable policies can be drafted to help them.

Financial inclusion is considered a fundamental tool for poor households who still lack access to dependable savings, and credit and insurance products. It also reduces high transaction costs of formal financial services, which are the greatest issue that development practitioners face in reaching out to the poor. In fact, the focus on costs often ignores other important factors that play into the uneven distribution of poverty and financial exclusion across different populations within the country. From this same perspective, it is argued that if poverty is analyzed from a "capability" perspective, it will be pertinent to note that the kind of poverty and severity of deprivations among the poorest of the poor will be of the same nature.

17.2 LITERATURE REVIEW

Globally, poverty is recognized as a big curse on the earth, generating several tough challenges for the well being of a marginalized section of society. It inhibits their freethinking toward the process of growth and development. It also develops the circle under which the marginalized section, by compulsion of their socio-economic conditions, are forced to live with the prevailing conditions. Now, the main concern is how to approach the circle of poverty, in order to break it. In this same orientation, the Indian planning commission's main thrust was always poverty reduction, and a variety of methods have been used to measure it. Many programs have been implemented to tackle poverty in India, mainly during the period 1947 until 2005, and even later on. Those programs aimed at addressing the issues related to poverty by providing banking services to farmers, via various schemes based on customers' needs and availability of credit with lower interest rates.

According to Chambers and Conway (1992), as well as Lont and Hospes (2004), to pay attention to sustainable livelihoods requires looking beyond work and income in order to in-depth understand poverty and vulnerability. From this perspective, Sebstad and Cohen (2000) state that "vulnerability to poverty is a risk of exposure of people to downward pressures and shocks and their ability to cope with the consequences of these risks". Furthermore, such vulnerability needs proper and good management, though it is a big challenge to include the population which is lying under these vulnerabilities. PFRC (2005) and Rutherford (1999) assert that saving behavior is best analyzed by the behavior of individuals and the working mechanism of institutions.

Thus, it will be important to identify the challenges of financial inclusion and related bottle necks which are responsible for financial exclusion and which keep people away from the formal banking System. In fact, Donald (1976) explained that poor people mostly take credit from informal sources, because of lower discrimination and easy to access to credit throughout the year. Consequently, friendly and flexible structures of credit and security requirements are the merits of Informal Credit Markets (ICM), and the informal sector is the real supporter of finance to the poor. Roe (1971) argued that ICM provided valuable services that weren't met by modern financial corporations. Formal institutions are providing services with a high discipline of banking regulations, which hardly covers the poor people of society. Pischke et al. (1983) added also that the informal sector was expected to be antidevelopment, exploitative, and prone to consumption rather than investment behavior and is incapable of expanding to provide an appropriate volume and range of financial services which have been proved wrong by the increasing framework of informal institutions.

Chandavarkar (1986) indicated that the informal sector plays a supportive role in the capacity development of the poor and their needs, which are not met by the formal financial sector. On the other hand, Bhat (1986) stated that informal credit markets operated largely on the basis of personal intimate information and knowledge and were in a better position to identify new opportunities in financial transactions. For Koning and Koch (1990), an unorganized financial market helps in exploring social welfare lost due to imperfections in the formal credit market and providing credit to the small informal sector which suffers from uncertainty and high transaction costs. For Floro (2019), formal banking institutes always treat poor people as unbankable and presenting higher risks for banks. Rahman (1992) outlined that ICM are required to support and nurture the sustaining of efficient and egalitarian development. However, Ghate (1992) proposed that the informal sector is not for interest rate regulation, credit allocation, and so on, because it leads to costs of lending which are lower than formal institutions.

Leyshon and Thrift (1993) reported on financial exclusion and specified that it was preceded by social exclusion and focused predominantly on the issue of geographical access to financial services, in particular banking outlets. According to Leyshon and Thrift (1995), financial exclusion simply refers to that group of persons who do not benefit from the banking system.

Ford and Rowlingson (1996), as well as Kempson and Whyley (1998), added, from this same perspective, that financial exclusion concerns all people who make little or no use of financial services. Financial exclusion is then related to the inability to have access to all banking services, which may arise from the unavailability of appropriate financial products, price differences, and much more.

From another perspective, Nair (1997) underlined that credit plays a supportive role for economic welfare, which pleads for reduction of poverty and avails us with economic welfare. However, UNDP (1997) advanced that for the urban poor, the informal credit sector works directly because of its flexible and easy services, and given the fact that the informal banking sector directly addresses the needs of clients and accepts them as more important than collateral. Slum dwellers lack access to basic survival means. The urban poor's poverty is further accentuated as more than 40% of the adult urban Indian population has not the authorization to get a bank account (this number would be greater if multiple person accounts are factored in), thereby depriving them of the formal financial system's deposits, credit, remittance, and other financial service facilities. Rupambara (2007), in her study on Urban Poor's Savings and Borrowing Behaviour, tried to understand the financial behavior of the urban poor and their correlation with financial exclusion. It was revealed from her research that the reasons for their low or no savings can be attributed to their poor earnings, larger family size with single earning member, and poor money management skills. In the same perspective Ananthi and Geethalakshmi (2017) carried out an investigation on migrants and their access to financial services and debts in Germany. It was found from their study that high unemployment rates, overcrowded accommodation, and low educational rates made migrants and ethnic groups vulnerable to financial exclusion. In their report on financial inclusion in the Mumbai Slums, Ramji (2009), Hirschler (2009), as well as Bhatia and Chatterjee (2010) found that many slum dwellers were excluded from formal banking services. Such people's financial literacy level has also been very weak. Hema Divya (2013), with reference to her report on the effect of financial inclusion, pushes for a more effective, large, and urban population to be brought under an inclusive financing.

17.2.1 Sustainable Finance: Future Needs

Growth and development complement the financial system. We experienced some of the most complex financial structures in the world in 2008, which spawned the worst global financial crisis seen in decades. When some economies crashed in some developed countries, others had eventually been pulled down in both developed and developing nations. Recognition has developed in the aftermath of this global financial crisis that the financial system must not only be sound and secure, but also sustainable in the way it makes the transition to a low-carbon and green economy. But achieving sustainable development would entail a realignment of the financial system with sustainable development goals (United Nations Environment Program, 2016). De Crombrugghe et al. (2008) used regression in Indian microfinance to explore factors influencing financial sustainability. The three metrics employed for the various aspects of sustainability were cost average by sales, repayment of loans, and cost control. The research was based on the database of 42 Microfinance Institutions (MFIs) in 2003 given by SADHAN (a non-profit organization for microfinance coordination and analysis). The findings of this study revealed that the cost-covering problems for small and partially unsecured loans can be addressed without raising the size of the loan itself. Without touching weak people, MFIs will achieve financial sustainability. Another research undertaken by Zerai and Rani (2012) showed that the majority of MFIs cannot cover their costs. Indeed, these scholars examined the technical efficiency of 19 Ethiopian microfinance institutions from the mix sector by using a stochastic border analysis. The findings revealed that an average efficiency score of 71.72% of Ethiopia's MFI assets, organizational sustainability, and scope of outreach have a major impact on performance. Empirical findings support the trade-off between Ethiopian MFIs' performance and outreach. In the same orientation, Cull, Demirguc-Kunt, & Morduch (2009) tested the effects of prudential supervision of MFIs' profitability outreach (by using data of 245 microfinance institutions) on small scale borrowers and women. Their findings suggested that benefits should drive the cost outreach to women and disadvantaged clients that are costly to meet.

Tehulu (2013) employed unbalanced panel data from 23 MFIs institutions located in East Africa between 2004 and 2009. To classify the factors influencing financial sustainability, the same scholar deployed models of regression for binary probit and ordinary probit. The regression of empirical findings showed that efficiency of management and at-risk portfolios are negative and significantly linked to financial sustainability. Inefficiency of management, risk portfolio, loan severity, and scale have then significant financial sustainability considerations. In his research entitled "A Comparative Analysis of the Financial Performance of Indian and Bangladesh Microfinance Institutions", Rai (2012) analyzed the financial performance of Indian MFIs and compared them with Bangladeshi MFIs from a variety of perspectives using different performance indicators and ratio Mann-Whitney U tests. His study covered the 2007–2008 period and reported that there is no significant difference between the Bangladeshi and Indian MFIs capital regarding operational self-sufficiency, gross loan portfolio yields, and asset returns. However, NBFC-MFIs are more viable financially and are highly valued. Kinde (2012) used a structured dataset of 14 MFIs over the 2002–2012 period to classify factors that influence MFIs' financial sustainability. His results showed that breath of outreach, deep outreach, dependency ratio, and cost and borrower are the major factors affecting the financial sustainability of MFIs in Ethiopia. Capital structure and employee productivity have a marginal impact on financial sustainability. Ali (2011) analyzed 30 MFIs based on financial sustainability and outreach and concluded that the diversification of branches, the number of active lenders, and the percentage of female borrowers have a greater influence on the financial viability of MFIs.

When it comes to India, it would be interesting to emphasize that the informal credit market is an important part of the financial market. This explains how the informal sector channels credit to small and poor borrowers in both urban and rural areas who serve all working capitals of all sizes.

In other words, the informal banking sector provides credit to the poorest people who are declared inefficient and unbankable by formal institutions. In line with this assumption, the informal sector is defined by considering two views: a traditional and a modern one. Besides, different studies claimed that positive and determined financial behavior could make us understand the efficiency of microfinance: on the one side, by assessing the role of microfinance institutions, and on the other side by analyzing the way by which such institutions don't accept the worth of financial services dedicated for the poorest of the poor. It appears so that microfinance has not achieved its developmental objective successfully due to neglect of the poorest of the poor. Thus, the principal hypothesis for our study could be suggested as follows:

Principal Hypothesis: The poorest households depend significantly on informal services dedicated for meeting their financial requirements.

17.3 METHODOLOGY

This study is purely based on primary survey data. Both qualitative and quantitative variables were examined. SPSS software was used for the analysis, and descriptive statistics were used.

The data was collected from a primary survey of 100 samples of slum households and 100 samples of beggars.

17.3.1 SAMPLING TECHNIQUE

Stratified simple random sampling was used in the case of slums and simple random sampling in the case of beggars.

17.3.2 DATA COLLECTION

A well structured questionnaire was employed for the data collection process. We formed strata based on economic activities related to slum dwellers, in order to make the population more representative and so that their economic activities remain heterogeneous. The samples were arranged at a proportion of equality and 25 samples were collected from four areas of the Lucknow city. Theses areas are: (1) Rajni Khand, (2) Mawaiyya, (3) SharamVihar Nagar, and (4) Daliganj-Pull. The samples sites of beggars are: (1) Dargah Khamman Peer Charbagh, Railway Station Lucknow, (2) Hanuman Mandhir, Charbagh, (3) Teen Lochan Hanumanji Mandhir, Charbagh, (4) Hanuman Setu, Mahanagar, and (5) Hanuman Setu Pull Mahanagar (Figure 17.1).

17.3.3 ANALYSIS METHOD

For the analysis, we followed a descriptive analysis, as a first step of our study.

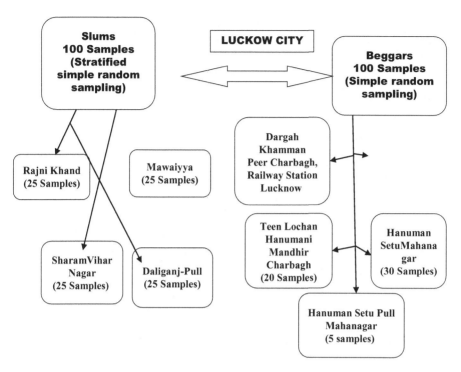

FIGURE 17.1 Flow chart of sampling methodology

17.4 PRESENTATION OF RESULTS AND ANALYSIS

Financial inclusion of poor and marginalized sections is a longstanding problem in India. In spite of the important efforts sustained by the government, this humanitarian/ economic issue is not yet resolved. Nevertheless, financial products are helpful instruments which can alleviate the problems related to poverty. Different types of financial services have been analyzed throughout the present study to understand the nature and needs of the poor. The saving service is one of the important financial services that plays a pivotal role. Data collection and financial information related to households have been collected in the field; and out of 200 samples of beggars and slums, the saving services of banks used by them represent respectively only 8% and 16%, which are low rates. More precisely, poor people mostly save their incomes for daily routine tasks and emergencies. However, credit is a financial instrument that helps both poor and non-poor. The credit service by banks is very poor; and only 2.5% of beggars and 7% of slums use such a service. More particularly, households mainly take credit for the purpose of family expenditures. Insurance as a security service also helps people during emergencies; and again a very small percentage of people are aware of how to get it via banks (5% for beggars and 5% for slums).

Banks accounts play a very important role in transferring money income from one place to another. From this perspective, the percentage of account holders from both beggars and slums are respectively 10% and 43%, which is very low for beggars and medium for slums. People are mostly not transferring their incomes because only 15% of beggars and 14% of slums use account transfers. However, microleasing services are considered as the new best originated services that are assumed to be very helpful in curbing the urban poor's problems. Indeed, the coverage of microleasing is quite low in the case of formal banking institutes, reflecting 1% of beggars and 9% of slums only. Lastly, pension schemes are employed by urban poor as a security for emergencies and old age; and the scheme has a low coverage of 12% and 49% respectively for beggars and slums. Financial services are then not the most efficient tool in providing benefits at a greater pace to poor persons. They do not cover the needs of poor people who can find more suitable services and target oriented implementations of already disposable schemes.

17.5 CONCLUSION

The financial inclusion of poor and marginalized sections remains a longstanding problem that India faces. In spite of the important efforts sustained by the government to tackle it, this humanitarian/economic issue is not yet resolved. Nevertheless, financial products might be helpful instruments and can alleviate the obstacles related to poverty. That is why different types of financial services have been analyzed throughout the present study in order to understand the nature and needs of the poor. The saving service is one of the important financial services that play a pivotal role in this context.

17.5.1 MAIN IMPLICATIONS OF THE STUDY

- Financial services for such marginal social groups, the urban poor, should be diversified and designed according to their needs and requirements.
- Slums and beggars need financial products mostly for short run purposes. Therefore, the design of these services should be comparatively economical.
- Poor urban people mostly take credit from friends and relatives without interest rates. Credit documentation should be simplified so that the urban poor can make use of financial services easily.
- The poorest of the poor should be provided with all the legal documentation of citizenship (i.e., voter ID cards, Adhar Cards, etc.) so that they have access to formal financial services.
- Poor people are ignorant about new financial services and government programs such as microleasing, insurance, and other pension schemes. Hence, they need proper awareness and training guidelines so that they can participate effectively.

The approach of Grameen Bank's business arrangements are more suitable to support the ultra poor. They issue loans on the bases of income generating activities in which they are successful across the globe. They have sanctioned these for over 100,000 beggars which has provided improvement in their lives by changing them from begging and to use business activities like selling door to door household items (Yunus, 2017).

17.5.2 LIMITATIONS OF THE STUDY

All undertaken researches have some limitations. For the current study, the limitations cited below may be taken into consideration:

- The study was restricted only to the city of Lucknow (in India).
- The population of the study is very volatile and inconsistent (which is reflected in their responses).
- Most of the respondents are illiterate and unaware about the issues and the financial problems they are facing.
- The sample size is too small to generalize the overall results. Some of these may be spurious due to inappropriate information given by the respondents.
- Analysis of the study is mostly based on sample observations and descriptive statistics. Because of the time constraint, we could not make use of advanced tools of statistics and econometrics, which may be done in further research for making the results of the research more reliable.

17.5.3 FUTURE RESEARCH DIRECTIONS

The present study might have greater scope by extending its domain from Lucknow to other Indian regions, in order to get better insights about vulnerable urban groups. It could also be expanded to other poverty factors so that the appropriate sustainable policy approach will be implemented to fight the problem of financial exclusion and inequality among the masses. Finally, the current research could gain in relevance by helping policymakers to understand the bottle necks behind urban poverty.

REFERENCES

Ali, K. H. (2011). The relationship between financial sustainability and outreach of microfinance institutions in Kenya.

Ananthi, M. R., and Geethalakshmi, R. (2017). Financial inclusion of urban poor in India–challenges and measures. competency mapping of employees working in hyper markets in India, 12.

Chambers, R., and Conway, G. (1992). *Sustainable rural livelihoods: practical concepts for the 21st century*. Institute of Development Studies (UK).

Chandavarkar, Anand (1986). The informal Financial Sector in Developing Countries; Analysis, Evidence and Policy Implication, *Resource Paper for Seacen Seminar*, Yogjakarta (Nov).

Cull, R., Demirguc-Kunt, A., & Morduch, J. (2009). *Does regulatory supervision curtail microfinance profitability and outreach?*. The World Bank.

De Crombrugghe, A., Tenikue, M., & Sureda, J. (2008). Performance analysis for a sample of microfinance institutions in India. *Annals of Public and Cooperative Economics*, 79(2), pp. 269–299.

Floro, S. L. (2019). *Informal credit markets and the new institutional economics: The case of Philippine agriculture*. Routledge Taylor & Francis.

Ghate, Prabhu. (1992). Interaction Between the Formal and Informal Financial Sectors: The Asian Experience, *World Development*, 20(6), pp.59–872.

Hanoch, Y., Barnes, A., and Rice, T. (2017). *Behavioral economics and healthy behaviors: Key concepts and current research*. Taylor & Francis Editions.

Hirschler, S. (2009). Migrant and Financial Exclusion, http://www.ecosocdoc.be/ static/ module/bibliographyDocument/document/001/700.

Kempson, E. and Whyley, C. (1998). *Access to current accounts*. London: British Bankers Association.

Kinde, B. A. (2012). Financial sustainability of microfinance institutions (MFIs) in Ethiopia. *European Journal of Business and Management*, 4(15), 1–10.

Leyshon, A. and Thrift, N. (1993). The restructuring of the UK financial services industry in the 1990s: a reversal of fortune? *Journal of Rural Studies*, 9(3), 223–241.

Leyshon, A. and Thrift, N. (1995). Geographies of financial exclusion: financial abandonment in Britain and the United States. *Transactions of the Institute of British Geographers, New Series*, 20, 312–341.

Lont H., and Hospes, O. (2004). *Livelihood and Microfinance: Anthropological and Sociological Perspectives on Savings and Debt*. Eburon Academic Publishers.

Nair, Tara'S (1997). *Paper Presented at National Seminar on informal Sector: Emerging Perspectives in Development*. Organized by IAMR and IHD December, New Delhi.

Navin, Bhatia, and Arnav Chatterjee, (2010). Financial Inclusion in the Slums of Mumbai, *Economic and Political Weekly*, 45(42), 23.

Patel, S. (2014). Community-Driven Solutions for Inclusive Urbanization. *Inclusive Urbanization: Rethinking Policy, Practice and Research in the Age of Climate Change*, 177.

PFRC. (2005). *Measuring financial capability: an exploratory study prepared for financial services authority by personal finance research centre*. Bristol: University of Bristol.

Pischke, Von J.D et al. (1983). A Penny Saved: Kenya's Co-operative Savings Scheme. In Von Pischke, Adam and Donald. pp. 302–307.

Rahman, A. (1992). The Informal Sector in Bangladesh: An Appraisal of it\ Role in Development. *Development and Change*. 23, pp. 147–168.

Rai, A. K. (2012). A comparative analysis of the financial performance of Micro Finance Institutions of India and Bangladesh.

Ramji, M. (2009). Financial Inclusion in Gulbarga: Finding Usage in Access. Institute of Financial Management and Research- Centre for Micro Finance, Working Paper Series No. 26, January

Rangarajan, C. and Dev, S. M. (2014). Counting the poor: Measurement and other issues (No. 2014–048).

Risbud, N. (2006). Changing Scenario of Housing for the Urban Poor in Indian Cities. Urban Governance and Management: Indian Initiatives, 199.

Roe, A. R. (1971). Some Theory concerning the role and failings of Financial Intermediation in Developing Countries. Capital Market Imperfections and Economic Development, World Bank Staff Working Paper, (338).

Rosenzweigh, K. and Binswanger, H. (1993). Wealth, Weather Risk and the Composition and Profit ability of Agriculture Investments, *Economic Journal* 103.

Rupambara. (2007). Financial Inclusion of the Urban Poor Issues and Options, *CAB CALLING*

Rutherford, S. (1999) *The poor and their money*, Institute for Development Policy and Management/University of Manchester.

Schreiner, M. and Sherraden, M. W. (2007). *Can the poor save?: saving & asset building in individual development accounts*. Transaction Publishers.

Sebstad, J. and Cohen, M. (2000). *Micro finance, Risk Management, and poverty*. AIMS

Sharma, K. (2012). Changing profile of urban poverty, A case study of Jharkhand (India). *Transcience*, 3(2), 37–50.

Tehulu, T. A. (2013). Determinants of financial sustainability of microfinance institutions in East Africa. *European Journal of Business and Management*, 5(17), 152–158.

United Nations Environment Programme. (2016). The Financial System We Need: Aligning the Financial System with Sustainable Development. UN.

World Bank (2000). *World Bank Estimates*. Washington: World Bank,

Yunus, M. (2017). *A world of three zeros: the new economics of zero poverty, zero unemployment, and zero net carbon emissions*. UK: Hachette.

Zerai, B. and Rani, L. (2012). Technical efficiency and its determinants of micro finance institutions in Ethiopia: a stochastic frontier approach. *African Journal of Accounting, Economics, Finance & Banking Research*, 8(8), pp.1–19.

Index

Page numbers in *italics* refer to content in *figures*; page numbers in **bold** refer to content in **tables**.

Printed in the United States
By Bookmasters